AF750754

Digital Twin Technology for Sustainable Agriculture

Arun Lal Srivastav • Naveen Kumar
Inga Zinicovscaia
Editors

Digital Twin Technology for Sustainable Agriculture

Applications, Implementation and Future Trends

Springer

Editors
Arun Lal Srivastav
Department of Applied Sciences
Chitkara University
Solan, Himachal Pradesh, India

Naveen Kumar
Department of Computer Science
Himachal Pradesh University
Shimla, Himachal Pradesh, India

Inga Zinicovscaia
Joint Institute for Nuclear Research
Dubna, Russia

ISBN 978-981-95-5914-5 ISBN 978-981-95-5915-2 (eBook)
https://doi.org/10.1007/978-981-95-5915-2

© The Editor(s) (if applicable) and The Author(s), under exclusive license to Springer Nature Singapore Pte Ltd. 2026

This work is subject to copyright. All rights are solely and exclusively licensed by the Publisher, whether the whole or part of the material is concerned, specifically the rights of translation, reprinting, reuse of illustrations, recitation, broadcasting, reproduction on microfilms or in any other physical way, and transmission or information storage and retrieval, electronic adaptation, computer software, or by similar or dissimilar methodology now known or hereafter developed.
The use of general descriptive names, registered names, trademarks, service marks, etc. in this publication does not imply, even in the absence of a specific statement, that such names are exempt from the relevant protective laws and regulations and therefore free for general use.
The publisher, the authors and the editors are safe to assume that the advice and information in this book are believed to be true and accurate at the date of publication. Neither the publisher nor the authors or the editors give a warranty, expressed or implied, with respect to the material contained herein or for any errors or omissions that may have been made. The publisher remains neutral with regard to jurisdictional claims in published maps and institutional affiliations.

This Springer imprint is published by the registered company Springer Nature Singapore Pte Ltd.
The registered company address is: 152 Beach Road, #21-01/04 Gateway East, Singapore 189721, Singapore

If disposing of this product, please recycle the paper.

To the farmers of the world who are the true innovators of the land, whose resilience, wisdom, and hard work sustain humanity. And also to the scientists and technologists for their continued efforts to nurture the harmony between nature and innovation.

Foreword

Digital twin skill is probable to renovate farming. By generating the simulated illustration of a physical unit, it supports nutrition fabricators in nursing, forecasting, and augmenting the production method remotely and even unconventionally. However, the growth in this field is comparatively gentler than in trades like manufacturing. An organized examination of agricultural digital twins' present position and growth is imperious.

Digital Twin Technology for Sustainable Agriculture: Applications, Implementation and Future Trends, this book particularized on the studies aiming farming digital twins from global trends, intensive areas (including domains, processes, and topics), reference architectures, and uncluttered questions, which could help researchers to scrutinize their research outline and sustenance the further growth of digital twins in agriculture.

Dr. Arun Lal Srivastav, Dr. Naveen Kumar, and Dr. Inga Zinicovscaia and the contributing researchers have vast expertise in this field that can narrate the gap and advancement in this technology and also incorporate the multidisciplinary approach. The book takes a thorough review from the theoretical aspects to the practical implementation to real-world applications. This book leverages to know the deep concepts about the digital twin process in agriculture but also gives the idea to make the agriculture more advanced with these modern tools.

It includes the practical implementation, different case studies, and other technological points to make it a very useful book for the persons who work in this field. It also motivates the readers to do more valuable things in the field of application of digital twin for sustainable farming.

I commend Dr. Srivastav, Dr. Kumar, and Dr. Zinicovscaia for thinking about the new technologies in the sustainable farming sector with digital twins' technology.

Dean Academic Affairs, Kurukshetra University
Kurukshetra, Haryana, India

Rakesh Kumar

Preface

The roots of cultivation can be drawn back to the initial antiquity of human civilization, lingering one of the universe's most imperative businesses even today. The declaration of fourth agricultural revolt or "Agriculture 4.0" in academia and business has carried out the ability of digitization, technological advancement, and increased efficiency

This book, *Digital Twin Technology for Sustainable Agriculture: Applications, Implementation and Future Trends*, is the collaborative efforts but the editors and the authors to apprehension alteration in farming and gives a view to the use of digital twin in the modern farming. Digital Twins means the simulation of physical units, systems, or processes in industries such as healthcare, production, and planning of urbanization. It also has vast applications in farming that opens the unprecedented opportunities to utilize the resources, improve the production, and maintain the sustainability of environment.

By the collaborative efforts from the experts, researchers, and academicians from this area, this edited book presents the theoretical insights, technological frameworks, and practical case studies that illustrate the power of digital twins in agriculture. Each chapter in this book explores the innovative pathways to monitoring the crop health, predicting the yields to managing the soil fertility.

Apart from discussing the immense potential of digital twins in farming, the book also points out the challenges that range from data privacy and other technical difficulties that are divided among farming groups. It is a true promise to address these challenges with informed strategies and inclusive innovation.

It is our hope that this book serves as a valuable resource for researchers, technologists, policymakers, and practitioners who are passionate about advancing sustainable agricultural practices. By bridging the gap between technology and traditional farming wisdom, *"Digital Twins"* aspires to contribute to a greener, smarter, and more resilient agricultural future.

Solan, Himachal Pradesh, India — Arun Lal Srivastav
Shimla, Himachal Pradesh, India — Naveen Kumar
Magurele, Romania — Inga Zinicovscaia

Contents

Editors and Contributors

About the Editors

Arun Lal Srivastav is an Associate Professor at Department of Applied Sciences, School of Engineering and Technology, Chitkara University, Himachal Pradesh, India. He earned his Ph.D. from the Indian Institute of Technology (BHU), Varanasi, in water treatment and completed post-doctoral research at National Chung Hsing University, Taiwan. His teaching expertise includes Environmental Science, Environmental Engineering, Disaster Management, and Design Thinking. His research interests span water quality surveillance, soil health, phytoremediation, and agro-waste management. He has published over 125 research papers in reputed journals such as Elsevier, Springer, and Taylor & Francis, Conferences, books and has edited multiple books with leading publishers. He also serves as a series editor for Elsevier and Nova on energy sustainability and e-waste management. A recipient of recognition among the world's top 2% scientists jointly by Stanford University and Elsevier (2023, 2024, 2025), he continues to advance sustainable environmental solutions.

Naveen Kumar is an Assistant Professor at Himachal Pradesh University, Shimla. He completed his Ph.D. from Himachal Pradesh University and has worked with the National Physical Laboratory, New Delhi, under the Ministry of Science and Technology. His areas of expertise include UI/UX design, web development, and software engineering. He has published more than 20 research papers and presented at over 15 national and international conferences. He also worked on research project under RUSA Himachal Pradesh University. His work reflects a strong commitment to advancing innovative technologies and enhancing digital education practices.

Inga Zinicovscaia is affiliated with the Horia Hulubei National Institute for R&D in Physics and Nuclear Engineering, Magurele, Romania. She earned her Doctor of Science degree from Moldova State University, Republic of Moldova, focusing on the environmental impact of metals determined through neutron activation analysis. Her research interests include water and soil remediation, air quality assessment, nanotoxicology, waste management, food quality evaluation, and radioecology. She has published over 175 research papers in prestigious ISI-indexed journals and authored several book chapters and monographs on wastewater bioremediation. She serves as a section editor for several journals and leads over 10 international collaborative projects between the Joint Institute for Nuclear Research and member states on various environmental issues.

Contributors

S. Aasha Nandhini Department of Electronics and Communication Engineering, Sri Sivasubramaniya Nadar College of Engineering, Kalavakkam, Tamil Nadu, India

Arti Sanatan Dharma College, Ambala Cantt, Kurukshetra University, Kurukshetra, India

Muskan Aggarwal Sanatan Dharma College, Ambala Cantt, Kurukshetra University, Kurukshetra, Haryana, India

Shalini Aggarwal S.U.S. Govt. College, Matak Majri (Indri), Karnal, Haryana, India

Monu Bhardwaj School of Management, Graphic Era Hill University, Dehradun, Uttarakhand, India

Vinod Kumar Chaudhary Department of Environmental Sciences, Dr. Rammanohar Lohia Avadh University, Ayodhya, Uttar Pradesh, India

Avnish Chauhan Department of Environmental Science, Graphic Era Hill University, Dehradun, Uttarakhand, India

Nafisa N. Chowdhury Centre for Climate Change and Environmental Research, BRAC University, Dhaka, Bangladesh

Sonu Dwivedi Department of Chemistry, D.B.S.(PG) College, Dehradun, Uttarakhand, India

Ashok Ganga Department of Electrical and Electronics Engineering, Aditya Institute of Technology and Management, Tekkali, Srikakulam, Andhra Pradesh, India

Kanwal Garg Department of Computer Science and Applications, Kurukshetra University, Kurukshetra, Haryana, India

Garima Sanatan Dharma College, Ambala Cantt, Kurukshetra University, Kurukshetra, India

Dipankar Ghosh Microbial Engineering and Algal Biotechnology Laboratory, Department of Biosciences, JIS University, Agarpara Kolkata, West Bengal, India

Girdhar Gopal Sanatan Dharma College, Ambala Cantt, Kurukshetra University, Kurukshetra, Haryana, India

Nguyen Thi Hoang Ha Faculty of Advanced Technologies and Engineering, VNU Vietnam Japan University, Hanoi, Vietnam

Uma Hapani Department of Environmental Studies, The Maharaja Sayajirao University of Baroda, Vadodara, Gujarat, India

Shama E. Haque North South University, Bashundhara, Dhaka, Bangladesh

Tran Thi Viet Ha Faculty of Advanced Technologies and Engineering, VNU Vietnam Japan University, Hanoi, Vietnam

Do Trung Hieu Faculty of Geography, VNU University of Science, Hanoi, Vietnam

Sabnam Hossain Microbial Engineering and Algal Biotechnology Laboratory, Department of Biosciences, JIS University, Agarpara Kolkata, West Bengal, India

Hurmat Chitkara University School of Pharmacy, Chitkara University, Baddi, Himachal Pradesh, India

Kanaka Raju Kalla Department of Electrical and Electronics Engineering, Aditya Institute of Technology and Management, Tekkali, Srikakulam, Andhra Pradesh, India

Santosh Kumar Karn Department of Biochemistry & Biotechnology, Sardar Bhagwan Singh University, Balawala, Dehradun, Uttarakhand, India

Ashok Kumar Kashyap Department of Computer Science, Himachal Pradesh University, Shimla, India

Balpreet Kaur Chitkara University School of Pharmacy, Chitkara University, Baddi, Himachal Pradesh, India

Navneet Kaur Department of Computer Science and Engineering, Chaudhary Devi Lal University, Sirsa, Haryana, India

D. Vijaya Kumar Department of Electrical and Electronics Engineering, Aditya Institute of Technology and Management, Tekkali, Srikakulam, Andhra Pradesh, India

Gobind Kumar Government Polytechnic College, Sirsa, Haryana, India

Akanksha Kumari Department of Food Technology and Nutrition (School of Agriculture), Lovely Professional University, Phagwara, Punjab, India

Kanchan Kumari Department of Biotechnology, Patna Women's College (Autonomous), Patna University, Patna, Bihar, India

Mukesh Kumar Department of Computer Science, IILM University, Greater Noida, Uttar Pradesh, India

Naveen Kumar Department of Computer Science, Himachal Pradesh University, Shimla, Himachal Pradesh, India

Pardeep Kumar Department of Computer Science and Applications, Kurukshetra University, Kurukshetra, Haryana, India

Pardeep Kumar Computer Science, Government College for Women, Sirsa, Haryana, India

Rakesh Kumar Department of Computer Science & Applications, Kurukshetra University, Kurukshetra, Haryana, India

Sandeep Kumar Department of Computer Science, Himachal Pradesh University, Shimla, India

Saurabh Kumar Department of Environmental Sciences, Dr. Rammanohar Lohia Avadh University, Ayodhya, Uttar Pradesh, India

Sunil Kumar Department of Electrical, Electronics and Communication Engineering, Galgotias University, Greater Noida, Uttar Pradesh, India

Sunny Kumar Department of Electrical Engineering, IIT Delhi, Delhi, India

R. Karthick Manoj Department of Electrical and Electronics Engineering, AMET Deemed to be University, Kanathur, Tamil Nadu, India

Anita Rani Mehta Department of Computer Science and Applications, Kurukshetra University, Kurukshetra, Haryana, India

Yogesh Mohan Department of Computer Science, Himachal Pradesh University, Shimla, India

Noha Nassar Faculty of Science, Cairo University, Giza, Egypt

Gaurav Pandey Forest Research Institute (FRI), Dehradun, Uttarakhand, India

Manish Pant Department of Chemical Engineering, Indian Institute of Science Education and Research, Bhopal, Madhya Pradesh, India

Veena Parihar Symbiosis Institute of Geoinformatics, Symbiosis International Deemed University, Pune, Maharashtra, India

Akansha Patel Department of Environmental Sciences, Dr. Rammanohar Lohia Avadh University, Ayodhya, Uttar Pradesh, India

Naveen Patel Department of Civil Engineering, Dr. Rammanohar Lohia Avadh University, Ayodhya, Uttar Pradesh, India

Braj Kishor Prasad Department of Computer Applications (MCA), Patna Women's College, Patna, Bihar, India

Le Kim Quy Faculty of Advanced Technologies and Engineering, VNU Vietnam Japan University, Hanoi, Vietnam

Vivek Raj Department of Biotechnology, Patna Women's College (Autonomous), Patna University, Patna, Bihar, India

Ritesh Rana Department of Computer Science, Himachal Pradesh University, Shimla, Himachal Pradesh, India

Lata Rani Chitkara University School of Pharmacy, Chitkara University, Baddi, Himachal Pradesh, India

Riddhi Rathva School of Environment and Sustainable Development, Central University of Gujarat, Gandhinagar, Gujarat, India

Kaiss Redouane Faculty of Economics and Management, Hassan 1st University, Settat, Morocco

Muneesh Sethi COER University, Roorkee, Uttarakhand, India

Anurag Singh Department of Mechanical Engineering, Dr. Rammanohar Lohia Avadh University, Ayodhya, Uttar Pradesh, India

Dhiraj Kumar Singh Amity School of Engineering and Technology, Amity University Haryana, Amity Education Valley, Gurugram, Manesar, Haryana, India

Gurwinder Singh Department of AIT-CSE, Chandigarh University, Mohali, Punjab, India

Harbinder Singh Department of Electronics and Communication Engineering, UCRD, Chandigarh University, Chandigarh, Punjab, India

Shivansh Singh School of Energy Systems, LUT University, Lappeenranta, Finland

Arun Lal Srivastav Department of Applied Sciences, Chitkara University, Solan, Himachal Pradesh, India

Om Prakash Suthar Department of Computer Engineering, Marwadi University, Rajkot, Gujarat, India

Jaganmohan Rao Tarra Department of Electrical and Electronics Engineering, Aditya Institute of Technology and Management, Tekkali, Srikakulam, Andhra Pradesh, India

Budumuru Giri Teja Department of CSE (AI & ML), Aditya Institute of Technology and Management, Tekkali, Srikakulam, Andhra Pradesh, India

Sanjay Tyagi Department of Computer Science and Applications, Kurukshetra University, Kurukshetra, Haryana, India

M. R. M. Veeramanickam Department of Computer Science & Engineering, Chandigarh University, Chandigarh, Punjab, India

Neeraj Kumar Verma Department of Data Sciencc, Manipal University, Jaipur, Rajasthan, India

Nguyen Minh Viet VNU Key Laboratory of Advanced Material for Green Growth, VNU University of Science, Hanoi, Vietnam

Introduction to Sustainable Farming 4.0 Practices

1

Tran Thi Viet Ha, Le Kim Quy, Nguyen Minh Viet, Nguyen Thi Hoang Ha, and Do Trung Hieu

Abstract

The shift to Sustainable Farming 4.0 signifies a significant advancement in agricultural practices by integrating digital technologies with agricultural operations and environmental management principles. This chapter examines the progression of agricultural technology from traditional practices to contemporary farming utilizing data-driven and intelligent technologies characteristic of Agriculture 4.0. Recent advancements, including Artificial Intelligence (AI), the Internet of Things (IoT), and notably Digital Twin technologies, are catalyzing significant transformations in agricultural operations by providing real-time modeling and forecasting data for crop and animal control. A special focus of this chapter is on Digital Twin applications in agriculture, which create virtual copies of physical systems (e.g., soil conditions, crop growth, greenhouse environments), then focusing on optimization of resource use, reducing risks, and enhancing productivity. The pressing need for sustainable agriculture in the face of climate change, soil degradation, water scarcity, and global food insecurity is also highlighted in this chapter. Based on the discussions, the chapter will emphasize how Farming 4.0 improves efficiency and supports long-term sustainability through precision input management, reduced environmental

T. T. V. Ha (✉) · L. K. Quy · N. T. H. Ha
Faculty of Advanced Technologies and Engineering, VNU Vietnam Japan University, Hanoi, Vietnam
e-mail: ttv.ha@vju.ac.vn

N. M. Viet
VNU Key Laboratory of Advanced Material for Green Growth, VNU University of Science, Hanoi, Vietnam

D. T. Hieu
Faculty of Geography, VNU University of Science, Hanoi, Vietnam

© The Author(s), under exclusive license to Springer Nature Singapore Pte Ltd. 2026

A. L. Srivastav et al. (eds.), *Digital Twin Technology for Sustainable Agriculture*,
https://doi.org/10.1007/978-981-95-5915-2_1

impact, and data-informed decision-making. By transitioning toward Farming 4.0, agriculture can evolve into a more adaptive and sustainable sector, aligned with the circular economy principles and the United Nations Sustainable Development Goals (SDGs).

Keywords

Agriculture 4.0 · Digital twin technology · Sustainable farming · Circular economy

1.1 Evolution of Agricultural Technology

From primitive humans to recently modern society, agriculture has always played a key role in human development. A long time ago, farming and animal domestication enabled primitive human communities to transition from nomadic lifestyles to settled civilizations, giving rise to culture, trade, and governance. After that, the transformations and advancements in agricultural activities have supported population growth and economic development through technological innovation. The integration of farming technologies has helped increase production efficiency and ensured sustainability, food security, and environmental protection. Over time, the key milestones in the development of agricultural technology can be identified, including: (i) conventional practices (before 1900), (ii) the rise of mechanical equipment and engines (1960s), (iii) the introduction of computers and IT (1980), (iv) the emergence of satellite and sensors (1990), (v) the use of mobile technologies and database software (2000s), and finally, (vi) the integration of the Internet of Things (IoT), Big Data, and Artificial Intelligence (AI) in the present era.

Before 1900, rudimentary tools such as ploughs, sickles, and hoes were often made from wood, stone, or metal in manual and conventional agriculture practices. Human and animal power, like buffalo, cow, horse, etc., were the primary energy sources. Farming techniques then relied on observation, traditional knowledge, and seasonal patterns. Thus, the agricultural activities at this time had low productivity, high labor demand, and high vulnerability due to natural climate conditions, pests, and diseases. Moreover, the lack of mechanization and scientific understanding limited the scale and efficiency of agricultural operations, often confining production to subsistence levels. As a result, food shortage and hunger were common occurrences, particularly during drought or crop failure.

Agricultural mechanization and chemical inputs were introduced in the 1950s and 1960s of the twentieth century. This period marked a significant turning point, the Green Revolution, significantly boosting agricultural productivity. Equipment like tractors or harvesters can help to reduce the reliance on human labor, allowing farmers to do the agricultural activities in much larger areas more efficiently. These advancements helped many countries achieve food security but brought new challenges, such as environmental degradation and increased dependence on chemical inputs. However, the overuse of synthetic fertilizers and pesticides at that time had led to several environmental issues such as soil degradation, contamination, and biodiversity loss.

The 1970s marked the beginning of digitalization in agriculture with the appearance of computers. With the development of computer technology and computer systems, the cost of agricultural activities was further reduced, and the efficiency of farming systems was significantly increased. Computer applications in agriculture are presented in three important areas: image analysis, crop models, and information technology (Usman 2024). By using computers, it was possible to make weather predictions and yield forecasts. At the same time, simulations helped to clearly understand the long-term impacts of different farming factors, such as crop rotation and water management. It was the solid background for the later precision agriculture, where decisions are made based on digital inputs and analytical models to optimize resource use and crop output.

The early 1990s marked a phase of integration between satellite and remote sensing technology in the agricultural sector. Crops experience biophysical and biochemical alterations in response to biotic and abiotic stressors. Awareness of the timing of such changes is crucial, hence requiring remote sensing in agriculture (Khan et al. 2023). The satellites can facilitate land monitoring and furnish essential data regarding meteorological conditions, soil quality, and vegetation health. Geographic Information Systems (GIS) facilitated the creation of intricate maps that depicted field circumstances, crop vitality, and productivity areas. This period marked a pivotal advancement in precision agriculture, enabling farmers to apply inputs such as water, fertilizers, pesticides, and other agricultural interventions solely as required, thereby minimizing waste and environmental impact while maintaining yield enhancement. Furthermore, these spatial technologies provide more precise and prompt decision-making at both field and regional levels. Variable Rate Technology (VRT) has also developed, enabling machines to modify input applied rates in real-time according to field data. These advances signified the shift from generalist managing farms to highly customized, site-specific agricultural treatments.

At the beginning of the new century, the 2000s, there was an explosion in mobile phone usage. At this period, everything is almost in the hands of the farmer, with just a mobile phone, especially later on, with powerful smartphones fully equipped with integrated features like weather forecasts, market prices, and expert advice via SMS or mobile apps. This digital accessibility empowered smallholder farmers in remote areas like islands, high mountains, etc. It helped them make more correct decisions based on the evidence and improve their productivity. It also provides digital agriculture platforms, where farmers can directly connect with customers, buyers, and service providers. For example, it was found that the impact of mobile phone and internet technology on rural farmers' income was impressive, with about 41% of agricultural profits and over 31% with support and non-agricultural profits (Khan et al. 2022). Mobile technologies also played a key role in helping farmers access loans from different sources, to have information on crop insurance, and to process digital payments, thus enhancing the financial situation for marginalized farming communities. Real-time information from mobile phones can help to reduce the unnecessary losses in harvesting by improving the logistics chain and timely market access at the suitable time. As the use of mobile phones continued to expand, the digital divide in agriculture began to narrow, contributing to more equitable and inclusive agricultural development.

The present-day agricultural sector is being significantly transformed with the Internet of Things (IoT), Big Data analytics, Artificial Intelligence (AI), and other high technologies etc., which can enable the "smart farming." For example, IoT devices such as different types of sensors, weather monitoring stations, and livestock trackers can easily collect real-time data from the field, and then provide continuous updates on crop and animal health. Big data analytics can process a huge amount of information to identify the trends, predict the outcomes, and support evidence-based decision-making. AI with machine learning can help to optimize the harvest conditions, forecast the weather, or detect the early signs of disease. The drones and autonomous machinery powered by AI are also used for tasks such as aerial imaging, seeding, spraying, and harvesting, significantly reducing human labor and enhancing efficiency. As a result, applying these technologies can significantly increase productivity in agriculture, thereby promoting sustainable natural resource usage by minimizing unnecessary waste and pollutants.

Furthermore, smart farming platforms integrate financial, weather, and market data, offering farmers actionable insights to make better business decisions. Blockchain is also being used to improve transparency and traceability in supply chains, increasing the level of trust of consumers and reducing supplier fraud. Overall, integrating these advanced digital technologies is opening a totally new era of precision smart agriculture that is data-driven, efficient, and resilient to climate and market uncertainties.

The evolution of agricultural technology is evidence of human intelligence and the efforts to overcome natural limitations. In summary, the evolution is shown in Fig. 1.1; each milestone improved productivity and reshaped agriculture's social, economic, and environmental aspects. Nowadays, we are in the digital agricultural

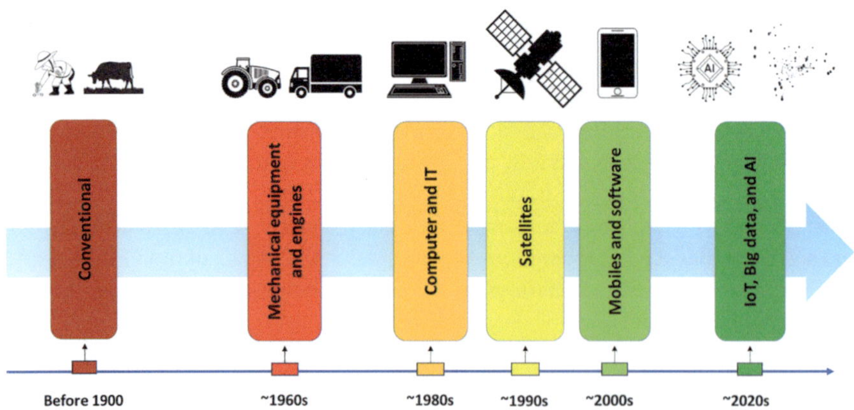

Fig. 1.1 Summary of the milestones in the evolution of agricultural technology

revolution with more efficient and environmentally friendly farming practices. However, for these innovations to be truly transformative and sustainable, equitable access, education, and supportive infrastructure should be ensured so that no farmer is left behind in this technological journey. Connecting the digital distance between developed and developing regions is still a big challenge that must be addressed through collaborative policy and investment. It is necessary for various stakeholders, like governments, the private sector, and research institutions, to work together to develop inclusive and scalable digital agriculture models. Enhancing capacity through training and combining local knowledge is important for successful technology adoption. Moreover, ensuring data privacy and ethical use of technology will significantly increase farmers' trust. Ultimately, the future of agriculture depends not only on technology but on how well it aligns with the broader goals of sustainable development. Advancing digital agriculture responsibly and inclusively directly contributes to several United Nations Sustainable Development Goals (SDGs), including SDG 2 (Zero Hunger), SDG 12 (Responsible Consumption and Production), SDG 13 (Climate Action), and SDG 9 (Industry, Innovation, and Infrastructure). By achieving the SDGs, agriculture's digital transformation can contribute to the sustainable global food system.

1.2 Defining Digital Twins in Agriculture

1.2.1 What Is the Digital Twin?

In 2003, Dr. Michael Grieves introduced the concept "virtual digital" at the University of Michigan's product lifecycle management course. It was initially described as a digital copy of one or a set of specific devices that can abstractly represent a real device and can be used as a basis for testing under real or simulated conditions. However, this concept was not taken deeply due to technical and cognitive limitations (Grieves 2005). Also, this concept was not yet named "digital twin" at that time. From 2003 to 2005, this concept was known as the "mirrored spaced model" (Grieves 2005); and from 2006 to 2010, it was known as the "information mirror model" (Grieves 2006). However, these conceptual models had all the elements of the digital twin model, which can be considered the initial form of the digital twin concept now. In 2011, Michael Grieves, in his book "Virtually Perfect: Driving Innovative and Lean Products through Product Lifecycle Management" (Grieves 2011), referenced the conceptual model of the "digital twin," a term described by his co-author John Vickers. This book describes digital twins as the virtual representations of real-world physical systems, whether living or non-living, that combine complex systems analysis, decision support, and technological integration.

The digital twin is not a totally new concept toward more comprehensive methods for Product Life Management; however, the digital twin provides a unique way to achieve this (Warren Purcell 2023). When traditional tools often rely on static models and historical data, digital twins allow for the use of continuous, real-time synchronization between physical assets and their digital counterparts. This dynamic interaction allows for more accurate simulations, predictive maintenance, and informed decision-making across the product lifecycle. Until now, digital twins have been increasingly adopted by several disciplines to address multidisciplinary problems, including agriculture.

Agriculture activities are the primary sources of global greenhouse gas emissions, land degradation, eutrophication, acidification, and broader social and environmental concerns. Therefore, achieving sustainable agricultural transformation in agricultural activities is critical to align with international policy frameworks and societal ambitions, such as the Green Deal, Paris Agreement, Sustainable Development Goals, and Farm-to-Fork Strategy. Using advanced technologies, digital twins support real-time observation, system modeling, and process automation via detailed digital replicas and continuous two-way data exchange. However, if not managed closely and carefully, the risk of system collapse or disruption is entirely possible, leading to disruption in agricultural production, especially in the current context of climate change.

1.2.2 Components of Agricultural Digital Twin

Agricultural digital twin typically involves the different components (Fig. 1.2). Digital twins in agriculture consist of two key parts: physical components and virtual components, which continuously interact through real-time data exchange. The basic idea of a digital twin is quite simple: linking physical and digital objects in an accurate and real-time manner. The physical components refer to the actual, on-the-ground elements of a farm, such as crops, livestock, soil, machinery, irrigation systems, and infrastructure like greenhouses or silos. Different sensors and IoT

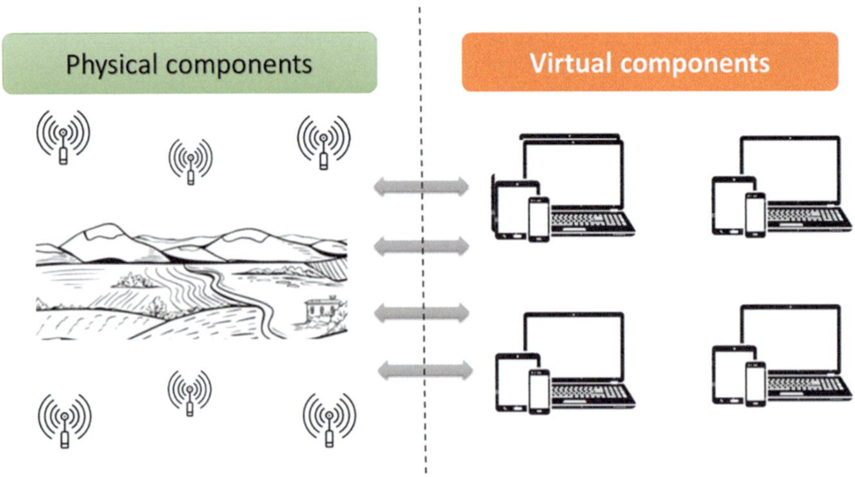

Fig. 1.2 The two main components of the digital twin in agricultural

(Internet of Things) devices can be equipped in order to collect data on environmental conditions, equipment status, soil moisture, crop health, weather patterns, and more. This continuous flow of data is transmitted to the virtual side of the digital twin system, where it is processed and visualized. The virtual component of the digital twin is mainly related to the digital technologies for data processing and analysis (Artificial Intelligence (AI)), Internet of Things (IoT)...), Communication Technologies (wireless, Low Power Wide Area Networks...), Cloud Computing, Virtual Representation, 3D Modeling, Simulation, Data Analytics, and Machine Learning, Application Programming Interfaces (APIs), and Physical Entity Model (Marc Escribà-Gelonch et al. 2024). Digital twin can make virtual copies or replicas of physical agricultural components, then simulate, predict, and optimize agricultural processes, enabling farmers to anticipate potential issues and make informed decisions proactively. Digital twins can integrate with tools like AI and big data to improve operational efficiency, reduce resource waste, and improve yield accuracy.

1.2.3 Applications of Digital Twins in Agriculture

The benefits of digital twins in agriculture include reduced production times and costs, hiding the complexity of integrating heterogeneous technologies, creating safer working environments, and establishing more environmentally sustainable operations. Digital twins can be tailored to different agricultural domains, such as crop farming, livestock farming, greenhouse and indoor farming, to support predictive maintenance, supply chain optimization, etc. For instance, simply speaking, digital twins allow farmers to see with their eyes various scenarios virtually before implementing them in reality by simulating a real-world agricultural system, thus reducing the risk and improving decision accuracy. They also facilitate real-time monitoring and feedback loops, allowing for dynamic adjustments to irrigation, fertilization, and pest control strategies. Digital twins can also combine with data on market information, contributing to more resilient and economically viable farming plans. In livestock farming, digital twins can monitor animal behavior and health, improving welfare and productivity through early anomaly detection. Digital twins are expected to play a central role in achieving precision, sustainability, and efficiency through the agricultural value chain as technology matures.

In crop cultivation, Xu et al. investigated the utilization of digital twins in agriculture via field tests and simulations of winter wheat growth (Xu et al. 2025). The study enhanced the management and monitoring protocols by integrating digital twin technology with crop growth models through a variety of experiments that simulated wheat cultivation. Drones and Internet of Things (IoT) sensors were employed to establish a digital twin ecosystem and consolidate data on wheat cultivation. The integration of digital twin technology and crop growth models may enable accurate simulation and intelligent management of wheat development processes. Skobelev et al. developed a Smart Plant Digital Twin (SPDT), an advanced software system for immediate monitoring and modeling of plant behavior, in accordance with actual plant development (Petr Skobelev et al. 2025). This system employed a multi-agent, multi-level model including crop physiology data and specialist expertise to forecast growth stages and resource requirements. An ontology structure facilitates diverse crops and genotypes, allowing for scalable modeling in the field or greenhouse. A broccoli plant in Taiwan was utilized, where agronomists employed sensor data and engaged with digital twins, validating the system's precision and practical utility in farming.

In cattle husbandry, the development of private animal digital twins has enabled the monitoring of health, nutrition, behavior, and environmental factors. Zhang et al. present a modality mapping completion network designed to translate imperfect sensor and video inputs, hence enhancing multimodal recognition of cow dairy activity in scenarios of modality loss (Yi Zhang et al. 2025). The method utilizes a multimodal behavior identification algorithm to recognize five distinct behaviors—drinking, feeding, reclining, standing, and walking—by mapping incomplete sensor or video data. The findings demonstrated enhanced reliability and versatility in recognizing cow behavior through multimodal data, especially in the presence of missing or partial information. It tackles tangible issues in creating digital twins for observing bovine behavior and provides pragmatic solutions for enhanced system resilience.

Digital twin technology additionally enhances greenhouse and indoor agriculture. Farmers can accurately replicate and change conditions for development by creating copies of numerous environmental factors such as CO_2 levels, humidity, and temperature in digital simulations, thereby enhancing yield while reducing energy and water consumption. This precision facilitates resource-efficient vertical agriculture and climate-regulated crop production. In contrast to traditional vertical farming systems that utilize static spectral recipes with predetermined illumination compositions, research by Alves et al. presents an Internet of Things-enabled smart vertical agricultural system that employs digital twin technology and a genetic algorithm (GA) to dynamically optimize lettuce growth by modulating RGB LED spectra during the crop cycle (Alves et al. 2025). In this study, the key environmental parameters within a growth tower, including temperature, humidity, and lighting, were monitored and controlled, and a digital twin that facilitates real-time data exchange between physical and virtual components was applied. At the same time, the GA iteratively refines the light composition, and the results pointed out the optimal RGB configuration. Another study developed a virtual acquisition method based on a virtual collection method for layer house temperature data based on the WOA-BiLSTM model when constructing a digital twin system (Yuchen Jia and Gao 2025).

Digital twins of farm equipment and machinery, such as tractors, harvesters, and irrigation pumps, are changing agricultural operations through smarter and more efficient maintenance and management (Cor Verdouw et al. 2021). It is possible for the farmers to continuously monitor mechanical performance, detect anomalies, and predict potential failures before they occur by creating virtual copies of each machine and then integrating with real-time data from onboard sensors. As a result, it can help to reduce sudden or unexpected breakdowns during critical farming periods, thereby minimizing downtime and ensuring smoother operations.

Additionally, digital twins allow for detailed tracking of equipment use, checking how many operating hours, fuel consumption, or how much remaining fuel is, providing insights into cost efficiency and the level of machinery wear. These data enable farmers to schedule more effectively and optimize the lifespan of each machine in their fields. Over time, this leads to significant cost savings and improved resource allocation. In large-scale farming where machinery costs and logistics have a significant impact, adopting digital twins ensures higher reliability, sustainability, and productivity. Ultimately, this technology supports smarter asset management, reducing both operational risk and long-term equipment depreciation.

The findings from released research and case analyses indicate that digital twins serve as a potent decision support tool in agricultural sectors, improving resource efficiency, early warning systems, and sustainability. Nevertheless, nearly all technologies are still in pilot or laboratory stages; enhancing infrastructure, standardizing data protocols, and scaling AI integration are essential for broad deployment. Digital twins have the potential to transform Agriculture 4.0 by providing intelligent, robust, climate-adaptive, and financially sustainable farming systems.

1.2.4 Challenges and Limitations

In the era of Smart Farming 4.0, new technologies, particularly digital twin technology, are poised to transform agricultural operations by improving decision-making, effectiveness, and sustainability. Within the larger framework of precision farming, digital twins have garnered interest for addressing difficulties including climate variability, resource scarcity, and supply chain inefficiencies. Their application encompasses soil health assessment, crop production forecasting, and machinery efficiency enhancement. However, despite these promising capabilities, the real practical implementation of digital twins in agriculture still remains several limitations, including high initial investment costs, data integration complexities, connectivity limitations in rural areas, and data security concerns.

The high initial investment cost is one of the most significant barriers that need to be considered to let digital twin technologies be widely applied in agricultural practices. As mentioned in the section below, the agricultural digital twin typically involves two components: physical and virtual. A combination of several sensors, edge devices, cloud infrastructure, and often customized software platforms is required to establish a functional digital twin, all of which demand substantial upfront capital. The expense of establishing a comprehensive digital twin model for a system for farming may provide a significant obstacle for small- and medium-sized farms that cannot embrace early (Cor Verdouw et al. 2021). Moreover, the expenses associated with hardwarc maintenance and ongoing software updates impose an economic cost for the ongoing utilization of the digital twin system. Thus, the integrated solutions between financial support mechanisms, public-private partnerships, and scalable solutions tailored to small farms are essential for inclusive adoption.

Data integration is another critical challenge. Agricultural components in the systems are heterogeneous, comprising various data sources coupled with different parameters, such as soil sensors, weather stations, drone imagery, satellite data, and IoT devices. Integrating these diverse data types into a unified digital twin framework is technically demanding and often hindered by the lack of standardized data formats and communication protocols. As emphasized by Jones et al., practical digital twin implementations require seamless integration across temporal and spatial data layers, which currently lacks industry-wide coordination in agriculture, and the integration between virtual entities is a problem that the digital twin will also have to address (David Jones et al. 2020).

Connectivity is another major challenge, especially in rural and remote agricultural areas. Reliable high-bandwidth internet connectivity is very important for real-time data collection, transfer, and synchronization between the physical and digital systems. However, many rural areas lack the necessary digital infrastructure to support these requirements of the digital twin system. Research by Wolfert et al. found that the infrastructure limitation often undermines the potential of digital agriculture. This problem is especially prevalent in developing countries or poor countries, where mobile or broadband networks are inconsistent or nonexistent (Sjaak Wolfert et al. 2017). In some regions, the precision and timeliness of transmitting information may be significantly compromised by delays or interruptions. Agriculturalists may encounter difficulties with sporadic signal coverage, hindering regular use of cloud-based platforms. Insufficient bandwidth and elevated latency can obstruct real-time applications, even in regions with fundamental connection. The absence of a dependable energy infrastructure exacerbates the implementation of sensor networks and IoT devices, hence constraining the feasibility and scalability of digital twin technologies in numerous farming contexts.

Data privacy and cybersecurity represent additional challenges that may hinder the implementation of digital twins in agriculture. Digital twins consolidate sensitive information, including proprietary agricultural procedures, yields, and financial data, making the preservation of data security and integrity a significant concern. Agriculturists continue to express concerns regarding the dissemination of their data to external service providers. The potential reasons include apprehensions regarding misuse, ambiguity in data ownership rights, and susceptibility to cyberattacks. Numerous farmers may feel uncertain regarding the ownership of data once it is published to external platforms. Concerns may arise when huge technology firms administer cloud-based digital twin systems, prompting inquiries over data monetization without consent. Furthermore, the intricacy of data-sharing contracts and the technical jargon employed in service contracts may deter adoption by non-expert consumers. The legal and ethical frameworks regulating data usage and ownership in agriculture remain underdeveloped in nearly all nations, rendering the trust difficulties justifiable and comprehensible.

The efforts on collaboration between stakeholders, such as policymakers, technology providers, and farmers, are necessary to overcome these challenges. While the benefits of digital twins in agriculture are well recognized, the main challenges must be systematically addressed to enable their widespread and equitable adoption in agricultural systems.

1.3 Importance of Sustainable Agriculture

Sustainable agriculture is increasingly recognized as a cornerstone for ensuring global food security, conserving environmental resources, and enhancing rural livelihoods. It is defined by the Food and Agriculture Organization (FAO) as the management and conservation of the natural resource base and the orientation of technological and institutional change in such a manner as to ensure the attainment and continued satisfaction of human needs for present and future generations ((FAO) 2014). Sustainable agriculture aims to balance environmental health, economic profitability, and social equity. The global population is predicted to reach nearly 10 billion by 2050 (Nations 2024), but the importance of sustainable agricultural systems has never been greater.

Unsustainable agricultural practices, for example, the excessive use of chemical fertilizers and pesticides in agriculture, the over-exploitation of water resources, and deforestation for land expansion, have caused significant environmental degradation and pollution. These practices have led to the degradation, water contamination, biodiversity loss, and increased greenhouse gas emissions. Soil degradation is a critical concern. Globally, an estimated 33% of soil is moderately to highly degraded due to erosion, nutrient depletion, acidification, and salinization ((FAO) 2015). Intensive tillage and chemical use reduce soil organic matter and microbial diversity, reducing soil fertility and productivity. It is reported that soil erosion due to unsustainable farming occurs at a much higher rate and far exceeds the natural soil regeneration rate, threatening the long-term viability of agriculture (Montgomery 2007).

Another prominent environmental effect is biodiversity loss. Modern industrial agriculture favors a narrow range of high-yielding crops, leading to genetic erosion. Furthermore, habitat destruction for agricultural expansion threatens native flora and fauna. It is highlighted that agricultural intensification contributes significantly to species decline, particularly in developing regions with rich biodiversity (Teja Tscharntke et al. 2012).

Water overuse and pollution also exemplify the negative externalities of unsustainable agriculture. It is reported that irrigated agriculture accounts for 70% of global freshwater withdrawals (Nations 2015). In many regions, this has led to the depletion of aquifers and surface water bodies. Additionally, nutrient runoff from fertilizers contributes to eutrophication in aquatic ecosystems, causing dead zones such as those in the Gulf of Mexico (Diaz and Rosenberg 2008).

Unsustainable agriculture not only harms the natural environment but also poses a risk to the socio-economic environment. Smallholder farmers, particularly in developing countries, are often trapped in cycles of debt due to reliance on expensive agrochemicals and high-yielding seeds. Land degradation reduces productivity, leading to lower incomes and food shortages.

In contrast, sustainable agriculture can promote economic resilience through diversified cropping, reduced input dependence, and improved resource efficiency. Agroecological practices such as crop rotation, intercropping, organic amendments, and integrated pest management contribute to long-term yield stability (Altieri et al.

2015). Importantly, the sustainable agriculture system also enhances resilience to climate variability, a critical feature in the context of increasing weather extremes.

The need for sustainable agriculture is well-articulated in international frameworks, including the United Nations Sustainable Development Goals (SDGs), particularly Goal 2 (Zero Hunger) and Goal 12 (Responsible Consumption and Production). To achieve these objectives, governments must facilitate the shift from industrialized monocultures to diversified, ecosystem-oriented strategies. Successful case studies serve as compelling evidence for this assertion. The ABC Program in Brazil assists farmers in implementing sustainable methods, including no-till farming, forestry, and pasture management. In Europe, the Common Agricultural Policy (CAP) has commenced the integration of "green architecture" initiatives, encompassing agricultural environmental actions, eco-schemes, and conditional subsidies to enhance biodiversity and resource efficiency.

Sustainable agriculture is not simply an environmental or economic choice; it is essential for safeguarding food systems in the twenty-first century. Sustainable agriculture, which mitigates ecological deterioration and fosters resilience, provides a pathway to a more equitable and prosperous future. Policy frameworks, scientific innovation, commercial incentives, and community engagement must converge to surmount existing obstacles and achieve the complete potential of sustainable agriculture systems.

Agricultural operations provide the main source of global greenhouse gas emissions, degradation of land, acidification, eutrophication, and various other social and economic issues. Recent technical breakthroughs have drastically modified agricultural systems, resulting in a considerable increase in production capacity. These innovations also play a crucial role in ensuring the supply of food and providing the raw materials essential for industrial growth. Achieving the sustainable transformation must align with international policy frameworks and societal ambitions, such as the Green Deal, Paris Agreement, Sustainable Development Goals, and the Farm-to-Fork Strategy. In modern agriculture, the advancement steadily reduces the need for a huge number of laborers by replacing traditional farming tools and practices with stronger and efficient types of machines. Moreover, technology has enabled the effective automation of various support functions in farming. The digital transformation of agriculture through approaches like smart farming, precision livestock farming, and, most recently, the adoption of digital twins offers practical tools to deal with the above-mentioned issues. Utilizing advanced technology, digital twins provide real-time monitoring, system modeling, and process control through intricate digital replicas and ongoing bidirectional data sharing. Digital technologies have undoubtedly integrated into the life of farmers throughout numerous places globally. It is essential to emphasize that if the extensive socio-ecological repercussions of these technological uses are not meticulously managed, the augmented automation linked to digital twins may exacerbate systemic fragility and render agricultural systems further vulnerable to disruptions, especially due to global warming.

Conventional agriculture using labor-intensive practices has undergone a significant transformation recently with the participation of modern technologies. This shift has introduced increasing levels of complexity, driven by the rapid adoption of advanced and innovative technologies, most notably precision agriculture and various digital farming tools (Athanasios Balafoutis et al. 2017). Despite the advantages, deploying such technologies often requires substantial investments in many parts, such as hardware, software, supporting infrastructure, and specialized technical expertise for effective utilization. Moreover, agricultural activities operate in a dynamic environment shaped by strong climate change, evolving consumer demands, and volatile market conditions, further challenging modern farmers. Agriculture faces climate change, resource depletion, population growth, and evolving consumer preferences, which threaten productivity and affect crop yields, livestock health, and food system stability. Nevertheless, there are always two sides to the problem, and these complexities simultaneously present many opportunities for innovation, knowledge generation, and sustainable growth.

1.4 Conclusion and Perspective

In this chapter, the evolution of agricultural practices has been explored in depth, illustrating how the sector has moved from manual labor-intensive practices to highly digitalized, intelligent farming systems. Every milestone in this revolution has enhanced efficiency and redefined the environmental, social, and economic underpinnings of agriculture. The transition to sustainable agriculture is accelerating due to advancements in digital technologies, signifying a pivotal moment in contemporary agricultural growth. Among these technologies, digital twins have emerged as one of the most promising advances, providing a new framework for modeling, simulating, and improving agricultural systems. Through the establishment of virtual replicas of physical entities in agricultural processes, encompassing soil, crops, cattle, and machinery, digital twins enable contemporary farmers to make informed decisions based on data, mitigate risks, and improve productivity while conserving natural resources.

The application of digital twins in agriculture supports multiple domains, such as crop modeling, greenhouse monitoring, livestock management, and machinery optimization, and so on. Despite these advancements, digital twin technology in agriculture is still in its early stages of deployment. Most systems remain in pilot phases, and several technical and socio-economic limitations constrain their scalability to broader, real-world applications. These include the high initial cost of implementation, the need for infrastructure, difficulties in integrating heterogeneous data sources, inconsistent connectivity, especially in rural areas, and persistent concerns regarding data privacy and ownership. Addressing these issues requires technical solutions, institutional reforms, and multi-sector collaboration.

Looking ahead, the future of digital twins in agriculture lies in their integration with emerging innovations such as edge computing, blockchain, machine learning, and next-generation communication networks (e.g., 5G and LPWAN). These developments could significantly reduce latency, enhance system scalability, and enable even more granular, real-time feedback. Multi-layered digital twins, incorporating genomic, environmental, and socio-economic data, may provide holistic insights for precision breeding, farm planning, and policy-making. Additionally, co-innovation with farmers, agronomists, and local communities will ensure that digital twin systems are designed to reflect on-the-ground realities and knowledge systems. For example, the participant or involvement approaches can be the solutions to local needs while fostering trust, ownership, and long-term engagement.

It is also important to emphasize that technological advancement alone is not enough to achieve sustainability in agriculture. Digital twins must be developed within inclusive, equitable, and ethically grounded frameworks. Furthermore, the in-depth evaluation of the long-term environmental and social impacts of this technology is necessary to prevent unintended consequences, such as over-reliance on automation or ecological oversimplification.

In conclusion, digital twin technology represents a transformative step forward in the journey toward Sustainable Farming 4.0. It offers a powerful decision-support framework capable of navigating the complexities of modern agriculture while optimizing productivity and sustainability. Yet, realizing the full potential of this technology will require systemic efforts to overcome technical, economic, and institutional barriers. With thoughtful implementation and inclusive governance, digital twins can help reimagine agriculture as a high tech, environmentally responsible, and socially just sector, capable of feeding the global population while protecting the planet for future generations.

References

Altieri MA, Nicholls CI, Henao A, Lana MA (2015) Agroecology and the design of climate change-resilient farming systems. Agron Sustain Dev 35:869–890

Alves RG, Lima F, Guedes ÍM, Gimenez SP (2025) Dynamic light optimization in vertical farming using an IoT-driven digital twin framework and artificial intelligence. Appl Soft Comput 174: 112985

Athanasios Balafoutis BB, Fountas S, Vangeyte J, Van Der Wal T, Soto I, Gómez-Barbero M, Barnes A, Eory V (2017) Precision agriculture technologies positively contributing to GHG emissions mitigation, farm productivity and economics. Sustainability 9(8):1339

Cor Verdouw BT, Beulens A, Wolfert S (2021) Digital twins in smart farming. Agric Syst 189: 103046

David Jones CS, Nassehi A, Yon J, Hicks B (2020) Characterising the digital twin: a systematic literature review. CIRP J Manuf Sci Technol 29:36–52

Diaz RJ, Rosenberg R (2008) Spreading dead zones and consequences for marine ecosystems. Science 321:926–929

(FAO), FAAO (2014) Building a common vision for sustainable food and agriculture

(FAO), FAAO (2015) Status of the world's soil resources

Grieves M (2005) Product lifecycle management: the new paradigm for enterprises. Int J Prod Dev 2(1/2):71–84

Grieves M (2006) Product lifecycle management: driving the next generation of lean thinking. McGraw Hill

Grieves M (2011) Virtually perfect: driving innovative and lean products through product lifecycle management. Space Coast Press

Khan MR, Crabbe RA, Malik NA, O'Meara L (2023) Applications of geospatial technologies for precision agriculture. In: Precision agriculture – evolution, insights and emerging trends. Academic

Marc Escribà-Gelonch SL, Van Schalkwyk P, Fisk I, Van Duc N, Long VH (2024) Digital twins in agriculture: orchestration and applications. J Agric Food Chem 72:10737–10752

Montgomery DR (2007) Dirt: the erosion of civilizations. University of California Press

Nations, DOEASA-U (2024) World Population Prospects 2024

Nations, U (2015) UN World Water Development Report 2015

Nawab Khan RLR, Zhang S, Osabuohien E, Ihtisham M (2022) Influence of mobile phone and internet technology on income of rural farmers: evidence from Khyber Pakhtunkhwa Province, Pakistan. Technol Soc 68:101866

Petr Skobelev AT, Simonova E, Zhuravel Y, Galitskaya A (2025) Multi-agent digital twin of broccoli: development and test evaluation. Procedia Comput Sci 252:674–683

Sjaak Wolfert LG, Verdouw C, Bogaardt M-J (2017) Big data in smart farming – a review. Agric Syst 153:69–80

Teja Tscharntke YC, Wanger TC, Jackson L, Motzke I, Perfecto I, Vandermeer J, Whitbread A (2012) Global food security, biodiversity conservation and the future of agricultural intensification. Biol Conserv 151:53–59

Usman S (2024) A review on computer applications in agricultural science. Comput Inform 4:65–79

Warren Purcell TN (2023) Digital twins in agriculture: a State-of-the-art review. Smart Agric Technol 3:100094

Xu X, Gao F, Xiong D, Fan Z, Xiong S, Dong P, Qiao H, Ma X (2025) Digital twin-based winter wheat growth simulation and optimization. Field Crop Res 329:109953

Yi Zhang YZ, Gao M, Wang X, Dai B, Shen W (2025) Multimodal behavior recognition for dairy cow digital twin construction under incomplete modalities: a modality mapping completion network approach. Artif Intell Agric 15:459–469

Yuchen Jia LL, Gao L (2025) Research on virtual collection method of layer house temperature for the construction requirements of digital twin system. Poult Sci 104:104771

Importance of Sustainable Agriculture: Focus on Climate-Smart Sustainable Agriculture

2

Nafisa N. Chowdhury and Shama E. Haque

Abstract

Sustainable agriculture is important in ensuring global food security. Unlike conventional farming, which often depends heavily on chemical inputs, machinery, and nonrenewable resources, sustainable practices prioritize sustainability, efficient resource use, ecological balance, and resilience against environmental stresses. Traditional farming methods contribute significantly to greenhouse gas emissions as well as a number of issues including deforestation, soil degradation, water contamination, and biodiversity loss. In contrast, sustainable agricultural systems promote crop rotation, organic soil management, and agroforestry, which restore ecosystem health. Moreover, sustainable agriculture emerges as a vital strategy to meet global food demands without depleting natural resources. This chapter examines the core principles and benefits of climate-smart sustainable agriculture, highlighting its contribution to the United Nations Sustainable Development Goals.

Keywords

Agricultural productivity · Climate-smart agriculture · Environmental sustainability · Resource conservation · Pollution reduction

N. N. Chowdhury
Centre for Climate Change and Environmental Research, BRAC University, Dhaka, Bangladesh

S. E. Haque (✉)
North South University, Bashundhara, Dhaka, Bangladesh
e-mail: shama.haque@northsouth.edu

© The Author(s), under exclusive license to Springer Nature Singapore Pte Ltd. 2026

A. L. Srivastav et al. (eds.), *Digital Twin Technology for Sustainable Agriculture*,
https://doi.org/10.1007/978-981-95-5915-2_2

2.1 Introduction

Modern agricultural practices, while highly effective at increasing crop yields, have also emerged as major drivers of environmental degradation. Their dependence on intensive chemical inputs and extensive land modification contributes to greenhouse gas (GHG) emissions, deforestation, soil fertility loss, water pollution, and declines in biodiversity (Li et al. 2025; Chowdhury et al. 2022; Tubiello 2019). In 2020, the global agri-food sector produced an estimated 16 billion tons of carbon dioxide equivalent (CO_2eq), indicating a 9% increase following 2000 (FAO 2022). Agriculture combined with land-use change is responsible for roughly 25% of worldwide emissions and in some developing nations, this figure exceeds 50% (Tubiello 2019). Recent climate data reinforce the urgency. The U.S. Environmental Protection Agency (2025) reported that from 1990 to 2023, the global warming potential of GHG emissions surged by 51%.

Additionally, the Intergovernmental Panel on Climate Change (2018) warns there is more than a 50% chance of global temperatures rising by at least 1.5 °C between 2021 and 2040 under all climate projection scenarios. Such warming has already intensified the frequency and intensity of natural disasters (Haque and Chowdhury 2025; Haque 2023), leading to reduced crop yields across several regions (Akpa 2024; Haque et al. 2023). In addition, climate change indirectly damages agriculture through the deterioration of infrastructure, farm equipment, and natural resources (Saleem et al. 2024; Charvériat 2012).

Bangladesh, one of the most climate-vulnerable nations in South Asia (Haque and Nahar 2023), demonstrates the magnitude of the challenge. By 2050, the country is projected to lose 17% of its land area owing to sea level rise, eliminating nearly 30% of its cultivable land and threatening food security (TDS 2024). Incorporating sustainable practices into national agriculture is therefore of utmost importance for both climate mitigation and adaptation. Sustainable agriculture promotes ecological balance, efficient use of resources, and resilience against climate shocks, all of which are essential for ensuring long-term food security, especially in vulnerable nations (Pandian et al. 2024; Zhang 2024; Çakmakçı et al. 2023; FAO 2018). This chapter focuses on the principles and advantages of sustainable and climate-smart agriculture, highlighting their role in advancing the United Nations Sustainable Development Goals (UN SDGs).

Section 2.1 of the present chapter outlines the background of the study and positions sustainable agriculture within the broader climate-smart agriculture (CSA) framework. Section 2.2 describes the research methodology, where Sect. 2.2.1 reviews insights from existing literature, and Sect. 2.2.2 presents details of the study area in climate-vulnerable Bangladesh. Section 2.3 presents environmental health and sustainable agricultural techniques, with Sects. 2.3.1 and 2.3.2 detailing soil health management and water-use efficiency, respectively. Next, Sect. 2.4 presents global sustainable agricultural techniques; Sects. 2.4.1 through 2.4.6 detail the various sustainable agriculture techniques including Integrated Pest Management (IPM), intercropping, and agroforestry, respectively. Section 2.5 discusses CSA practices of Bangladesh, and Sect. 2.6 outlines the limitations of sustainable agricultural methods. Next, Sect. 2.7 discusses the role of sustainable agriculture in achieving the UN SDGs and policies. Lastly, Sect. 2.8 draws the conclusions of the study.

2.2 Methodology

The study uses a mixed-method approach, blending both qualitative and quantitative techniques to understand the recent advances in sustainable agriculture. The methodology is structured around two key components—(i) literature review and (ii) field surveys.

2.2.1 Literature Review

In this chapter, a systematic review identified 86 scholarly literatures from various databases (e.g., Google Scholar, Web of Science, ScienceDirect, and Scopus). Keywords including 'agricultural practices', 'climate-resilient agriculture techniques', 'sustainable agricultural', 'climate-smart', and 'environmental pollution' were utilized, focusing on highly cited research published between 1982 and 2025 to ensure advancements and emerging trends. Table 2.1 provides a summary of selected recent studies conducted on sustainable agricultural issues between 2017 and 2025.

Table 2.1 Summary table of key findings of recent articles on sustainable agriculture

Author(s)	Area of research	Key findings
Kabato et al. (2025)	Focuses on CSA as a strategy to improve agricultural productivity, resilience to climate change, and curb GHG emissions	The findings indicate that combining multiple CSA practices is more effective than using single approaches. However, adoption is limited by financial, technical, and institutional barriers, especially in developing countries. Custom-made, site-specific strategies and stronger policy support are crucial for broader implementation.
Mabhaudhi et al. (2025)	Explores the integration of Weather and Climate Information Services (WCIS) with CSA practices, particularly in the context of agricultural water management among farmers with a focus on Africa.	The study concluded that tailored climate information can enhance the timing and effectiveness of CSA decisions. The findings further indicate that weather smart practices were highly associated with drought, temperature, and precipitation patterns, whereas nitrogen smart practices were connected to the end of precipitation and temperature.
Al-Imran et al. (2025)	Investigated impacts of floating bed agriculture versus traditional soil-based farming on the growth and yield of chili pepper varieties in waterlogged regions of Bangladesh, under varying seasonal conditions.	Determined that positive effects of floating beds are linked to improved moisture retention, nutrient availability, and pH stability. Floating beds significantly improved shoot morphology, root development, and reproductive traits of chili plants during the winter season, nonetheless had minimal impact in summer months.

(continued)

Table 2.1 (continued)

Author(s)	Area of research	Key findings
Hiywotu (2025)	Focuses on advancing sustainable agriculture for UN SDG 2 through practices, policies, and technologies.	The study emphasizes the necessity for integrated strategies in agriculture, exploring emerging technologies such as precision agriculture, agroforestry, and digital platforms to lessen environmental impacts while promoting sustainability.
Pandey and Mishra (2024)	Focuses on the importance of artificial intelligence (AI) in improving food production and distribution efficiency	The findings indicate that AI offers transformative potential in agriculture by enhancing precision farming, crop monitoring, and disease detection, while optimizing supply chains, storage, and curbing post-harvest losses.
Bouri et al. (2023)	Explores climate-smart pest management and its role in addressing weather-driven phytosanitary issues.	CSA emphasizes early detection and management of climate-related crop threats, with disease control central to sustainability. Despite challenges in implementation and coordination, strengthening research, surveillance, and capacity building enhance climate-smart pest management's effectiveness.
Coulibaly et al. (2021)	Focuses on adoption of sustainable agricultural practices	The findings suggest farmers' adoption of sustainable agricultural practices is influenced through various factors, including self-identity, moral customs, and risk perception. Farmers may intend to adopt sustainable agricultural practices but often lack intention to take action.
Sarvade et al. (2019)	Discusses soil and water conservation techniques that are crucial for promoting sustainable agriculture in India.	India faces severe natural resource degradation, with 146.8 Mha of land affected, primarily owing to soil erosion causing 5.3 billion tons of annual loss and reducing reservoir capacity. Intensive farming coupled with over-extraction of groundwater has led to substantial depletion, impacting crop productivity. The findings indicate that addressing the issue requires integrated soil and water management, improved practices along with supportive conservation policies.
Blaser et al. (2018)	Discusses the role of agroforestry in balancing production with sustainability goals.	In West Africa, low-to-intermediate shade cocoa systems maintain yields while enhancing climate and biodiversity benefits. Beyond ~30% canopy cover, these benefits decline, highlighting the need to optimize shade levels.
Sikka et al. (2018)	Focused on the role of climate-smart land and water management practices to achieve sustainable agriculture in India.	The influences of changing climate considerably impact agriculture and future food security in India. Low-cost, participatory land and water interventions have shown promise in building resilience, highlighting the need for scalable solutions backed by effective policies.

2.2.2 Field Survey

A field survey was conducted in the unions of Char Duanti and Patharghata Sadar under Patharghata Upazila (between 21°58′ and 22°14′ N latitudes and between 89° 53′ and 90°05′ E longitude; Fig. 2.1), a coastal subdistrict in Barguna District, Bangladesh. Patharghata Upazila occupies an area of 387.36 km^2 including 37.2 km^2 forest areas (Rahim et al. 2018). Patharghata has low-lying coastal plains geography as a result the area is particularly susceptible to climate-induced natural disasters. Specifically, the land outside the embankments becomes non-arable owing to elevated levels of salinity, and arable land within the embankments suffers during flash flood events due to saltwater infiltration (Rahaman 2023). The coastal community comprises of ultra-poor residents who heavily rely on agriculture for livelihood that is severely impacted by frequent and intense climatic disasters (Tiller 2019; Rahim et al. 2018).

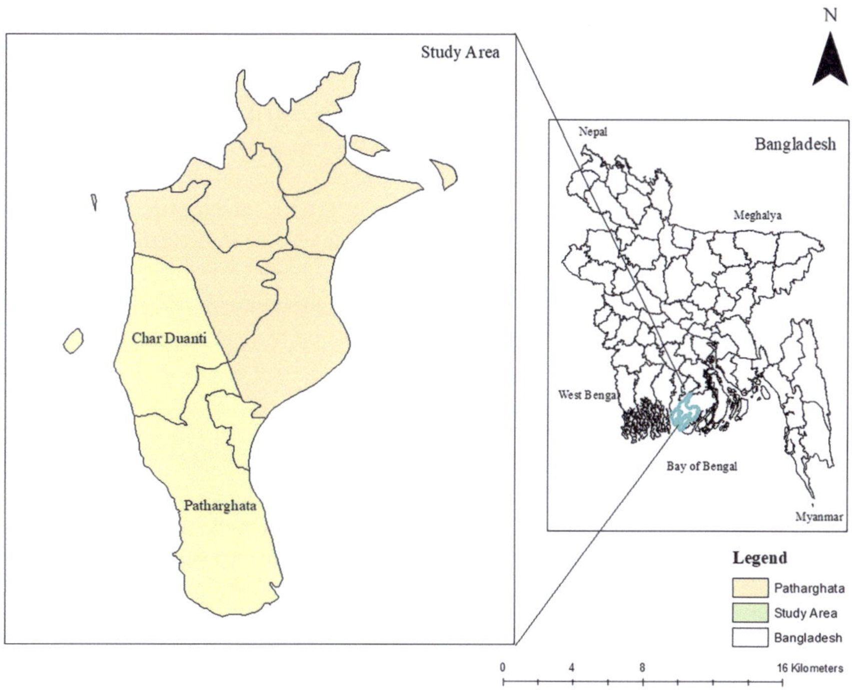

Fig. 2.1 Map of Bangladesh showing location of the study area in Barguna District

2.3 Environmental Health and Sustainable Agricultural Techniques

2.3.1 Soil Health Management

An important strategy in sustainable agricultural techniques is soil health management through crop rotation, diverse crop cultivation, organic nutrient management, and agroforestry (Barman et al. 2022; Tahat et al. 2020). Crop rotation involves the sequential cultivation of various crop species on the same piece of land, which reduces soil erosion, enhances soil fertility, and improves overall crop productivity. Crop diversification involves growing more than one crop in an area that results in pest, disease, and weed control, and promotes soil fertility (Volsi et al. 2022). Conservation tillage retains crop residues on the surface of the soil to minimize soil susceptibility to erosion by the action of water and wind. Cover cropping between cash crop cycles involves growing any crop to cover the soil. Later, the cover crop may be added to the soil for enrichment. These practices enhance organic matter content and microbial activity that are key for nutrient cycling and long-term productivity (Srivastava et al. 2024).

2.3.2 Water-Use Efficiency in Sustainable Agriculture

Global food production must rise by 70% in order to meet the demand for food by 2050 (Velasco-Muñoz et al. 2018; Wu and Ma 2015), which will require either expanded cultivation, intense use of existing farmland, or integrated management techniques (Velasco-Muñoz et al. 2018; Wu and Ma 2015). This may result in a 53% increase in water usage and a 38% expansion in agricultural land in low-output scenarios. Additionally, irrigation demand may rise by 50% in developing and 16% in developed countries (De Fraiture and Wichelns 2010; Fischer et al. 2007). Such intensification risks biodiversity loss, soil degradation, and ecosystem disruption. Additionally, the influences of changing climate threaten water availability and disrupt agricultural practices through altered precipitation, extreme weather events as well as altered soil condition and surface runoff (Yuan et al. 2024). Past research indicates that drip irrigation, rainwater harvesting, and mulching are promising agricultural practices that can substantially reduce water consumption quantities (Zhang et al. 2021; Nouri et al. 2019). For instance, in the arid regions of northwest China, rainwater harvesting increased soil moisture in the 0–200 cm profile by 13.7%, whereas advanced drip irrigation techniques (e.g., partial root-zone drying and infiltration pipes) increased water usage efficiency by 17.2%.

2.4 Global Sustainable Agricultural Techniques

2.4.1 Climate-Smart Agriculture

Food security is threatened by the influences of global climate change. Such threats can be minimized by increasing the adaptive capacities of farmers, and increasing resilience and resource use efficiency in agricultural production systems. Climate-smart agriculture is an approach that aims to transition and reorient agricultural systems to support food security for an estimated nine billion people by 2050 under the various scenarios of changing global climate. Note that multiple definitions of CSA are available in published literature. For example, according to Steenwerth et al. (2014), CSA tackles the problem of supplying the increasing demand for food, fiber, and fuel, regardless of the effects of climate change and the limited opportunities to increase the amount of arable land that is now available. A review of 13 CSA-related documents by Chandra et al. (2018) indicates that CSA is primarily centered on three pillars: (i) mitigation; (ii) adaptation; and (iii) food security. However, the findings indicate that the approach can be interpreted differently across global, developing, and developed contexts. These scientists found that CSA-related research increasingly focuses on smallholder agriculture in developing countries, while studies from developed nations remain limited.

2.4.1.1 Climate-Smart Agriculture in Bangladesh

In Bangladesh, CSA strategies are important in managing the impacts of saltwater intrusion, which is projected to reach 8 km inland by 2030 (The CIAT; World Bank 2017). Currently, the country has CSA technologies available that are actively applied to promote sustainable agricultural production and increase resilience against the influences of changing climate (Chaki et al. 2023). Locally, CSA systems include innovative practices such as vertical and rooftop gardens, floating gardening, and homestead cultivation. Specifically, floating gardening, utilizing aquatic plants as the medium for plant growth, is a traditional/indigenous cultivation system in the south and southwestern parts of the country (Al-Imran et al. 2023; Irfanullah et al. 2008). Furthermore, effective interventions include traditional methods such as the Sorjan system, which utilizes a series of raised beds for upland crops and ditches for wetland crops, enabling simultaneous cultivation (Domingo and Hagerman 1982).

2.4.2 Integrated Pest Management

Integrated Pest Management is a relatively new dimension in the development of sustainable agricultural system that minimizes pest impact by combining methods such as biological control, habitat alteration, cultural practices, and use of resistant crop varieties, thereby reducing reliance on chemical pesticides. Romeh (2018) indicates that while biological control is vital to IPM, employing these methods can cut pesticide use by up to 50% compared to routine spraying. Pest-resistant varieties offer added benefits such as low cost, reduced insecticide use, improved safety, support for natural enemies, and ease of adoption in the field. In India, currently, 35 Central IPM Centers have been established across 28 states and 2 Union Territories to promote the adoption of IPM (CEEW 2025). Here, 3–5% (4.2–7 million hectares; CEEW 2025) of the country's total cultivated area is under IPM, which employ a range of approaches such as cultural practices, host plant resistance, mechanical and physical methods, biological control, behavioral strategies, regulatory measures along with need based chemical interventions (Karuppuchamy and Venugopal 2016).

2.4.3 Agroecological Crop Protection

Agroecological Crop Protection offers a sustainable, ecology-based approach by combining various approaches and disciplines, including ecology, agroecology, organic farming, permaculture (i.e., permanent agriculture), and IPM (Yarahmadi and Rajabpour 2024; Deguine et al. 2023). This approach involves optimizing interactions among plant, animal, and microbial communities both above and below ground, within and neighboring agroecosystems. Agroecological Crop Protection represents a major departure from conventional practices as it considers broader goals such as sustainable food systems and the interconnections between soil, plant, animal, and human health, while contributing to several SDGs.

2.4.4 Intercropping and Agroforestry

As the influences of climate change intensifies, it may surpass the adaptive capacity of existing systems. Adaptive capacity refers to an ecosystem's ability to withstand climate change impacts with minimal disruption to its functions and to adjust its resilience in response to environmental changes (Angeler et al. 2019). Intercropping, the relay or simultaneous cultivation of multiple crops, and agroforestry-based farming method, which integrates trees on at least 10% of farmland, are sustainable approaches that increase productivity as well as agricultural resilience (Burgess et al. 2022). Both intercropping and agroforestry offer practical alternatives to monoculture, reduce reliance on synthetic chemicals, promote biodiversity, and improve system resilience. Agroforestry is an afforestation activity and an intensive land management technique. This technique increases the advantages from the biological interactions resulting from intentional combination of trees and/or shrubs with crops and/or livestock (Association for Temperate Agroforestry 2021). Additionally, agroforestry is known to promote soil resilience to drought by increasing soil organic carbon, fixing nitrogen, and encouraging greater diversity and abundance of soil microbial communities (Muchane et al. 2020).

2.4.5 Precision Agriculture

Precision agriculture marks a major transition from conventional agricultural systems owing to its utilization of modern techniques (e.g., Global Positioning Systems, Geographic Information Systems, various sensors, yield monitors, and targeted management to enhance crop yield) while reducing consumption of natural resources from the planning to post-harvest processing (Karunathilake et al. 2023). This approach enables farmers to evaluate changes in crop health, soil composition, and variation in soil moisture levels to make informed decisions in planting, fertilizing, and harvesting, promoting more sustainable and profitable farming (Sanyaolu and Sadowski 2024; Case-Cohen 2018).

Afroj et al. (2016) indicate that a developing country such as Bangladesh can benefit from precision agriculture due to its high propensity to be adversely impacted by the influences of climate change. However, financial limitation may apply for precision farming practices in the country as some of the technologies are relatively expensive and advanced to follow through for the existing set-up.

2.4.6 Sustainable Intensification

Sustainable intensification aims to improve resilience to stresses, reduce input dependency, enhance biodiversity, and sustain ecosystem services without expanding further into non-agricultural land (Garibaldi et al. 2017; Pretty and Bharucha 2014). Petersen and Snapp (2015) conducted a study where 30 agricultural experts found that sustainable intensification is often misunderstood and seen as similar to current practices. Concerns include the vague definition of sustainable and insufficient focus on biological processes. On the other hand, Kamau et al. (2023) note that though there are ongoing questions over the specific practices and conditions that make sustainable intensification effective, diversification stands out as a promising approach.

2.5 Climate Smart Agriculture in Bangladesh

2.5.1 Climate Resilient and Sustainable Agriculture Techniques in Bangladesh

Bangladesh, due to its diverse geography and susceptibility to multiple climate hazards, is categorized into 11 distinct climate stress areas, as identified in the country's National Adaptation Plan (Table 2.2). Each of these regions experiences different climatic threats, ranging from cyclones and storm surges to floods, droughts, and salinity intrusion. As a result, farmers and local communities have developed and adopted sustainable, climate-resilient practices that are specifically tailored to the environmental pressures and agro-ecological conditions of their respective regions.

Table 2.2 Climate stress areas of Bangladesh

Climate stress area	Districts	Prominence of climate hazards
South-western coastal area and Sundarbans (SWM)	Satkhira, Khulna, Bagherhat, Pirojpur, Barguna, Barisal, Patuakhali, Jhalokhathi, Bhola, Shariatpur, Gopalganj, Jashore	Rainfall variability, river floods, sea level rise, salinity, tropical cyclone, storm surges, drought, extreme heat waves, extreme cold, riverbank erosion, and lightning
South-east and eastern coastal area (SEE)	Noakhali, Feni, Lakshmipur, Chattogram, Cox's Bazar, Chandpur	Rainfall variability, river floods, sea level rise, salinity, tropical cyclone, storm surges, drought, extreme heat waves, extreme cold, riverbank erosion, lightning, and landslides
Chattogram Hill Tracts (CHT)	Rangamati, Khagrachari, Bandarban	Rainfall variability, flash floods, tropical cyclone, storm surges, drought extreme heat waves, extreme cold, lightning, and landslides
Rivers, floodplains, and erosion-prone areas (FPE)	Nilphamari, Kurigram, Lalmonirhat, Gaibandha, Rangpur, Bogura, Sirajganj, Pabna, Rajshahi, Jamalpur, Tangail, Manikganj, Dhaka, Munshiganj, Mymensingh, Sunamganj, Netrokona, Habiganj, Kishorganj, Sylhet, Brahmanbaria, Narsingdi, Narayanganj, Rajbari, Faridpur, Madaripur, Gopalganj, Narail, Sariatpur, Barisal, Patuakhali, Bhola, Jhalokathi, Khulna, Chandpur, Cumilla, Noakhali, Lakshmipur, Cox's Bazar	Rainfall variability, river floods, tropical cyclones, tornado, extreme heat waves, extreme cold, riverbank erosion, and lightning
Haor and flash flood areas (HFF)	Sunamganj, Netrokona, Habiganj, Kishorganj, Sylhet, Maulvibazar, Brahmanbaria	Rainfall variability, flash floods, tropical cyclone, tornado, extreme heat waves, intense cold, riverbank erosion, lightning, and landslides
Drought-prone and Barind areas (DBA)	Naogaon, Chapai Nawabganj, Rajshahi, Bogura, Joypurhat, Rangpur, Dinajpur, Meherpur, Chudanga, Kushtia, Jashore, Magura, Jhenaidah	Rainfall variability, tropical cyclone, tornado drought, extreme heat waves, extreme cold, and lightning
Northern, north-western region (NNW)	Panchagarh, Thakurgaon, Nilphamari, Lalmonirhat, Rangpur, Kurigram, Dinajpur	Rainfall variability, river floods, flash floods, tropical cyclone, tornado, drought, extreme heat waves, extreme cold, riverbank erosion, lightning, and landslides

(continued)

Table 2.2 (continued)

Climate stress area	Districts	Prominence of climate hazards
Chalan beel and low-lying area of the north-western region (CBL)	Pabna, Natore, Sirajganj	Rainfall variability, river floods, tropical cyclone, tornado, extreme heat waves, extreme cold, riverbank erosion, and lightning
Char and Islands (CHI)	Nilphamari, Lalmonirhat, Kurigram, Gaibandha, Sirajganj, Jamalpur, Mymensingh, Manikganj, Munshiganj, Shariatpur, Chandpur, Bhola, Patuakhali, Feni, Noakhali, Lakshmipur, Chattogram, Cox's Bazar	Rainfall variability, river floods, sea level rise, salinity, tropical cyclone, tornado, storm surges, extreme heat waves, extreme cold, river bank erosion, lightning, higher sea surface temperature, and ocean acidification
Bay of Bengal and ocean (BoB)	Bay of Bengal (maritime boundary)	Rainfall variability, sea-level rise, tropical cyclone, tornado, storm surges, extreme heat waves, lightning, higher sea surface temperature, hypoxia, and ocean acidification
Urban areas (URB)	43 cities	Rainfall variability, urban floods, sea level rise, salinity, tropical cyclone, storm surges, drought, extreme urban heat waves, extreme cold, and lightning

National Adaptation Plan, Bangladesh

Recurrent cyclones and storm surges are prominent in the southwestern coastal zone that in turn intensify salinity intrusion, severely affecting crop productivity. To tackle these challenges, farmers have shifted to salt-tolerant rice varieties such as Bangladesh Rice Research Institute (BRRI) dhan47, Binadhan-10, and BRRI dhan61, which deliver higher yields in saline soils (Rabbani et al. 2013; Haque et al. 2019). Farmers have also diversified into non-traditional crops such as sunflowers, which are well-suited to saline conditions and require less water. Sunflower cultivation has become particularly attractive during the winter season, when many staple crops perform poorly, offering households a reliable source of food and income (Mongabay 2024).

With the Integrated Floating Cage Aquageoponics System, farmers can simultaneously produce fish and vegetables in shaded or waterlogged ponds. This system has specifically improved household nutrition and diversified incomes in areas of southwestern Bangladesh (Haque et al. 2015). Field observation indicates that in the study area, farmers have transitioned to duck farming, which is effective as pest control and well-adapted to the wet conditions left by the aftermath of cyclones (Fig. 2.2; Momin 2022). Additionally, in flood-prone regions, such as Sunamganj and Kishoreganj, that have been severely impacted by flash flood events (Haque et al. 2024; Sultana 2024), farmers have adopted floating gardens that require

Fig. 2.2 Current CSA practices in Patharghata and Char Duanti Unions

constructing organic beds from water hyacinth, bamboo, and composted material to grow vegetables on water surfaces. The beds allow cultivation during flood periods, and contribute to soil regeneration and biodiversity enhancement.

Alternate Wetting and Drying (AWD) is a deficit irrigation method that is practiced in such dry areas. While traditional rice cultivation method requires significant amount of water, in this process, the soil is allowed to dry out for a period of one to several days (Lampayan et al. 2015). During this stage, even though the fields appear dry on the surface, roots of rice crops still receive enough moisture to grow. The BRRI first introduced this AWD technique in 2005 (Pearson et al. 2018). More recently, Rana et al. (2023) tested automated AWD systems using soil moisture sensors and digital timers, showing that they can reduce irrigation water use by up to 20% compared to traditional manual methods. Beyond conserving water, AWD also offers significant environmental benefits: it can lower greenhouse gas (GHG) emissions and reduce arsenic buildup in rice grains by nearly 50% (Runkle et al. 2018), a crucial advantage given that much of Bangladesh's groundwater is already contaminated with arsenic (Haque 2018).

2.5.2 Technological Advancements in Bangladesh's Agricultural Sector

Bangladesh's agricultural sector has seen a significant transformation over the last few years with the incorporation of advanced technologies. These are aimed at boosting productivity, efficiency, and climate resilience. Machine learning, GIS, remote sensing, early warning systems, and AI-driven platforms are increasingly being utilized to collect and interpret agricultural datasets, enabling smallholder farmers to make better decisions in the face of climate change.

GIS-based Land Resources Information Systems enabled more targeted interventions in regions affected by salinity, drought, and waterlogging by providing detailed spatial maps of land suitability, soil fertility, and cropping patterns (Miah and Islam 2011). Additionally, remote sensing also acts as a valuable tool for long-term planning. Nath et al. (2022) used Landsat satellite data to track shifting cropping patterns in southeastern Bangladesh, showing how spatial and temporal monitoring can directly inform climate adaptation strategies. Moreover, machine learning and satellite imagery are advancing precision in crop monitoring. Tiwari et al. (2024) and Son et al. (2024) developed frameworks using Sentinel-1 and Sentinel-2 data for highly accurate in-season rice field classification. Integrated into platforms like Google Earth Engine, these tools allow for large-scale, real-time monitoring. Using this approach, Tiwari and Son mapped more than 90,000 km^2 of Boro rice fields annually between 2014 and 2018, capturing variations in planting dates, crop health, and cultivated areas. Such frameworks are particularly valuable for tracking agricultural shifts, responding quickly to climate shocks such as floods or droughts, and supporting evidence-based policies on food security, crop insurance, and adaptive planning (Ibrahim and Balzter 2024). In addition, the rollout of early warning systems in the haor basins and coastal regions has provided farmers with advance alerts about extreme weather events, giving them crucial time to protect their crops and livelihoods.

The CARE (a leading international humanitarian organization)-led SUFAL project, and the Department of Agricultural Extension (DAE) have played primary roles in expanding access to localized agrometeorological services, helping farmers adjust planting dates, safeguard crops, and minimize post-harvest losses (CARE Bangladesh 2024). Complementing these systems is the rapid growth of mobile-based and AI-powered applications, which enable farmers to diagnose plant diseases by analyzing images of symptomatic crops. Moreover, platforms such as *Khamari* and *Geopotato* provide customized agro-advisory services based on satellite and field data analytics, such as pest-outbreak notifications, fertilizer suggestions, and water management advice (Table 2.3).

Table 2.3 Key technological agricultural advancements in Bangladesh

Technology	Example use case	Benefits
GIS & Remote Sensing	Agro Ecological Zoning, Crop Suitability Mapping	Location-specific crop planning and sustainable land use
Satellite Crop Mapping	Sentinel 1/2 for area estimation	Real-time, high-accuracy crop acreage and health monitoring
Early Warning Systems	Disseminating flood/cyclone alerts to farmers	Reduction in crop losses and timely harvest
AI in Pest Management	Image-based crop disease detection tools	Early and accurate disease identification and timely treatment
Digital Climate Services	DAE weather-based advisories for planting schedules	Optimization of planting times; reduction of climate-induced crop failure risk

2.6 Limitations of Sustainable Agricultural Methods

While sustainable agriculture offers a crucial pathway toward environmentally responsible farming, its adoption is not without challenges, particularly in developing countries. One of the biggest difficulties is the high upfront cost of transitioning to sustainable practices. Investments in specialized equipment, organic inputs, and certification processes can be burdensome, especially for smallholder farmers who already face severe financial constraints. The transition toward sustainability also necessitates abandoning farming traditions, many of which have been passed down for generations. Beyond these financial and cultural barriers, infrastructure limitations further complicate adoption. Poor road networks, limited market access, and inadequate storage facilities frequently lead to post-harvest losses, discouraging farmers from embracing new methods.

2.7 Achieving UN SDGs

Sustainable agriculture is significant in achieving UN SDG 1 (No Poverty) and SDG 2 (Zero Hunger) by enhancing farmer resilience and food security through diverse, sustainable practices. Socially, sustainable agriculture promotes SDG 3 (Good Health and Well-being) by curbing pesticide application, SDG 5 (Gender Equality) through women's empowerment, and SDG 10 (Reduced Inequality) via fair trade and equitable resource access. Environmentally, sustainable agriculture contributes to SDG 6 (Clean Water and Sanitation), SDG 13 (Climate Action), and SDG 15 (Life on Land) by encouraging water efficiency, climate-resilient approaches, and biodiversity conservation. Finally, it supports SDG 17 (Partnerships) through cross-sector collaboration to scale sustainable practices.

2.8 Conclusions

The sustainable agriculture practices explored in this chapter highlight that sustainable agriculture is highly effective when grounded in local adaptation, ecological principles, and innovative technologies. Particularly, in climate-vulnerable Bangladesh, the alignment between climate-smart agriculture practices and national policy such as the National Adaptation Plan 2022–2050, which promotes localized adaptation strategies, such as saline-tolerant crop varieties, integrated water resource management, and digital agrometeorological services, is crucial. Further, the findings suggest that traditional agricultural practices, when included in national plans as proven examples, can help shape broader strategies to increase the resilience of agriculture. However, concerted efforts are necessary to make the whole transition to climate-smart and sustainable agriculture. For sustainable agriculture to be both equitable and scalable, grassroots ideas must be supported by public-private partnerships, capacity-building initiatives, and easily accessible financial sources.

References

Afroj M, Kazal MMH, Rahman MM (2016) Precision agriculture in the World and its prospect in Bangladesh. Res Agric Livest Fish 3(1):1–14

Akpa AF (2024) The effects of climate extreme events on selected food crop yields in Sub-Saharan Africa. Heliyon 10(9):e30796

Al-Imran M, Mitra S, Rahman MA, Das SK (2023) Effect of different macrophytes on crop cultivation under floating agriculture system for climate change adaptation in Bangladesh. Aquat Bot 189:103699

Al-Imran M, Islam ATMR, Karmaker D, Mitra S, Haider I, Rahman MA, Das SK (2025) Evaluating chili pepper (Capsicum annuum L.) varieties in floating agriculture systems: a climate-smart agriculture approach for Bangladesh's wetlands. J Crop Improv 39(1):19–42

Angeler DG, Fried-Petersen HB, Allen CR, Garmestani A, Twidwell D, Chuang WC, Donovan VM, Eason T, Roberts CP, Sundstrom SM, Wonkka CL (2019) Adaptive capacity in ecosystems. In: Advances in ecological research, vol 60. Academic, pp 1–24

Association for Temperate Agroforestry (AFTA) (2021) Association for temperate agroforestry [online]. Available at: https://aftaweb.org/. Accessed 11 May 2025

Barman A, Saha P, Patel S, Bera A (2022) Crop diversification an effective strategy for sustainable agriculture development. In: Sustainable crop production-recent advances. IntechOpen

Blaser WJ, Oppong J, Hart SP et al (2018) Climate-smart sustainable agriculture in low-to-intermediate shade agroforests. Nat Sustain 1:234–239. https://doi.org/10.1038/s41893-018-0062-8

Bouri M, Arslan KS, Şahin F (2023) Climate-smart pest management in sustainable agriculture: promises and challenges. Sustainability 15(5):4592. https://doi.org/10.3390/su15054592

Burgess AJ, Cano MEC, Parkes B (2022) The deployment of intercropping and agroforestry as adaptation to climate change. Crop Environ 1(2):145–160

Çakmakçı R, Salık MA, Çakmakçı S (2023) Assessment and principles of environmentally sustainable food and agriculture systems. Agriculture 13(5):1073

CARE Bangladesh (2024) Scaling-up Flood Forecast-based Action and Learning in Bangladesh (SUFAL-II). Available at: https://www.carebangladesh.org/project-details/54. Accessed 29 May 2025

Case-Cohen AL (2018) Understanding the soil health knowledge of farmers in the Yakima valley. Doctoral dissertation, Evergreen State College

Chaki A, Zahan T, Hossain MD, Ferdous Z, Islam MA et al (2023) Climate-smart agriculture technologies and practices in Bangladesh. SAARC Agriculture Centre, South Asian Association for Regional Cooperation, BARC Complex, Farmgate, New Airport Road, Dhaka – 1215, Bangladesh

Chandra A, McNamara KE, Dargusch P (2018) Climate-smart agriculture: perspectives and framings. Clim Pol 18(4):526–541

Charvériat C (2012) Natural disasters in Latin America and the Caribbean: an overview of risk. SSRN Electron J. https://doi.org/10.2139/ssrn.1817233

Chowdhury S, Khan S, Sarker MFH, Islam MK, Tamal MA, Khan NA (2022) Does agricultural ecology cause environmental degradation? Empirical evidence from Bangladesh [online]. Heliyon 8(6):e09750

CIAT; World Bank (2017) Climate-smart agriculture in Bangladesh. CSA country profiles for Asia series. International Center for Tropical Agriculture (CIAT); World Bank, Washington, DC, p 28

Coulibaly TP, Du J, Diakité D (2021) Sustainable agricultural practices adoption. Agriculture 67(4): 166–176

Council on Energy, Environment and Water (CEEW) (2025) Integrated pest management in India [online]. Available at: https://www.ceew.in/publications/sustainable-agriculture-india/integrated-pest-management. Accessed 11 May 2025

De Fraiture C, Wichelns D (2010) Satisfying future water demands for agriculture. Agric Water Manag 97:502–511

Deguine JP, Aubertot JN, Bellon S, Côte F, Lauri PE, Lescourret F, Ratnadass A, Scopel E, Andrieu N, Bàrberi P, Becker N (2023) Agroecological crop protection for sustainable agriculture. Adv Agron 178:1–59

Domingo AA, Hagerman HH (1982) Sorjan cropping system trial in irrigated wet land conditions. Philipp J Crop Sci 7:154–161

FAO (2018) Transforming food and agriculture to achieve the SDGs. Food and Agriculture Organization of the United Nations. http://www.fao.org/3/I9900EN/i9900en.pdf. Accessed 06 May 2025

FAO (2022) Greenhouse gas emissions from agrifood systems. Global, egional and country trends, 2000–2020. FAOSTAT Analytical Brief Series No. 50. Rome, FAO. chrome-extension://efaidnbmnnnibpcajpcglclefindmkaj/https://openknowledge.fao.org/server/api/core/bitstreams/121cc613-3d0f-431c-b083-cc2031dd8826/content. Accessed 11 May 2025

Fischer G, Tubiello FN, van Velthuizen H, Wiberg DA (2007) Climate change impacts on irrigation water requirements: effects of mitigation, 1990–2080. Technol Forecast Soc Chang 74:1083–1107

Garibaldi LA, Gemmill-Herren B, D'Annolfo R, Graeub BE, Cunningham SA, Breeze TD (2017) Farming approaches for greater biodiversity, livelihoods, and food security. Trends Ecol Evol 32(1):68–80

Haque SE (2018) An overview of groundwater quality in Bangladesh. In: Mukherjee A (ed) Groundwater of South Asia. Springer hydrogeology. Springer, Singapore. https://doi.org/10.1007/978-981-10-3889-1_13

Haque SE (2023) The effects of climate variability on Florida's major water resources. Sustainability 15:11364. https://doi.org/10.3390/su151411364

Haque SE, Chowdhury NN (2025) Bangladesh plastic consumption trend: current scenario. Water Air Soil Pollut 236:218. https://doi.org/10.1007/s11270-025-07862-x

Haque SE, Nahar N (2023) Bangladesh: climate change issues, mitigation and adaptation in the water sector. ACS ES&T Water 3(6):1484–1501. https://doi.org/10.1021/acsestwater.2c00450

Haque MM, Alam MR, Alam MM, Basak B, Sumi KR, Belton B, Murshed-E-Jahan K (2015) Integrated floating cage aquageoponics system (IFCAS): an innovation in fish and vegetable production for shaded ponds in Bangladesh. Aquacult Rep 2:1–9

Haque MR, Sultana R, Rahman MH, Islam S, Hasan AK (2019) Impact of adopting salt tolerant rice varieties in the coastal areas of Bangladesh. Fundam Appl Agric 4(2):823–828

Haque SE, Gazi-khan L, Rashid H (2023) Climate change mitigation and adaptation strategies, the environment, and impacts of the COVID-19 pandemic: a review of the literature. In: Role of green chemistry in ecosystem restoration to achieve environmental sustainability. Elsevier. https://doi.org/10.1016/B978-0-443-15291-7.00006-7

Haque SE, Nahar N, Chowdhury NN, Sayanno TK, Haque MS (2024) Geomorphological changes of river Surma due to climate change. Int J Energy Water Resour 9(1):113–132

Hiywotu AM (2025) Advancing sustainable agriculture for goal 2: zero hunger – a comprehensive overview of practices, policies, and technologies. Agroecol Sustain Food Syst 49(7): 1027–1055. https://doi.org/10.1080/21683565.2025.2451344

Ibrahim SA, Balzter H (2024) Evaluating flood damage to paddy rice fields using PlanetScope and Sentinel-1 data in North-Western Nigeria: towards potential climate adaptation strategies. Remote Sens 16(19):3657

IPCC (2018) Summary for policymakers. In: Masson-Delmotte V, Zhai P, Pörtner H-O, Roberts D, Skea J, Shukla PR, Pirani A, Moufouma-Okia W, Péan C, Pidcock R, Connors S, Matthews JBR, Chen Y, Zhou X, Gomis MI, Lonnoy E, Maycock T, Tignor M, and Waterfield T (eds) Global warming of 1.5°C. An IPCC special report on the impacts of global warming of 1.5°C above pre-industrial levels and related global greenhouse gas emission pathways, in the context of strengthening the global response to the threat of climate change, sustainable development, and efforts to eradicate poverty. Cambridge University Press, Cambridge/New York, pp 3–24. https://doi.org/10.1017/9781009157940.001

Irfanullah HM, Adrika A, Ghani A, Khan ZA, Rashid MA (2008) Introduction of floating gardening in the north-eastern wetlands of Bangladesh for nutritional security and sustainable livelihood. Renewable Agric Food Syst 23(2):89–96

Kabato W, Getnet GT, Sinore T, Nemeth A, Molnár Z (2025) Towards climate-smart agriculture: strategies for sustainable agricultural production, food security, and greenhouse gas reduction. Agronomy 15(3):565

Kamau H, Roman S, Biber-Freudenberger L (2023) Nearly half of the world is suitable for diversified farming for sustainable intensification. Commun Earth Environ 4:446. https://doi.org/10.1038/s43247-023-01062-3

Karunathilake EM, Le-Anh T, Heo S, Chung YS, Mansoor S (2023) The path to smart farming: innovations and opportunities in precision agriculture. Agriculture 13(8):1593. https://doi.org/10.3390/agriculture13081593

Karuppuchamy P, Venugopal S (2016) Integrated pest management. In: Ecofriendly pest management for food security. Academic, pp 651–684

Lampayan RM, Samoy-Pascual KC, Sibayan EB, Ella VB, Jayag OP, Cabangon RJ, Bouman BAM (2015) Effects of alternate wetting and drying threshold level and plant seedling age on crop performance, water input, and water productivity of transplanted rice in Central Luzon, Philippines. Paddy Water Environ 13:215–227. https://doi.org/10.1007/s10333-014-0423-5

Li L, Awada T, Shi Y, Jin VL, Kaiser M (2025) Global greenhouse gas emissions from agriculture: pathways to sustainable reductions. Glob Chang Biol 31(1):e70015. https://doi.org/10.1111/gcb.70015. PMID: 39740017; PMCID: PMC11683860

Mabhaudhi T, Dirwai TL, Taguta C, Senzanje A, Abera W, Govid A, Dossou-Yovo ER, Aynekulu E, Chimonyo VGP (2025) Linking weather and climate information services (WCIS) to Climate-Smart Agriculture (CSA) practices. Clim Serv 37:100529

Miah MG, Islam MA (2011) Development of GIS-based Land Resources Information System (LRIS) for land suitability assessment in Bangladesh. Soil Resource Development Institute (SRDI), Ministry of Agriculture, Dhaka

Momin MR (2022) Effectiveness of climate-resilient adaptation strategies in saline-prone areas of Bangladesh. Available at: https://ccdbclimatecentre.com/wp-content/uploads/2023/08/EFFECTIVENESS-OF-CLIMATE-RESILIENT-ADAPTATION-STRATEGIES-IN-SALINE-PRONE-AREAS-OF-BANGLADESH.pdf. Accessed 12 June 2025

Mongabay (2024) In Bangladesh, sunflower grows where other crops don't amid increasing salinity. Mongabay. Available at: https://news.mongabay.com/2024/01/in-bangladesh-sunflower-grows-where-other-crops-dont-amid-increasing-salinity/. Accessed 27 May 2025

Muchane MN, Sileshi GW, Gripenberg S, Jonsson M, Pumariño L, Barrios E (2020) Agroforestry boosts soil health in the humid and sub-humid tropics: a meta-analysis. Agric Ecosyst Environ 295:106899

Nath B, Hossain M, Majumder SC (2022) Agricultural crop pattern mapping and change analysis at a sub-district level using landsat satellite data: evidence from South-Eastern region, Bangladesh. J Agric Environ Int Dev 116(2):5–38

Nouri H, Stokvis B, Galindo A, Blatchford M, Hoekstra AY (2019) Water scarcity alleviation through water footprint reduction in agriculture: the effect of soil mulching and drip irrigation. Sci Total Environ 653:241–252

Pandey DK, Mishra R (2024) Towards sustainable agriculture: harnessing AI for global food security. Artif Intell Agric 12:72

Pandian K, Mustaffa MRAF, Mahalingam G, Paramasivam A, Prince AJ, Gajendiren M, Mohammad ARR, Varanasi ST (2024) Synergistic conservation approaches for nurturing soil, food security and human health towards sustainable development goals. J Hazard Mater Adv 16: 100479

Pearson KA, Millar GM, Norton GJ, Price AH (2018) Alternate wetting and drying in Bangladesh: water-saving farming practice and the socioeconomic barriers to its adoption. Food Energy Secur 7(4):e00149

Petersen B, Snapp S (2015) What is sustainable intensification? views from experts. Land Use Policy 46:1–10

Pretty J, Bharucha ZP (2014) Sustainable intensification in agricultural systems. Ann Bot 114(8): 1571–1596

Rabbani G, Rahman A, Mainuddin K (2013) Salinity-induced loss and damage to farming households in coastal Bangladesh. Int J Global Warming 5(4):400–415

Rahaman MM (2023) Exploring the community-based adaptation strategies: an anthropological study on local knowledge in Patharghata Upazila, Bangladesh. South Asian J Soc Sci Humanit 4(3):39–53

Rahim MA, Siddiqua A, Nur MNB, Zaman AM (2018) Community perception on adverse effects of natural hazards on livelihood and enhancing livelihood resiliency: a case study at Patharghata Upazila, Barguna. Procedia Eng 212:149–156

Rana MM, Rahman MM, Oliver MMH, Miah MY (2023) Development and evaluation of the first automated irrigation system for alternate wetting and drying technique in rice cultivation in Bangladesh. Smart Agric Technol 6:100348

Romeh AA (2018) Integrated pest management for sustainable agriculture. In: Negm A, Abu-hashim M (eds) Sustainability of agricultural environment in Egypt: part II. The handbook of environmental chemistry, vol 77. Springer, Cham. https://doi.org/10.1007/698_2018_267

Runkle BR, Suvočarev K, Reba ML, Reavis CW, Smith SF, Chiu YL, Fong B (2018) Methane emission reductions from the alternate wetting and drying of rice fields detected using the eddy covariance method. Environ Sci Technol 53(2):671–681

Saleem A, Anwar S, Nawaz T et al (2024) Securing a sustainable future: the climate change threat to agriculture, food security, and sustainable development goals. J Umm Al-Qura Univ Appll Sci 11:595–611. https://doi.org/10.1007/s43994-024-00177-3

Sanyaolu M, Sadowski A (2024) The role of precision agriculture technologies in enhancing sustainable agriculture. Sustainability 16(15):6668

Sarvade S, Upadhyay VB, Kumar M, Imran Khan M (2019) Soil and water conservation techniques for sustainable agriculture. In: Jhariya M, Banerjee A, Meena R, Yadav D (eds) Sustainable agriculture, forest and environmental management. Springer, Singapore. https://doi.org/10.1007/978-981-13-6830-1_5

Sikka AK, Islam A, Rao KV (2018) Climate-smart land and water management for sustainable agriculture. Irrig Drain 67(1):72–81

Son NT, Chen CF, Lin HS, Cheng YS, Chen CR, Syu CH, Zhang YT, Liu TS, Toscano P, Chen SL, Chen SH (2024) Machine learning-based early prediction of rice-growing fields using multi-temporal Sentinel-1 synthetic aperture radar and Sentinel-2 multispectral data. J Appl Remote Sens 18(3):038503–038503
Srivastava RK, Purohit S, Alam E, Islam MK (2024) Advancements in soil management: optimizing crop production through interdisciplinary approaches. J Agric Food Res 18:101528
Steenwerth KL, Hodson AK, Bloom AJ et al (2014) Climate-smart agriculture global research agenda: scientific basis for action. Agric Food Secur 3:11. https://doi.org/10.1186/2048-7010-3-11
Sultana T (2024) Our farmers are leading the fight against climate change. The Daily Star. Available at: https://www.thedailystar.net/opinion/views/news/our-farmers-are-leading-the-fight-against-climate-change-3902056. Accessed 27 May 2025
Tahat MM, Alananbeh KM, Othman YA, Leskovar DI (2020) Soil health and sustainable agriculture. Sustainability 12(12):4859
TDS (The Daliy Star; 2024) Adaptation strategies for climate change impacts in Bangladesh [online]. Available at: https://www.thedailystar.net/opinion/views/news/adaptation-strategies-climate-change-impacts-bangladesh-3722791. Accessed 11 May 2025
Tiller (2019) Final Socio-Economic Survey Report: pack 3—Barguna & Patharghata. Payra-Kuakata Comprehensive Plan (PKCP) [online]. Available at: http://pkcp.gov.bd/report/Other%20Reports/Final%20Socio%20economic%20survey%20report%20_Pack%203_%20Barguna%20&%20Patharghata.pdf. Accessed 1 June 2025
Tiwari V, Tulbure MG, Caineta J, Gaines MD, Perin V, Kamal M, Krupnik TJ, Aziz MA, Islam AT (2024) Automated in-season rice crop mapping using Sentinel time-series data and Google Earth Engine: a case study in climate-risk prone Bangladesh. Journal of Environmental Management, 351, 119615, (2017). Trends Ecol Evol (Amst.) 32(1):68–80
Tubiello FN (2019) Greenhouse gas emissions due to agriculture. In: Encyclopedia of food security and sustainability. Elsevier, pp 196–205
U.S. Environmental Protection Agency (EPA) (2025) Climate change indicators: greenhouse gases [online]. Available at: https://www.epa.gov/climate-indicators/greenhouse-gases. Accessed 12 June 2025
Velasco-Muñoz JF, Aznar-Sánchez JA, Belmonte-Ureña LJ, Román-Sánchez IM (2018) Sustainable water use in agriculture: a review of Worldwide research. Sustainability 10(4):1084. https://doi.org/10.3390/su10041084
Volsi B, Higashi GE, Bordin I et al (2022) The diversification of species in crop rotation increases the profitability of grain production systems. Sci Rep 12:19849. https://doi.org/10.1038/s41598-022-23718-4
Wu W, Ma B (2015) Integrated nutrient management (INM) for sustaining crop productivity and reducing environmental impact: a review. Sci Total Environ 512–513:415–427
Yarahmadi F, Rajabpour A (2024) Insecticides and natural enemies: applications in integrated pest management programs—challenges, criteria, and evaluation for recommendations. In: In Insecticides in pest control – impact, challenges and strategies pp. 1–19. (Agricultural Sciences). InTechOpen. https://doi.org/10.5772/intechopen.1005830
Yuan X, Li S, Chen J, Yu H, Yang T, Wang C, Huang S, Chen H, Ao X (2024) Impacts of global climate change on agricultural production: a comprehensive review. Agronomy 14(7):1360. https://doi.org/10.3390/agronomy14071360
Zhang QF (2024) From sustainable agriculture to sustainable agrifood systems: a comparative review of alternative models. Sustainability 16(22):9675
Zhang W, Sheng J, Li Z, Weindorf DC, Hu G, Xuan J, Zhao H (2021) Integrating rainwater harvesting and drip irrigation for water use efficiency improvements in apple orchards of northwest China. Sci Hortic 275:109728

Emerging Digital Technologies in Agriculture: Innovations for a Sustainable Future

3

Sabnam Hossain and Dipankar Ghosh

Abstract

Digital technologies are transforming agriculture, providing critical solutions to global concerns such as food security and climate change. This abstract investigates developing digital technologies, long-term future trends, and innovations that are generating a more resilient agricultural landscape. Artificial intelligence and machine learning are critical technologies for predictive analytics and automated operations, while the Internet of Things enables real-time monitoring and smart resource management. Drones and robotics improve precision agriculture, while Big Data analytics enables more informed decision-making. Blockchain ensures supply chain transparency, while Controlled Environment Agriculture maximizes resource use. These improvements help to improve resource efficiency, minimize environmental impact, and increase climate resilience, eventually empowering farmers and assuring a more secure and sustainable food future.

Keywords

Digital agriculture · Precision farming · Sustainable agriculture · Artificial intelligence (AI) · Food security

S. Hossain · D. Ghosh (✉)
Microbial Engineering and Algal Biotechnology Laboratory Department of Biosciences, JIS University, Agarpara Kolkata, West Bengal, India

© The Author(s), under exclusive license to Springer Nature Singapore Pte Ltd. 2026

A. L. Srivastav et al. (eds.), *Digital Twin Technology for Sustainable Agriculture*,
https://doi.org/10.1007/978-981-95-5915-2_3

3.1 Introduction

Agriculture in the twenty-first century has worldwide challenges due to a rapidly growing population, so ensuring food security has become the top concern for many countries (Aryal et al. 2022). Additionally, the agricultural sector continues to be most affected by climate change, which disrupts traditional cropping patterns, water availability, and soil fertility. High temperatures, irregular rainfall, and extreme weather like floods and droughts have a significant impact on crop productivity, particularly in economically weak nations (Change 2016). The Green Revolution of the 1960s and 1980s brought tremendous advances in agricultural practices and significantly improved food production at the farm level, but it did not resolve the overall food security dilemma (Beddington 2010). Although persistent efforts are made in sufficient food production and lowering farming costs, the fundamental question raised by early thinkers such as Charlotte Perkins Gilman is whether these methods can sustainably feed the world still remains unresolved (Polefrone 2021).

Also adding to these concerns, the issue of resource scarcity particularly arable land, freshwater, and manpower has threatened the sustainability of conventional food production systems. According to FAO, 2022 agriculture sector utilizes nearly 70% of global freshwater, which is a major for water-scarce regions, along with availability of arable land per capita is declining continuously due to urbanization, land degradation, and soil erosion (Kay et al. 2022). Also, the migration of youth to non-agricultural sector for employment has created labour shortages and declined productivity in many developing countries (Yeboah and Flynn 2021). In order to meet the growing food demands, the traditional agricultural method must undergo a transformation that is not only efficient but also resilient, sustainable, and adaptive. Also, lack of real-time data, inefficient resource utilization, and limited scalability of traditional systems have contributed to lower yield, post-harvest losses, and increased vulnerability to climate-related disruptions.

The challenges posed by traditional farming are solved to a major extent by incorporating technology in agriculture. The growth of agricultural technology is witnessed between 1784 and 1870, for soil tillage, planting, weeding, irrigation, and harvesting, which resulted in increase in food output. This period is known as "Agriculture 2.0." "Precision Farming", or Agriculture 3.0, emerged as a result of advances in electronics and computing technology (Dayıoğlu and Turker 2021). These developments led to reduced chemical use in the field as well as saving in energy and water. Precision agriculture, according to the International Society of Precision Agriculture (ISPA), is "a management strategy that gathers, processes and analyses temporal, spatial and individual data and combines it with other information to support management decisions according to estimated variability for improved resource use efficiency, productivity, quality, profitability and sustainability of agricultural production" (https://ispag.org/about/definition). Agriculture 4.0 refers to the use of advanced digital technologies such as AI, ML, Blockchain, IoT, Robotics, and Big Data to modernize agricultural operations (Dayıoğlu and Turker 2021). The introduction of these digital tools has helped shift practices from intuition-based decision-making to data-driven precision farming. The AI-ML system can detect nutrient deficiencies, insect outbreaks, and disease onset in crop production. A convolutional neural network can detect crop diseases using picture analysis (Ali et al. 2025a, b). The Internet of Things improves the operation of AI-ML systems by incorporating sensors that continuously monitor soil moisture, temperature, and humidity surrounding growing crops, contributing to precision management (Ahmed 2024). To enhance resource efficiency smart irrigation system is used that reduces water usage up to 40% (Lakhiar et al. 2024). Drones and robotics have enabled field monitoring, targeted pesticide application, and automated harvesting that contributes to reduction of labour cost. The integration of satellite imaging, drones, and predictive analytics allows for the prediction of droughts, pest outbreaks, and fertilizer needs. The use of digital twins in agriculture is increasing, where a virtual model of a farm system could replicate crop production and resource utilization. Big Data analytics integrates information from market data and weather forecasts for effective plantation schedules, crop rotation, and supply chain logistics (Delgado et al. 2019). Mobile systems offer weather forecasts, crop advisory services, and market prices to smallholder farmers, which improves decision-making accuracy and income. Blockchain ensures traceability of product which improves food safety in supply chains for global trade (Tripoli and Schmidhuber 2018). Controlled Environment Agriculture such as hydroponics, aquaponics, and vertical farming systems utilizes automation and climate sensors to grow crops with minimal usage of water, space, and chemicals (Cowan et al. 2022). Digital agriculture promotes Sustainable Agriculture, which aims to boost productivity, improve resilience, and reduce emissions.

Digital transformation also aligns with global Sustainable Development Goals, including SDG 2 (Zero Hunger), SDG 9 (Industry, Innovation, and Infrastructure), and SDG 13 (Climate Action) (Varriale et al. 2025). Digitalization of agriculture enhances productivity and climate resilience and provides strategic solution to combat poverty, food insecurity, and environmental degradation.

To this purpose, the current chapter highlights the emerging digital technologies that are contributing to a more resilient and sustainable agricultural production system. These technologies aim to address food security, climate change adaptation, efficient resource utilization, and explore the innovations to combat the challenges of traditional farming system. The goal is to create a technologically empowered food system capable of meeting the demands of present and future generations (Fig. 3.1).

3.2 The Imperative for Digital Transformation in Agriculture

With increasing wealth and changing dietary patterns of people in emerging nations, food demand is anticipated to increase by roughly 60% relative to 2010 levels in the year 2050 (Valin et al. 2021). Food insecurity affects mostly the vulnerable and disadvantaged groups in these countries, such as women, widows, people with disabilities, and low-income families living in areas of risk. This imbalance is expected to worsen in the coming decades due to unequal access to resources. Major economic shocks, like as the 2007–08 global food price surge and the

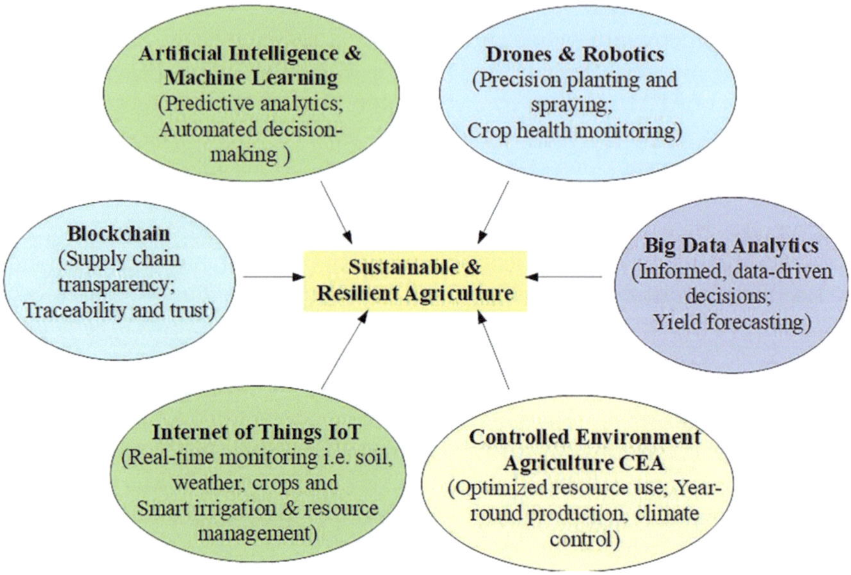

Fig. 3.1 A conceptual representation of how emerging digital technologies collectively contribute to a resilient, climate-smart, and sustainable agricultural system

COVID-19 outbreak, have hindered efforts of ensuring nutritional and food security under the SDGs' goal of Zero Hunger and Poverty. COVID-19 demonstrates the current gap in access to health systems and emphasizes the critical necessity for a well-designed public safety in larger areas to deal with any unanticipated shocks in the future (Aryal et al. 2022). High-value crops like fruits, vegetables, and oilseeds are increasing in demand due to rising awareness of nutrition and wellness among developing nations. This requires agricultural productivity to be increased and diversified without a proportionate rise in arable land. Crop modelling techniques (such as DSSAT and APSIM) can predict nutritional production (e.g., Vitamin A, protein) from a variety of biofortified or underused crops, which helps with nutritional planning (Gavasso-Rita et al. 2024).

Digital consulting platforms help farmers transition from monoculture to climate-resilient and nutrient-dense crops (such as millet and pulses). Traditional methods to increase productivity such as excess application of fertilizers have already degraded arable land and caused ecological damage (Hossain et al. 2020). Climate change has severe impact on agricultural production, mainly in developing countries where farmers depend on weather conditions. Rise in temperature has shorten the growing season that causes early crop maturation resulting in reduced yield. Variations in rainfall patterns cause waterlogging, soil erosion, and drought that makes agricultural planning difficult affecting food security and farmers' livelihoods (Change 2016). Agro-ecological zoning, which uses geospatial and digital models, aids in matching crops with appropriate climate zones, hence improving production stability. Climate forecasting systems are able to use historical weather, satellite data, and

AI algorithms to anticipate the likelihood of flooding, droughts, or extreme weather, allowing for better planting decisions. Warmer and wetter climates help many pests to survive and spread more easily. Changes in pest populations and disease transmission create new challenges for crop health and productivity. Water scarcity worsens in arid and semi-arid regions, which are home to some of the most food-insecure populations (Mdemu 2021). Agricultural practices account for 30% of global greenhouse gas emissions (Tubiello et al. 2013). Greenhouse gas emissions have serious effects for the environment, human health, and the economy. Reducing greenhouse gas emissions, such as carbon dioxide, is critical to addressing climate-adverse weather conditions, ocean acidification, and potential disturbances to ecosystems and food security. This creates a vicious cycle where agriculture influences climate change and is also affected by it. Availability of resources is finite and reducing consumption, reusing products, and recycling are critical for preserving resources for future generations. This includes conserving water and electricity, as well as avoiding waste. A reduced environmental impact helps to preserve biodiversity and ecosystems as well. Pollution, deforestation, and habitat degradation are all harmful to the ecosystem and the animals that rely on it.

Arable land per capita is declining due to rapid industrialization and the expansion of urban settlements. Agriculture uses 70% of the world's freshwater for irrigation which is expected to increase more than double by 2050 (Kay et al. 2022). Water scarcity is found in major agricultural regions like China and India due to groundwater depletion and variations in rainfall patterns (Mdemu 2021). The aging of rural populations and the continuous migration of youth to urban sector for non-agricultural employment opportunities reduce the availability of manpower. This demographic shift results in a significant loss of traditional knowledge and contributes to labour shortages in peak agricultural seasons. Limited labour supply increases the production costs which in turn discourages smallholder farmers from expanding operations. Digitalization in agriculture has the ability to empower rural youth and women by providing inclusive mobile platforms, e-extension services, and rural e-commerce. To foster entrepreneurship, Africa and South Asia are promoting women-led digital innovation hubs (Lukuyu et al. 2021).

Sustainable agricultural practices are important to reduce negative impacts on biodiversity, soil, water, and air while maintaining productivity (Rehman et al. 2022). Precision agriculture tools, AI-driven nutrient management systems, and smart irrigation systems are examples of digital tools that help farmers use resources more efficiently (Ali et al. 2025a, b). Digital platforms assist farmers in converting waste (such as crop residue and manure) into resources (biofertilizer and bioenergy), thereby encouraging circular agriculture and lowering greenhouse gas emissions. Understanding the environmental impact of farming requires tracking carbon footprints which is made available by digitalization. Remote sensing platforms enable monitoring of soil health, including organic matter, pH, and moisture content. Water conservation is also aided by intelligent irrigation systems (Adeyemi et al. 2017).

Traditional farming knowledge can be preserved through the integration of digital databases with indigenous knowledge. To lessen their reliance on middlemen, blockchain-enabled agri-wallets can be used to obtain microcredit and subsidies. Minimizing environmental effect is a moral and financial commitment for long-term natural resource sustainability and future food security.

3.3 Key Emerging Digital Technologies and Their Applications in Agriculture

Industry 4.0 has the ability to change the production capacities of whole industries, including agriculture. IoT technology is becoming more integrated into agricultural equipment, making connectivity a crucial aspect of this revolution. Technologies like Edge Computing, AI, Robotics, 5G connectivity, Blockchain, and supercomputing can improve farm efficiency, sustainability, and competitiveness (Dayıoğlu and Turker 2021). The European Commission defines advanced digital technology as follows:

3.3.1 Artificial Intelligence (AI) and Machine Learning (ML) in Agriculture

Artificial Intelligence(AI) analyses historical and current data to predict crop productivity and disease outbreaks (Pokhariyal et al. 2023). This system helps farmers to choose the required hybrid seeds, determine plantation schedule for optimized output (Shaikh et al. 2022). Automatic tractors, intelligent irrigation systems, and robotic harvesters reduce the need of manpower. Herbicides are identified and sprayed selectively by weed detection systems, reducing environmental harm (Huang et al. 2025). When necessary, an intelligent irrigation system will turn on watering systems after determining the soil's moisture content (Liao et al. 2021). By processing data from sensors, satellites, and drones, artificial intelligence (AI) facilitates real-time decision-making for crop selection and soil nutrient management (Patel et al. 2025). AI helps agricultural stakeholders reduce lending risks by forecasting harvest results. Banks use AI-driven platforms to assess land productivity based on historical and current performance data, improving transparency and financial inclusivity (Avickson and Ogunola 2024).

3.3.2 Internet of Things (IoT) in Agriculture

IoT-based smart sensors monitor environmental variables such as temperature, moisture content, nutrient levels, and crop health (Shahab et al. 2025). These systems optimize growth conditions, especially for delicate crops like orchids, by utilizing cloud platforms, Raspberry Pi-based controllers, and wireless sensor networks (WSNs). IoT architecture has the data collection layer, where environmental and crop data are acquired using in-field sensors or remote sensing devices like drones and UAVs, and then uploaded to the cloud over the internet (Zhang et al.

2018). In field traps, IoT can automatically capture, count, and classify insects for real-time pest detection (Farooq et al. 2019). In order to maximize farming practices across crop and livestock systems, smart agriculture incorporates a variety of digital technologies. Wearable Internet of Things (IoT) systems are utilized in the Smart Livestock Farming (SLF) industry to continuously monitor animal health, welfare, and production metrics using a variety of sensor types classified as "At Cow," "From Cow," and "Near Cow" sensors. These tools help with early disease detection and better herd management by measuring things like heart rate, body temperature, salivary pH, and even behavioural patterns. IoT systems use Crop Water Stress Index (CWSI) and adjust irrigation schedules accordingly (King and Shellie 2023). To further enhance field-level automation, a streamlined irrigation system architecture utilizing mobile interfaces, solenoid valves, and soil moisture sensors has been suggested. IoT technologies detect site-specific nutrient requirements using remote sensing for applying fertilizer uniformly (Gorai et al. 2021). Geographic Information Systems (GIS) allows machines to treat each individual plot with precision, for example, autonomous drones for aerial spraying where needed. Real-time crop health data is visualized by sensors for early detection of threats such as blight or rust (Shafi et al. 2022) and Variable Rate Technology (VRT) sprayers or robotics apply treatments (Taseer and Han 2024).

3.3.3 Drones and Robotics in Agriculture

Drones used in agriculture offer high-resolution and real-time data on crop health to evaluate factors like disease symptoms, water stress, and nutrient deficiencies. By detecting plant density and growth stage, multispectral technologies enable early intervention prior to the onset of visible symptoms (Peng et al. 2022). The European Union has also made investments in a number of robotics projects, such as MARS, GRAPE, VINEROBOT, ROMI, and Flourish. These projects use Unmanned Ground Vehicles (UGVs) and Unmanned Aerial Vehicles (UAVs) to automate weed control, disease prediction, grape monitoring, and crop health assessment for precise field mapping. Automated pest detection is made possible by deep learning techniques in drones, which identify minute variations in crop coloration and canopy structure. Robotic manipulators and GPS navigation can automate the processes of seeding, harvesting, and weeding (Ali 2024). Weeding robots can distinguish between crop and weed species using vision systems and apply target-specific treatment. In vast rangelands, drones monitor the movement, grazing habits, and spatial distribution of herds. To monitor abnormal body temperatures for illness or injury, drones are equipped with thermal imaging systems (Rietz et al. 2023). Manipulators are being developed for specific agricultural tasks that collaborate with human workers and achieve energy-efficient operation. These initiatives show how precision agriculture technologies can boost sustainability and productivity in contemporary farming systems while drastically lowering labour, pesticide use, and energy input. Commercial adoption of drones and robotics is still limited and technical advancements are in progress for mainstream agricultural application in global food security.

3.3.4 Big Data Analytics in Agriculture

The collection, integration, and analysis of large and varied agricultural datasets are all part of big data analytics in agriculture. These datasets come from a variety of sources, such as market databases, weather stations, satellites, IoT devices, and remote sensing systems (Delgado et al. 2019). Using strong machine learning and data mining techniques, these datasets are transformed into insightful information that guides the best possible agricultural decision-making. The main objective of big data analysis is to increase production, resource use, lowering risks, and facilitate sustainable agricultural practices.

Advanced Sensor Technology Systems monitor temporal and spatial variability in agricultural environments using sensors, robots, and remote sensing to make precise recommendations for farm treatments. Through the analysis of climate-soil-yield relationships, these technologies offer site-specific solutions that reduce production cost. Risk management in agriculture through big data system is concerned with controlling hazards like crop failure by adverse weather, degraded soil, or presence of pests by integrating multiple datasets, such as yield records, weather forecasts, and satellite imagery. ICT platforms are used by agricultural management systems to provide all-inclusive farm management solutions. To connect farmers, suppliers, markets, and stakeholders, they make use of cloud-based technologies and wireless connectivity (Tantalaki et al. 2019).

3.3.5 Blockchain for Transparency and Trust in Agriculture

According to the CDC, food contamination causes 48 million illnesses and 3000 fatalities in the United States alone each year. Blockchain contributes to food safety by recording detailed hygiene and handling conditions at every stage and detecting mislabelled product (Freund 2023). Walmart and Kroger were among the first to track the origin of products such as pork and mangoes using blockchain and obtained results instantly than the traditional methods which take about a week (Priya et al. 2024). Integration with IoT enables real-time monitoring of cold-chain integrity for alerting stakeholders instantly about deviations in storage conditions. Distributed ledger technology (DLTs) eliminates uncertainty and facilitates value transfers through a shared ledger, creating a digital institution of trust. DLTs provide a secure mechanism for storing transactions in a digital database, eliminating the need for third-party intermediaries. This decreases transaction costs, allows for speedier and even real-time transactions, ensures immutable data, and offers access to the database for all network participants (Tripoli and Schmidhuber 2018). This system increases supply chain transparency, traceability (farm-to-fork), and promote fair trade and consumer trust. In developing regions, AgriLedger has built trust among farmers by recording all activities and payments digitally (Baker 2021). Farm Share leverages blockchain's equity model to build a community-based agricultural system (Kim and Laskowski 2018). Also, contract-based insurance compensates farmers during calamities.

CEA techniques like vertical farming and smart greenhouse system maximize production through controlled environment conditions including light, water, and CO_2 levels (Cowan et al. 2022). By combining CEA with Blockchain, traceability of product is possible. Energy consumption, nutrient levels, and irrigation volume are tracked on blockchain and accessible to the customers via mobile apps or QR codes. This promotes soil conservation and carbon footprint tracking, contributing to the Sustainable Development Goals (SDGs).

3.3.6 Genomics and Gene Editing in Agriculture

With adverse weather conditions comes the need for creating stress-tolerant crops which has been made possible by genomics. Whole-genome sequencing, transcriptomics, and marker-assisted selection are used for identifying stress-tolerant genes that enable drought tolerance, heat resistance, and salinity adaptation. Gene editing technologies like CRISPR-Cas9 allow for targeted modifications in crop genomes without introducing foreign DNA (Arora and Narula 2017). This includes editing of genes involved in stomatal regulation, root architecture, and osmotic balance that enhance the plant's ability to withstand abiotic stress. For example, editing the ARGOS8 gene in maize resulted in stable yields under limited water conditions. Gene editing facilitates the biofortification of crops by modifying specific genes that regulate micronutrient pathways. Researchers developed varieties of enriched food products like in golden rice with vitamin A and also others with iron, zinc, and folate, which are able to address the issues like malnutrition (Dubock 2025). Genomics and gene editing empower plant breeders to develop high variety crops that are nutrient-dense and also climate-resilient.

In order to identify plant traits in dynamic stress environments, modern breeding programs are combining transcriptomics and phenomics. Gene activation is becoming easier without causing double-strand breaks thanks to the development of CRISPR-Cas12 and base-editing tools, which also lessen off-target effects (Arora and Narula 2017). To alter gene expression in a way that is both reversible and heritable, researchers are investigating epigenome editing. This opens up new possibilities for stress adaptation. Proteomics and metabolomics help us understand how plants react to abiotic stress. These innovations facilitate the development of crop varieties that are more resilient to biotic stressors like pests and diseases and have improved water and nutrient use efficiency.

Additionally, by recording growth parameters in real time, digital phenotyping tools that use AI and computer vision are speeding up trait selection. The breeding cycle is shortened, and the cost of creating climate-resilient crops is decreased.

3.4 Sustainable Future Trends and Innovations to Attain Sustainable Agriculture

The FAO outlines five principles for boosting productivity and sustainability, creating a production system that benefits the ecosystem and meets human needs, and achieving 2030 SDG targets. FAO's five core principles include increasing production, employment, and value addition in food systems, protecting natural resources, improving livelihoods, fostering equitable economic growth, strengthening resilience, and adapting to new challenges (Dayıoğlu and Turker 2021).

3.4.1 Enhanced Resource Efficiency

Sustainable agriculture is dependent on technological and systemic innovations that enhance resource efficiency. Smart irrigation systems include sensors, weather data, and automation to apply only the required amount of water to the field. This approach minimizes water usage for plant hydration in water-scarce regions. AI-based irrigation models can improve water savings by 40–50% in arid zones (Alharbi et al. 2024). Drip fertigation systems, for example, aid in synchronizing nitrogen administration with irrigation time, increasing yield and lowering runoff. Variable rate technology (VRT) allows for site-specific delivery of fertilizers and pesticides using real-time data from sensors, drones, and satellite imaging, lowering environmental contamination and input costs (Taseer and Han 2024). VRT direct use of lime or micronutrients to specific field zones which helps in improving pH balance and nutrient use efficiency of soil. When combined with UAVs (drones) and hyperspectral imaging, VRT enhances early disease identification and spatial crop performance analysis.

3.4.2 Minimized Environmental Impact

Practices such as composting organic waste, reusing livestock manure as biofertilizer, and converting agricultural residues into bioenergy reduce landfill dependence supporting zero-waste goals. Sustainable agricultural practices help in reducing primary greenhouse emissions such as carbon dioxide, methane, and nitrous oxide (Ali et al. 2025a, b). This includes techniques such as nutrient management, manure handling, and the use of renewable source of energy to preserve soil structure, increase the amount of organic matter in the soil, and prevent soil erosion. Plant growth-promoting rhizobacteria (PGPR) and mycorrhizal fungi reduce reliance on chemical fertilizers and also promote soil-biodiversity.

Agro-ecological methods promote intercropping, the production of a variety of crops, and conservation of natural habitats. It helps in nutrient cycling through pollinators, beneficial insects, and natural predators. Agroecology preserves traditional farming knowledge and promotes community seed banks.

3.4.3 Increased Climate Resilience

Adaptation to extreme weather events requires sustainable agricultural systems to incorporate techniques that enhance soil structure, prevent erosion, and increase organic matter. Food production in adverse climatic conditions is ensured by the development of crop varieties that can withstand heat and drought and flourish under abiotic stress conditions. Adopting techniques like improved soil management, efficient irrigation, and integrated pest management can boost crop production and minimize losses during storage and transportation (Mariani and Ferrante 2017).

Logistics and distribution systems become more efficient with the application of digital agriculture including sensors that monitor field conditions in real time, facilitating timely harvest, storage, and timely delivery, reducing delays ensuring that food reaches consumers with minimal spoilage. Establishing systems to provide timely information about potential climate hazards, like droughts or floods, allows farmers to take preventative measures and minimize losses. Creating transparent and traceable supply chains builds trust and ensures that food safety standards are met, contributing to a more reliable and secure food system. Smart irrigation, renewable energy systems, agroecological intensification, and digital extension platforms help to achieve SDG 2 (Zero Hunger), SGG 6 (Clean Water), and SDG 13 (Climate Action).

3.4.4 Farmer Empowerment and Livelihoods

Farmers require timely information of weather reports such as weather patterns, crop disease outbreaks, and market data to make informed decisions about what to grow, when to sell, and where to sell, and financial services for access to loans, insurance, and other financial products critical to invest in new technologies and risk management, which has been made possible by digital platforms and mobile applications (Caine et al. 2015).

Farmers can employ digital technologies to improve resource use (water, fertilizer, etc.), resulting in decreased input costs and increased efficiency. E-commerce platforms and other market linkages can assist farmers in reaching larger customers, perhaps resulting to higher prices for their produce and reducing their dependency on middlemen. Farmers must receive training in financial management, planning, and investment in order to efficiently manage their finances and make informed decisions (Adnan et al. 2018). They can benefit from training programs that teach them about new technologies, sustainable farming techniques, and other pertinent skills that will increase their productivity and resilience.

3.4.5 Integration with Renewable Energy

Farmers adopt *solar-powered irrigation systems*, which provide a sustainable alternative to diesel or electric pumps and reduce operational costs, improve energy access in remote regions, and lower greenhouse gas emissions. Agricultural waste like husks, stems, and manure can be recycled into biogas through anaerobic digestion, bioplastics which is made from starch and compost for soil improvement. Utilizing biodegradable bioplastics derived from agricultural waste can lessen reliance on fossil fuels and enhance soil health by reducing microplastics pollution (Yu et al. 2018). In order to apply fertilizer precisely, sensor-based soil health cards are being used to map the macro- and micronutrient status of farmland. Pest detection systems also reduce crop damage and pesticide use by sending alerts about potential infestations. Alley cropping and agroforestry increase carbon sequestration, incorporate trees into croplands, and produce additional income from fruit, lumber, and fodder. Blockchain provides eco-certification and traceability systems making it easier for farmers to use sustainable methods to access markets. Solar-powered agricultural technologies are in line with SDG 7 (Affordable and Clean Energy), while circular bioeconomy practices also support SDG 12 (Responsible Consumption). These technologies support rural innovative ecosystems, reduce poverty (SDG 1), and foster inclusivity (SDG 5) (Pandey and Pandey 2023).

Implementation of sustainable practices should include biodiversity and environmental conservation, quality and healthy food, agricultural innovation, and aligned economic and social goals. The focus will shift from customer requirements to those of the entire ecosystem, including farmers, producers from other industries, legislators, state governments, and other stakeholders. Advanced algorithms and information will be utilized to manage agricultural activities (Hrustek 2020). Therefore, sustainable agriculture in the digital era is not only a productivity concern it is a holistic growth path connected with the global 2030 Agenda (Bajpai and Biberman 2021).

3.5 Challenges and Considerations

While Agriculture 4.0 promises significant benefits, there are still difficulties to overcome for a successful digital transition. Agriculture 4.0 presents issues as follows:

3.5.1 Cost of Adoption

Smallholder farmers continue to face significant challenges due to the high initial investment costs. Many digital technologies, such as precision agriculture instruments, sensors, drones, and controlled irrigation systems, require significant upfront investment. Despite assertions that not all digital technologies are expensive, price remains a relative issue. Even low-cost alternatives may be unattainable to small farms with thin margins. Other expenses include maintenance, software upgrades, and cloud service subscription (e.g., AI decision assistance), contributing to financial unpredictability. Public-private partnerships (PPPs) are critical for developing countries like Kenya and India which took initiative like introduction of digital voucher systems for marginal farmers. Large agribusinesses have more resources to invest in digital transformation than small-scale farmers. The concern is not only about financial costs, but also about risk investing in novel technologies with no definite returns (Caine et al. 2015).

3.5.2 Digital Divide

Access to digital infrastructure, such as high-speed internet and smart devices, limits reach and benefits. This separation is not just between rural and urban regions, but also across farm sizes and locations. Farmers in areas with low internet connectivity cannot access real-time data, cloud-based technologies or market-access platform. Digital gap is worsened by socioeconomic disparities and varied degrees of digital competence (Helsper 2021). As a result, certain groups are still excluded from transformative potential, which undermines the inclusive ideal of smart agriculture. The digital difference between genders remains large. Women in agriculture are less likely to hold mobile internet in LMICs, limiting their access to e-extension and market platforms. Rural areas lack connectivity and also digital infrastructure, such as mobile towers, inexpensive smartphones, and electricity, expanding inequality even more.

3.5.3 Data Security and Privacy

With the advent of big data and smart farming, there are growing worries about data ownership, data sovereignty, and the exploitation of personal or farm-level data. Farmers are concerned about who owns the data created in their fields, how it is preserved, and who profits from it. There is no widely agreed notion of data sovereignty in agriculture and regulatory frameworks sometimes fall behind. While initiatives like the 2018 Code of Conduct seek to set ethical norms, they are non-binding and provide limited protection (Koščík and Myška 2018). Farmers regularly express concern about technology suppliers that may misuse raw data for commercial gain without ensuring transparency or shared benefits. Smallholders frequently struggle to understand data contracts, resulting in unbalanced power dynamics.

3.5.4 Skill Gap

Digital revolution raises the risk of expanding the skills gap in agriculture. Farmers are increasingly required to use complex technologies, understand big datasets, and make decisions based on algorithms. This transition for conventional farmer, often with no formal education or assistance is difficult (Pham and Stack 2018). There are growing concerns about "digital land grabs," in which agribusinesses acquire comprehensive geospatial and farm-level data in order to establish market dominance.

The lack of blockchain connectivity makes contracts more vulnerable to data tampering and unfair manipulation, especially in insurance and supply chain transactions. Recent case studies (e.g., in Brazil and India) reveal farmers inadvertently signing off on data ownership transfers in e-agreements due to a lack of legal awareness. Without adequate investment in education and extension services, adoption of new technology may be superficial or ineffectual. The demand for capacity-building and agricultural digital literacy is especially urgent in underdeveloped countries where traditional knowledge dominates farming techniques (Caine et al. 2015). Generational shifts in farming populations mean that many elderly farmers have cognitive difficulties while using digital tools, leading to resistance to adoption.

3.5.5 Interoperability

Interoperability makes the integration of digital platform with various datasets (López-Morales et al. 2020). Because of the current lack of standardization in the range of data sources, collection formats, and technical systems, it is difficult to integrate them into a single agriculture management system. Farmers have to manage numerous incompatible systems, which increases their workload and hinders the exchange of knowledge and seamless decision-making. Data-driven agriculture's full potential remains limited in the absence of common protocols for data processing and device connectivity. Innovation and scalability are hampered by persistent interoperability gaps.

3.5.6 Policy and Regulatory Frameworks

The rapid expansion of digital agriculture far exceeds governments' capacity to establish effective policy and regulatory solutions. Existing regulatory structures are frequently too inflexible, antiquated to manage the intricate and dynamic issues of digital agriculture (Martens and Zscheischler 2022). This policy lag causes confusion among stakeholders and may discourage investment. Data ownership governance also creates disagreement on whether agricultural data should be recognized as a private asset or common resource. There is no international agreement on digital agriculture governance which has resulted in imbalanced standards for data use, cross-border technology sharing, and farmer rights. Due to the absence of legal provisions, farmers encounter unfair contracts by data aggregators or technology businesses. To address these issues, UN agencies are developing Digital Agriculture Code of Ethics, focusing on AI, big data, and sensor-based surveillance (Härtel 2022).

For the proper establishment of digital agriculture strict regulatory system is needed which should include data-rights rules, open licensing policies, and ethical use of AI and digital systems. Government should create a collaborative system with global institutions, farmer cooperatives, and technology entrepreneurs to develop a trustable, scalable, and resilient controlling system.

3.6 Conclusion

The agriculture sector is shifting to a new paradigm known as Agriculture 4.0, which includes a number of technologies to improve agricultural production. Today, when traditional farming methods are no longer sufficient, countries must focus on modern practices to increase productivity while decreasing expenses. A digital twin in agriculture gives the virtual depiction of a farm that has the potential to increase productivity with efficacy while reducing energy consumption and other losses. Smart Agriculture, also known as Agriculture 4.0, refers to the digitalization of production, similar to Industry 4.0. Indeed, modern technology exists to make agriculture more environmentally friendly and productive. Technical advances in the agriculture industry not only help to increase production efficiency but also improve quality, reduce carbon footprint, and mitigate production-related risks. Artificial Intelligence (AI), Machine Learning (ML), Internet of Things (IoT), drones, robotics, and blockchain technologies are changing traditional agriculture into data-driven, efficient, and accurate systems (Dayıoğlu and Turker 2021). The need for digitally resilient food systems is highlighted by their capability of mitigating concerns such as climate change, resource scarcity, labour shortages, and food insecurity. Precision agriculture method increases productivity with minimal waste production. Controlled environment agriculture and gene-editing technology provide climatically resilient crop production which is also able to fulfil the growing nutritional requirements. Food safety is ensured by blockchain that provides traceability and transparency throughout the supply chain. With the increasing population and climatic concerns, digital agriculture provides a way for more inclusive growth and food security. It supports global sustainability goals such as SDG 2 (zero hunger), SDG 13 (climate action), and SDG 9 (industry, innovation, and infrastructure) (Pandey and Pandey 2023).

Despite these promising developments, limitations such as access to internet connectivity, electricity, and digital literacy in rural areas continue to persist which can be tackled with policy frameworks. As digital platforms collect massive amounts of agricultural data, ownership, privacy, and benefit-sharing become controversial problems. Open data initiatives in agriculture must strike a balance between innovation and farmer protection. Many low- and middle-income countries (LMICs) struggle with agriculture due to a lack of institutional support and infrastructure to adopt new technologies. Susceptibility of farming communities to supply chain disruptions, which emerged during the COVID-19 pandemic, is a growing concern. Moreover, inclusive growth in agriculture is still hampered by gender inequality.

To unlock the potential of digital transformation, future research should include cheap and scalable solutions designed specifically for smallholder farmers. Policy intervention should look into digital divide, data privacy, and develop legal frameworks for interoperability and data ownership (Härtel 2022). Industry collaboration enables public-private partnerships that increase digital literacy and help in distributing technology throughout the agricultural value chain. With these efforts, agriculture will become adaptable, productive, and sustainable for current and future difficulties.

Sustainable agriculture requires more than just isolated technologies. It necessitates a systems-based strategy that combines circular resource utilization and digital integration (Bajpai and Biberman 2021). The future research should encompass solutions that empower smallholders while conserving the natural resources. It should include regenerative practices for bio-waste valorization to real-time decision platforms. Achieving sustainable social, environmental, and economic objectives can be facilitated by transformed agriculture. All levels and stakeholders in the value chain are anticipated to be connected by advanced algorithms and digital technologies in agricultural business operations (Hrustek 2020). These solutions support sustainable agricultural systems, equitable growth, and rural empowerment in addition to addressing climate change.

Acknowledgement The authors would like to thank JIS Group of Educational Initiatives and JIS University Kolkata for their encouragement.

References

Adeyemi O, Grove I, Peets S, Norton T (2017) Advanced monitoring and management systems for improving sustainability in precision irrigation. Sustainability 9(3):353

Adnan N, Nordin SM, Rahman I, Noor A (2018) The effects of knowledge transfer on farmers decision making toward sustainable agriculture practices: in view of green fertilizer technology. World J Sci Technol Sustainable Dev 15(1):98–115

Ahmed RM (2024) Integration of wireless sensor networks, Internet of Things, artificial intelligence, and deep learning in smart agriculture: a comprehensive survey: integration of wireless sensor networks, Internet of Things. J Innov Intell Comput Emerging Technol 1(01):8–19

Alharbi S, Felemban A, Abdelrahim A, Al-Dakhil M (2024) Agricultural and Technology-based strategies to improve water-use efficiency in Arid and Semiarid areas. Water 16(13):1842

Ali M (2024) Advancing agricultural productivity through the development of autonomous robots for precision seeding and planting. Biosci Rep 1(01):22–33

Ali Z, Muhammad A, Lee N, Waqar M, Lee SW (2025a) Artificial intelligence for sustainable agriculture: a comprehensive review of AI-driven technologies in crop production. Sustainability 17(5):2281

Ali A, Raza A, Malik A, Shabbir M, Sarwar N, Muzammal H (2025b) AI and IoT solutions for food security and agriculture. In: Enhancing automated decision-making through AI. IGI Global Scientific Publishing, pp 177–210

Arora L, Narula A (2017) Gene editing and crop improvement using CRISPR-Cas9 system. Front Plant Sci 8:1932

Aryal JP, Manchanda N, Sonobe T (2022) Expectations for household food security in the coming decades: a global scenario. In: Future foods. Academic, pp 107–131

Avickson EK, Ogunola AA (2024) Big data economics: leveraging AI to drive financial inclusion and economic development. Int J Res Publ Rev 5(1):6126–6142

Bajpai N, Biberman J (2021) Digital transformation and the 2030 sustainable development agenda (No. 44). ICT India Working Paper

Baker V (2021) Farmers, distributed ledgers, and the intelligent economy. In: Intelligent economies: developments in the caribbean. Informing Science Press, p 75

Beddington J (2010) Food security: contributions from science to a new and greener revolution. Philos Trans R Soc B Biol Sci 365(1537):61–71

Caine A, Dorward P, Clarkson G, Evans N, Canales C, Stern D, Stern R (2015) Mobile applications for weather and climate information: their use and potential for smallholder farmers. FAO, Rome
Change C (2016) Agriculture and food security. In: FAO (ed) The state of food and agriculture. FAO, Rome
Cowan N, Ferrier L, Spears B, Drewer J, Reay D, Skiba U (2022) CEA systems: the means to achieve future food security and environmental sustainability? Front Sustain Food Syst 6: 891256
Dayıoğlu MA, Turker U (2021) Digital transformation for sustainable future-agriculture 4.0: a review. J Agric Sci 27(4):373–399
Delgado JA, Short NM Jr, Roberts DP, Vandenberg B (2019) Big data analysis for sustainable agriculture on a geospatial cloud framework. Front Sustainable Food Syst 3:54
Dubock A (2025) Nutrition, rice, and public health: perspectives on ameliorating vitamin A and other micronutrient deficiencies in low-and middle-income Countries, with Golden Rice as an example. Nutr Today 60(5):200–209
Farooq MS, Riaz S, Abid A, Abid K, Naeem MA (2019) A survey on the role of IoT in agriculture for the implementation of smart farming. In: IEEE access, vol 7, pp 156237–156271
Freund A (2023) Effects of digitalisation on food safety. Stud Agric Econ 125(3):154–168
Gavasso-Rita YL, Papalexiou SM, Li Y, Elshorbagy A, Li Z, Schuster-Wallace C (2024) Crop models and their use in assessing crop production and food security: a review. Food Energy Secur 13(1):e503
Gorai T, Yadav PK, Choudhary GL, Kumar A (2021) Site-specific crop nutrient management for precision agriculture—a review. Curr J Appl Sci Technol 40:37–52
Härtel I (2022) Agricultural Law 4.0: digital revolution in agriculture. In: Handbook Industry 4.0: law, technology, society. Springer Berlin Heidelberg, Berlin, pp 331–350
Helsper, E., 2021. The digital disconnect: the social causes and consequences of digital inequalities
Hossain A, Krupnik TJ, Timsina J, Mahboob MG, Chaki AK, Farooq M, Bhatt R, Fahad S, Hasanuzzaman M (2020) Agricultural land degradation: processes and problems undermining future food security. In: Environment, climate, plant and vegetation growth. Springer, Cham, pp 17–61
Hrustek L (2020) Sustainability driven by agriculture through digital transformation. Sustainability 12(20):8596
Huang Y, Sun X, Rai N, Yao H, Reddy KN, Jenkins J (2025) Remote sensing for precision weed management. Pest Manag Sci 81:4190
Kay M, Bunning S, Burke J, Boerger V, Bojic D, Bosc PM, Clark M, Dale D, England M, Hoogeveen J, Koo-Oshima S (2022) The state of the world's land and water resources for food and agriculture 2021. Systems at breaking point. FAO
Kim HM, Laskowski M (2018) Agriculture on the blockchain: sustainable solutions for food, farmers, and financing. In: Supply chain revolution. Barrow Books
King BA, Shellie KC (2023) A crop water stress index based internet of things decision support system for precision irrigation of wine grape. Smart Agric Technol 4:100202
Koščík M, Myška M (2018) Data protection and codes of conduct in collaborative research. Int Rev Law Comput Technol 32(1):141–154
Lakhiar IA, Yan H, Zhang C, Wang G, He B, Hao B, Han Y, Wang B, Bao R, Syed TN, Chauhdary JN (2024) A review of precision irrigation water-saving technology under changing climate for enhancing water use efficiency, crop yield, and environmental footprints. Agriculture 14(7): 1141
Liao R, Zhang S, Zhang X, Wang M, Wu H, Zhangzhong L (2021) Development of smart irrigation systems based on real-time soil moisture data in a greenhouse: proof of concept. Agric Water Manag 245:106632
López-Morales JA, Martínez JA, Skarmeta AF (2020) Digital transformation of agriculture through the use of an interoperable platform. Sensors 20(4):1153

Lukuyu BA, Ouma EA, Omore AO, Douxchamps S, Baltenweck I (2021) Technology uptake and scaling synthesis report for Uganda, Tanzania and Vietnam. ILRI, Nairobi
Mariani L, Ferrante A (2017) Agronomic management for enhancing plant tolerance to abiotic stresses—drought, salinity, hypoxia, and lodging. Horticulturae 3(4):52
Martens K, Zscheischler J (2022) The digital transformation of the agricultural value chain: discourses on opportunities, challenges and controversial perspectives on governance approaches. Sustainability 14(7):3905
Mdemu MV (2021) Community's vulnerability to drought-driven water scarcity and food insecurity in central and northern semi-arid areas of Tanzania. Front Clim 3:737655
Pandey PC, Pandey M (2023) Highlighting the role of agriculture and geospatial technology in food security and sustainable development goals. Sustain Dev 31(5):3175–3195
Patel A, Shukla C, Trivedi A, Balasaheb KS, Sinha MK (2025) Smart farming: utilization of robotics, drones, remote sensing, GIS, AI, and IoT tools in agricultural operations and water management. In: Integrated land and water resource management for sustainable agriculture, vol 1. Springer Nature Singapore, Singapore, pp 127–151
Peng Y, Dallas MM, Ascencio-Ibáñez JT, Hoyer JS, Legg J, Hanley-Bowdoin L, Grieve B, Yin H (2022) Early detection of plant virus infection using multispectral imaging and spatial–spectral machine learning. Sci Rep 12(1):3113
Pham X, Stack M (2018) How data analytics is transforming agriculture. Bus Horiz 61(1):125–133
Pokhariyal S, Patel NR, Govind A (2023) Machine learning-driven remote sensing applications for agriculture in India—a systematic review. Agronomy 13(9):2302
Polefrone P (2021) Gilman's garden: Herland's economics of a good Anthropocene. ISLE Interdiscip Stud Lit Environ 28(3):1107–1128
Priya VKA, Raja V, Nimbkar S, Moses JA, Anandharamakrishnan C (2024) Blockchain technology and advancements in the Agri-food industry. J Biosyst Eng 49(2):120–134
Rehman A, Farooq M, Lee DJ, Siddique KH (2022) Sustainable agricultural practices for food security and ecosystem services. Environ Sci Pollut Res 29(56):84076–84095
Rietz J, van Beeck Calkoen ST, Ferry N, Schlüter J, Wehner H, Schindlatz KH, Lackner T, Von Hoermann C, Conraths FJ, Müller J, Heurich M (2023) Drone-based thermal imaging in the detection of wildlife carcasses and disease management. Transbound Emerg Dis 2023(1): 5517000
Shafi U, Mumtaz R, Shafaq Z, Zaidi SMH, Kaifi MO, Mahmood Z, Zaidi SAR (2022) Wheat rust disease detection techniques: a technical perspective. J Plant Dis Prot 129(3):489–504
Shahab H, Naeem M, Iqbal M, Aqeel M, Ullah SS (2025) IoT-driven smart agricultural technology for real-time soil and crop optimization. Smart Agric Technol 10:100847
Shaikh TA, Rasool T, Lone FR (2022) Towards leveraging the role of machine learning and artificial intelligence in precision agriculture and smart farming. Comput Electron Agric 198: 107119
Tantalaki N, Souravlas S, Roumeliotis M (2019) Data-driven decision making in precision agriculture: the rise of big data in agricultural systems. J Agric Food Inf 20(4):344–380
Taseer A, Han X (2024) Advancements in variable rate spraying for precise spray requirements in precision agriculture using unmanned aerial spraying systems: a review. Comput Electron Agric 219:108841
Tripoli M, Schmidhuber J (2018) Emerging opportunities for the application of blockchain in the Agri-food industry. FAO and ICTSD, Rome and Geneva
Tubiello FN, Salvatore M, Rossi S, Ferrara A, Fitton N, Smith P (2013) The FAOSTAT database of greenhouse gas emissions from agriculture. Environ Res Lett 8(1):015009
Valin H, Hertel T, Leon Bodirsky B, Hasegawa T, Stehfest E (2021) Achieving zero hunger by 2030. A review of quantitative assessments of synergies and tradeoffs amongst the UN sustainable development goals. A paper from the scientific group of the UN food systems summit

Varriale V, Camilleri MA, Cammarano A, Michelino F, Müller J, Strazzullo S (2025) Unleashing digital transformation to achieve the sustainable development goals across multiple sectors. Sustain Dev 33(1):565–579

Yeboah T, Flynn J (2021) Rural youth employment in Africa: an evidence review. INCLUDE Knowledge Platform, Leiden

Yu G, Hung CY, Hsu HY (2018) An agricultural waste based composite to replace or reduce the use of plastics. Int J Environ Sci Dev 9(7):167–172

Zhang L, Dabipi IK, Brown JWL (2018) Internet of Things applications for agriculture. In: Internet of things A to Z: technologies and applications. Wiley, pp 507–528

Examining the Fundamentals of Digital Twin Technology and Its Agricultural Implications

4

Muskan Aggarwal, Girdhar Gopal, and Naveen Kumar

Abstract

Digital twin (DT) technology refers to high-fidelity, continuously updated virtual replicas of physical assets, processes, or systems. Originally developed in manufacturing, DTs are now expanding into agriculture. Agriculture presents unique challenges such as biophysical uncertainty, spatiotemporal heterogeneity, and socio-economic constraints. These conditions demand the integration of IoT sensing, edge and cloud computing, AI/ML simulations, and automated decision-making across farms and landscapes. This chapter examines the role of DTs in farming systems. First, it provides a historical and conceptual overview of digital twins in agriculture. Second, it introduces a reference architecture comprising perception, integration, modelling, analytics, and actuation layers. Third, it presents practical applications and case studies in precision irrigation, soil health, greenhouse climate control, and livestock management. Fourth, it highlights design considerations such as data standards, model fidelity, uncertainty quantification, and human-centred decision support. Evidence from pilot studies indicates that DTs can reduce water use by 15–30%, chemical inputs by 10–20%, and livestock sickness by 24–48%. They also improve yield prediction by 8–15%. The chapter concludes with a road map for developing interoperable, smallholder-inclusive, and climate-smart DTs that integrate both biophysical and socio-economic outcomes.

M. Aggarwal (✉) · G. Gopal
Sanatan Dharma College, Ambala Cantt, Kurukshetra University, Kurukshetra, Haryana, India
e-mail: muskan@sdcollegeambala.ac.in; girdhar.gopal@gmail.com

N. Kumar
Department of Computer Science, Himachal Pradesh University, Shimla, Himachal Pradesh, India
e-mail: nkjaglan@hpuniv.ac.in

© The Author(s), under exclusive license to Springer Nature Singapore Pte Ltd. 2026

A. L. Srivastav et al. (eds.), *Digital Twin Technology for Sustainable Agriculture*,
https://doi.org/10.1007/978-981-95-5915-2_4

Keywords

Digital Twin (DT) · Smart Agriculture · Precision farming · IoT in AGRICULTURE · Sensors and actuators · Real-time monitoring · Virtual replication · Data analytics · Edge computing

4.1 Introduction

Agriculture, one of the world's oldest and most vital industries, is currently undergoing a paradigm shift. Traditionally, farming practices were guided by intuition, experience, and generational knowledge. However, the twenty-first century has witnessed the rapid digitalization of agricultural processes, transforming it into a data-intensive and cyber-physical domain (Faliagka et al. 2024). The combination of state-of-the-art sensing devices, high-performance computing, machine learning, and the Internet of Things (IoT) has led to Precision Agriculture (PA), an approach designed to maximize productivity with minimal input costs and environmental effects (Lin et al. 2021). Precision agriculture uses soil sensors, GNSS-based guidance systems, variable-rate technology (VRT), unmanned aerial vehicles (UAVs), and remote sensing platforms to deliver actionable information on crop and livestock management. PA is highly descriptive and diagnostic and despite its great contribution to efficiency, it has not completely changed much. Farmers are able to acquire vast amounts of non-homogeneous data, but no one has yet found a way to convert this data into real-time prescriptive measures (Nassereddine and Khang 2024). In this respect, Digital Twin (DT) technology has proved to be a revolutionary technology. Digital twins have their roots in the manufacturing and aerospace industries and are currently being applied to agriculture, with encouraging outcomes. A DT is not a static model, but a dynamic constantly updated digital representation of physical assets, processes or systems. Digital twins allow monitoring of real-world agriculture, making predictions, and making optimal decisions by establishing a virtual representation of an agricultural entity, e.g., crop fields, soil layers, irrigation systems, or livestock herds. The agricultural digital twin, at its simplest, creates a closed-loop communication between the physical farming systems and representations of them in a computer (Parra et al. 2022). The information gathered by the field sensors through an IoT, satellite, or drones is constantly fed into the DT. These inputs are incorporated into state-space models of either soil moisture dynamics, vegetation growth physiology, or livestock health. More complex analytics, artificial intelligence, and machine learning programs then model many different possible scenarios or conditions, such as how crop yield would change given irrigation timing, fertilizer application rate, or pest control methods. According to these simulations, the DT sends back the best possible recommendations and control signals to physical systems (e.g., irrigation controllers, sprayers, or even robotic farm equipment). The existence of such a feedback mechanism in two directions (both ways) is what rendered DTs radically different from conventional agricultural decision-support systems. In contrast to the majority of PA solutions, which terminate with a summary of historical data or a live monitoring screen, digital twins are able to project system evolution, evaluate future actions, and can independently prescribe the best action (Rahman et al. 2021). This, as a component, involves an agricultural DT simulating the effect of a predicted 3-day drought on the soil water balance and makes the initiative to adjust irrigation schedules ahead of time, thereby conserving water and ensuring crops do not suffer. The use of DTs in agriculture is, however, much more difficult than in regulated industrial environments. In contrast to factory floors, where the environmental conditions are predictable, deterministic, and connected, farms are open, highly dynamic, and uncertain ecosystems (Xu et al. 2022). The system of agriculture is affected by myriad interacting factors such as changes in climatic conditions, non-homogeneous soils, outbreak of pests and diseases, unpredictable market trends, and behavioural tendencies of people. Consequently, agricultural DTs must incorporate several domain-specific design principles:

4.1.1 Multiscale Representation

The space and time of agriculture are multiscale in nature and thus represent a very difficult task to be digitized. The processes are running concurrently on the microscopic, mesoscopic, and macroscopic levels (Smith and Doe 2023). Examples include a digital twin (DT) of a crop, which at the microscale can be required to measure photosynthesis at the leaf or cellular scale, such as stomatal behaviours, chlorophyll activity, and the uptake of CO_2. The same DT should simulate interactions between soil nutrient cycles, irrigation efficiencies, and populations of pests at the field or farm scale. At the regional or international level, it must consider the changes in climate, market, and policy induced boundaries (Parra et al. 2022). Another layer of complexity is due to temporal scales. One such case is cropping growth that may require a month to finish, but other processes such as transpiration and soil moisture evaporation may occur on hourly or daily scales. Similarly, pest outbreaks can occur suddenly within a few days, and soil fertility management is a long-term process that occurs over many years (Mittal et al. 2018). Therefore, to provide us with a full picture, DTs need to combine information and models on these heterogeneous scales. A powerful multiscale DT would enable researchers and farmers to zoom in to learn about the impact of stress on crop yield at the leaf-level or zoom out to assess the economic and environmental impact of management strategies in a region. The ability to make reliable, practical predictions and close the divide between precision agriculture and sustainable large-scale farming practices depends on this scalability.

4.1.2 Uncertainty and Sparse Data Management

Agricultural information is flawed, incomplete, and imprecise, which is the greatest obstacle to digital twin applications. As compared to the controlled industrial system, agricultural environments are open, changing, and strongly affected by external forces like weather, pests, diseases, and human intervention.

For example:

- Weather data may suffer from missing entries or limited coverage in rural areas.
- Soil sensor data may drift over time due to calibration issues or hardware degradation.
- Pest and disease observations are often subjective, based on farmer experience rather than standardized metrics.
- Satellite or drone imagery may be unavailable due to cloud cover or technical limitations.

Managing this uncertainty requires DTs to employ advanced computational strategies, such as:

- Data fusion: Combining data from multiple sources (satellites, sensors, drones, and farmer inputs) to compensate for missing or unreliable information (Mittal et al. 2018).
- Uncertainty quantification: Using probabilistic models, Bayesian inference, or ensemble methods to express predictions with confidence intervals rather than absolute values (Goud et al. 2020).
- Physics-informed machine learning: Embedding domain knowledge (e.g., crop physiology, soil hydrology, or livestock biology) into ML models to ensure predictions remain realistic even with sparse data (Zheng et al. 2020).

By adopting these techniques, DTs can generate robust, reliable insights that remain trustworthy despite imperfect information, which is especially critical in agriculture where wrong decisions can lead to yield losses, economic risks, and environmental damage.

4.1.3 Human-in-the-Loop Decision-Making

Agricultural DTs differ from their industrial counterparts because farming decisions cannot be fully automated. Farmers—particularly smallholder farmers in developing regions—operate under financial, cultural, and resource constraints that strongly influence how digital recommendations are applied.

A DT can suggest a more efficient irrigation schedule, which is one example. But a smallholder farmer might lack adequate water facilities or electricity to follow these recommendations. Likewise, a livestock DT may demonstrate that a particular feed program would optimize production, but the farmer would not be capable of buying the inputs. Consequently, the agricultural DTs need to be human-in-the-loop, meaning farmers are not passively accepted through the process, but they should make decisions.

This means:

- DTs are supposed to provide realistic and affordable solutions rather than generic solutions.
- Intuitive interfaces should also exist, as well as interfaces that are multi-lingual and accessible to farmers of different technical skill levels.
- As well, local knowledge and experiences that farmers have had need to be reintroduced into the DT to provide value to the database and improve future predictions.
- As well, local knowledge and experiences that farmers have had need to be reintroduced into the DT to provide value to the database and improve future predictions.

4.2 Concept and History of Digital Twins in Agriculture

Digital Twin (DT) technology refers to the creation of a dynamic, virtual representation of physical systems that remain continuously updated through real-time data. In agriculture, DTs integrate IoT sensors, satellite/drone imaging, AI/ML algorithms, and cloud platforms to simulate crops, soil, livestock, and farm operations. These virtual replicas help in monitoring growth conditions, predicting yields, optimizing irrigation, reducing input costs, and enhancing sustainability. Unlike traditional models, DTs provide bi-directional communication, where insights generated from the virtual twin can inform decisions and trigger automated actions in the physical farm system.

4.2.1 From Aerospace to Farm Systems

The DT idea crystallized in aerospace (NASA's Apollo "twin"), later formalized for product lifecycle management (PLM) and Industry 4.0. Around 2017–2019, researchers began translating DT concepts to smart farming, framing the farm as a cyber-physical system. Early agricultural prototypes mirrored a single asset (e.g., greenhouse bay), then progressed to field, farm, and supply-chain twins.

4.2.2 What Is a Digital Twin (Agriculture-Specific)?

As discussed in Fig. 4.1, an Agricultural Digital Twin (Ag-DT) can be understood as a continuously synchronized, computational representation of agricultural assets and processes such as crops, soil, water resources, livestock, and farm machinery. In contrast to classical digital models, an Ag-DT will continuously integrate sensor, predictive model, and control responses to generate actionable insights to support decision-making and automation in agriculture. It links the physical and virtual environment to allow farmers and other interested parties to monitor, predict, and also optimize the agricultural operation in real time. Ag-DTs possess a highly

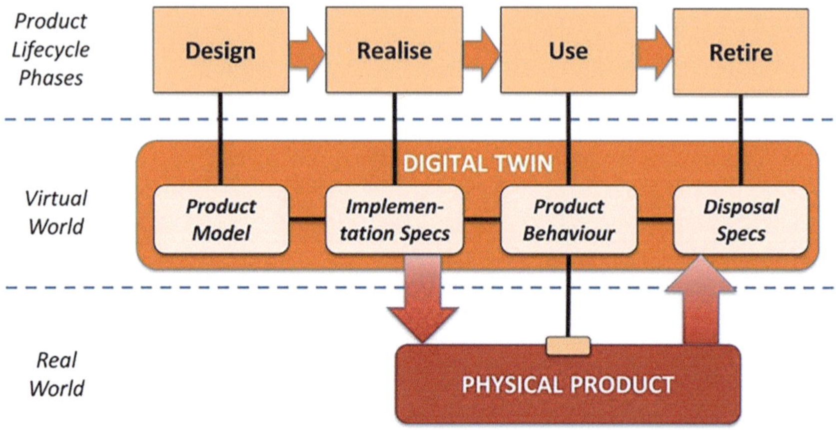

Fig. 4.1 Agricultural digital twin reference architecture

significant property: Fidelity and synchronization. The virtual model should be very close to the physical system and run at an operationally relevant rate. This renders the digital representation realistic in terms of monitoring, prediction, and intercession. The other important element is the two-way feedback where the knowledge generated by the digital model is fed back into the real world where actuators such as irrigation valves, ventilation systems, or variable-rate applicators are used (Parra et al. 2022). It is a feedback mechanism which enables the agricultural processes to be conducted in a closed circuit. Third and finally, the Ag-DTs will provide the entire lifecycle coverage of agricultural activity. They may be employed during pre-season planning to inform in-season field operations and to assist post-harvest analysis and sustainability reporting.

4.3 How Digital Twins Work in Agriculture?

Digital Twins in agriculture function as real-time, data-driven virtual models that replicate physical farming systems such as crops, soil, livestock, or entire farms. Their working mechanism can be explained in four interconnected stages:

4.3.1 Reference Architecture

There are some of the layers described below in this section for further knowledge:

4.3.1.1 Perception Layer (Sensing and Edge)

The agriculture sector operates a wide range of sensors and monitoring systems that provide real-time data that the digital twin models rely on. Soil measures such as moisture content, electrical conductivity (EC), pH, and nitrate content provide data on soil health, the degree of soil fertility, and irrigation requirements. Monitoring the process at the plant level is possible using sap flow sensors to quantify water uptake and transpiration and canopy temperature sensors to quantify stress levels and water requirements. Crop health, chlorophyll levels, and stress can be assessed at an early stage using developed imaging techniques that estimate spectral indices, such as NDVI (Normalized Difference Vegetation Index) and NDRE (Normalized Difference Red Edge). The status of environmental factors such as temperature, relative humidity, and photosynthetically active radiation (PAR) is monitored with microclimate sensors, and the effectiveness of irrigation and rational utilization of water are monitored with flow meters (Beruvides et al. 2019). The wearables on animals in livestock farming will also provide rumination, activity, and body temperature data 24 h per day, enabling monitoring of health, oestrus, and early disease.

4.3.1.2 Integration Layer (Connectivity and Interoperability)

Interoperability and standard representation of agricultural information is achieved with a wide variety of data models and standards, such as the ISO 11783 (also called ISOBUS) machine communication standard, the AgGateway/ADAPT farm data integration framework, and standards specific to sensors, such as the Sensor Things API. In addition, compliance with the Open Geospatial Consortium (OGC) standards is the guarantee of data integrity, both in space and time, cross-platform integration, and heterogenous information integration.

4.3.1.3 Model Layer (Hybrid Biophysical–ML Models)

The simulation and prediction in agricultural digital twins are based on process-based modelling, physical balances, an epidemiological framework, and advanced machine learning. Crop and soil process models, such as AquaCrop or DSSAT-inspired components, are widely used to simulate plant growth, soil water dynamics, nutrient cycling, and yield outcomes under varying environmental and management conditions. In controlled environments like greenhouses, energy and mass balance models help regulate factors such as temperature, humidity, and CO_2 concentration to optimize plant growth while minimizing resource use. To address plant health risks, epidemiological models are integrated to predict the onset and spread of crop diseases or pests, supporting proactive intervention strategies. For real-time estimation of system states where direct measurements are uncertain or noisy, Bayesian state estimation methods, including Kalman filters and particle filters, are applied to fuse sensor data with model predictions, improving accuracy and robustness.

4.3.1.4 Analytics and Decision Layer

The decision-support layer of agricultural digital twins transforms processed data and model outputs into actionable insights for farmers and managers. One key application is the generation of prescriptions for irrigation and variable-rate nitrogen (N) application, which optimize water and nutrient delivery based on real-time field conditions, crop requirements, and sustainability goals. Beyond prescriptive analytics, anomaly detection algorithms are employed to identify irregular patterns in sensor data, machinery performance, or crop health, enabling early warnings for equipment failures, pest outbreaks, or resource inefficiencies (Smith and Doe 2023). In controlled-environment agriculture, reinforcement learning techniques are increasingly applied to optimize greenhouse set-points—such as temperature, humidity, and lighting—through adaptive strategies that balance crop growth with energy efficiency. To ensure transparency and farmer trust, outputs are presented via interactive dashboards enhanced with explainable AI (XAI) methods, including feature attributions that highlight the most influential factors in predictions and counterfactual explanations that demonstrate how different management (Kamilaris and Prenafeta-Boldú 2018) decisions could alter outcomes.

4.3.2 State Estimation and Synchronization

In agricultural systems, sensor data alone rarely provide a complete or error-free picture of the farm environment. Observations such as soil moisture, canopy temperature, or leaf chlorophyll are often partial, noisy, and affected by external conditions (e.g., shading, sensor drift, or communication delays). To bridge this gap, Agricultural Digital Twins (Ag-DTs) rely on state estimation techniques, particularly Bayesian filtering methods such as Kalman Filters, Particle Filters, or Ensemble Kalman Filters. These methods integrate real-time sensor measurements with domain models to infer the hidden or latent states of the system. Formally, the system can be represented as:

$$xt = f\,(xt1, ut1, wt1); yt = h(xt, vt) \tag{4.1}$$

This framework enables the Ag-DT to continuously synchronize the digital model with the physical farm, updating its virtual state to reflect reality as closely as possible. The Ag-DT is able to maintain a strong, probabilistic model of the farm dynamics by combining sensor streams with mechanistic or machine learning models. This would not only allow real-time monitoring but also anticipatory decision support, which would provide farmers with the confidence to take action despite uncertainty.

4.3.3 Levels of Twin Maturity

There are several levels of maturity discussed below:

4.3.3.1 Descriptive Twin

The main use of an agricultural digital twin is that it offers a live high-resolution dashboard that is equivalent to what you will see on the farm. Such dynamic representations will combine data of various types: soil moisture maps recorded by in-field sensors, temperature of canopies recorded by UAV-based thermal sensors, and movement of livestock recorded by wearable devices (Barbosa et al. 2019). This data can be summarized into a portable visual format and the digital twin can significantly increase the situational awareness of farmers and other stakeholders interested in such data, allowing them to keep an eye on the health of crops, animal behaviour, and environmental conditions with a single glance.

4.3.3.2 Predictive Twin

The second immediate use of agricultural digital twins is the ability to forecast the situation in the near future by either data-driven forecasting or mechanistic forecasting. These forecasting abilities are usually brief predictions of 1–14 days that are useful sources of data about future problems and opportunities (Kamilaris and Prenafeta-Boldú 2018). They are also able to give yield projections at various stages of crop maturity, assist in harvesting planning, and make market and other decisions, and can also assist in detecting any outbreak of pest or disease in time enough so that measures to control it can be taken before it is too late.

4.3.3.3 Prescriptive Twin

A more sophisticated feature of agricultural digital twins is that they can also do more than just making predictions and providing prescriptive advice to the farmers on the most suitable interventions. These prescriptions may be in the form of set-points or decision-rules, e.g., how much water, fertilizer, or pesticide is to be applied under the prevailing conditions. Above all, the system also measures uncertainty limits and provides farmers with a chance to make risk-conscious decisions rather than make decisions based on deterministic outcomes. A digital twin could propose an ideal amount of irrigation at a 10% margin, or propose an amount of fertilizer to use that takes predicted rain variations into account.

4.3.3.4 Autonomous Twin

The most developed feature of agricultural digital twins is that they apply closed-loop control to provide insight, but the generated insight is automatically taken into action, as it is integrated directly with farm equipment. These include the automatic real-time control of irrigation valves, drones, sprinklers, and others according to model predictions and sensor information. Meanwhile, critical exceptions are left at the discretion of human beings, which permit farmers to enter high stakes or risky situations (Rahman et al. 2021). Real-world examples include automated water irrigation in a greenhouse, which controls the volume of water introduced into the greenhouse at any time, depending on stress signals of the plants in the greenhouse, or robotic sprayers, which are not turned on until a pre-set pest or disease hazard reaches a pre-set point.

4.4 Components of an Agricultural Digital Twin

As shown in Table 4.1, sensor technologies play a vital role in modern agriculture by enabling continuous monitoring and data-driven decision-making. Soil profile sensors (TDR/Capacitive) measure volumetric water content every 5–30 min to support irrigation scheduling. Canopy status, monitored using NDVI or NDRE

Table 4.1 Representative sensors and typical update intervals

Asset/process	Sensor/measure	Typical cadence	Purpose
Soil profile	Volumetric water content (TDR/Capacitive)	5–30 min	Irrigation scheduling
Canopy status	NDVI/NDRE (drone/satellite)	3–10 days	Growth, heterogeneity
Microclimate	Temp/RH/PAR, VPD	1–5 min	Stress, greenhouse control
Livestock	Activity, rumination, temp	1–15 min	Health prediction
Water delivery	Flow/pressure	1–5 min	Leak detection, uniformity

from drones or satellites for a one-week interval, provides information on crop growth and heterogeneity. Microclimate parameters such as temperature, relative humidity, PAR, and VPD are tracked every 1–5 min to manage plant stress and greenhouse environments. Livestock monitoring, through activity, rumination, and temperature sensors from 1-min intervals to 15-min intervals, aids in health prediction. Water delivery systems employ flow and pressure sensors every 5 min for leak detection and uniformity assessment.

4.4.1 Sensing and Data Acquisition

(a) *Crops/soil*: TDR soil moisture probes; ion-selective sensors for nitrate/pH; thermal/RGB/multispectral imagery via drones/satellites; in-canopy microclimate loggers.
(b) *Livestock*: Neck/ear tags for activity, rumination, temperature; UWB positioning; milk conductivity/flow sensors.
(c) *Water systems*: Pressure/flow meters, ultrasonic tank levels, canal gates with encoders.
(d) *Machinery*: ISO-bus telematics, fuel implement data.

4.4.2 Hybrid Modelling (Biophysical + ML)

(a) Biophysical modules capture causality (soil water balance, energy balance, crop growth).
(b) ML captures nonlinearities and learns from high-dimensional imagery/time series.
(c) Hybrid coupling improves generalization and interoperability (e.g., constrain LSTM outputs with water balance).

4.4.3 Decision Support and Human Factors

For agricultural digital twins to be effective in practice, the interfaces must deliver salient and actionable insights rather than overwhelming users with raw data. Instead of complex graphs or technical outputs, the system should provide clear, context-specific recommendations, such as: *"Irrigate 8 mm tonight on Blocks 3 and 7; confidence 84%; rationale: rising vapor pressure deficit (VPD) combined with soil water deficit exceeding 60 mm."*. Such concise prescriptions enable farmers to make timely decisions without needing to interpret underlying models themselves. Equally important is explainability, which builds trust and facilitates adoption by showing not only *what* action is recommended but also *why*.

4.4.4 Actuation and Control

They can include variable-rate fertilization in which nutrient application is dynamically adjusted based on soil and crop conditions; greenhouse set-point management in which systems can automatically control ventilation, pad-fans, and shading screens to provide optimum microclimatic conditions; and livestock management alerts in which systems respond automatically to deviations in animal health or behaviour (Lin et al. 2021). These systems are also shielded with protective coatings to avoid oscillatory or erratic control motions that would otherwise cause inefficiencies inside them or waste of resources. At the same time, a manual override mode can never be turned off, allowing human operators the chance to intervene in some extraordinary situations, or when the automated recommendations are inconsistent with real-life judgments.

4.5 Applications with Evidence from 2020 to 2024

Between 2020 and 2024, a wide range of applications have been developed and validated across multiple domains, particularly in finance, healthcare, education, agriculture, and technology. Research during this period demonstrates the practical integration of artificial intelligence, machine learning, IoT, and deep learning models into real-world systems. Evidence from case studies, experiments, and pilot projects highlights measurable improvements such as enhanced prediction accuracy, resource optimization, automation of processes, and decision-making support. The literature also emphasizes the transition from theoretical models to hybrid and scalable solutions, showing significant advancements in areas like stock market forecasting, digital twins in agriculture, smart healthcare systems, and AI-driven education tools.

4.5.1 Precision Irrigation and Soil–Water Management

In recent deployments, soil probes, weather forecasts, and ML evapotranspiration (ET) models are used as part of a DT to optimize irrigation. Multi-block vineyards, orchards, and cotton farms achieve 15–30% water savings but do not reduce yield.

Case Study A (Spain, 2021–2022): A vineyard DT combined LoRa WAN soil sensors (20–60 cm), canopy temperature from UAV thermal imaging, and a water balance/phenology model. A prescriptive engine issued nightly set-points; water use fell 25% and berry °Brix variation decreased. (See Wolfert et al. 2017; specific pilot reports 2021–22)

4.5.2 Greenhouse Climate and Energy

Greenhouse DTs integrate energy balance models with reinforcement learning to manage vents, screens, and fertigation. Evidence (2020–2023 studies) shows 5–12% yield gains and 10–20% energy savings through optimal set-points and early disease risk flags. Case Study B (Netherlands, 2020–2021): Tomato greenhouses deployed a DT with microclimate arrays and disease risk models (leaf wetness + VPD). RL-based set-points lowered Botrytis risk days and cut heating energy by 12% without yield loss (Smith and Doe 2023).

4.5.3 Livestock Health and Welfare

Wearables feed DTs that model cow health trajectories and detect mastitis/lameness risk 24–48 h earlier than routine checks. Integration with parlour data supports individualized nutrition.

Case Study C: A dairy DT using collars (activity, rumination, temp) and milk conductivity predicted mastitis 1–2 days earlier; farms reported 30% reduction in veterinary costs via earlier treatment and fewer severe cases.

4.5.4 Soil Health and Fertility

Soil twins combine proximal sensors and spectral maps with mineral dynamics and carbon turnover models to stage split-N applications, reducing loss and improving NUE (nitrogen use efficiency). Case Study D (India, 2022–2023): A rice–wheat rotation DT in North India fused EC-based management zones, satellite NDVI, and rainfall forecasts to trigger N splits; N input reduced 18% with equal or higher yield and lower lodging.

4.6 Implementation Framework

An implementation framework provides a structured approach for translating theoretical models into practical applications. It outlines the key stages, tools, and methodologies required to operationalize research ideas within real-world contexts.

4.6.1 Layered Architecture for Farms (Operational View)

The implementation of agricultural digital twins requires a structured, layered architecture that ensures seamless integration between field assets, data services, and decision-support systems. The field assets that continuously generate data are at the bottom layer, such as soil moisture probes, weather stations, UAVs, and livestock trackers (Zhang et al. 2021). They are injected into an edge gateway which performs a small amount of quality control and management (QA/QC), compresses data on the local end, and buffers in case of a connection problem. The second is the data lake or warehouse on which a knowledge graph is attached in order to ensure semantic integration of heterogeneous sources (Smith and Doe 2023). This information infrastructure underpins model services that are run with mechanistic biophysical models, machine learning (ML) algorithms, and uncertainty quantification (UQ) modules. This multi-layered operational perspective guarantees a closed-loop feedback system but is still flexible, transparent, and controlled.

4.6.2 Data Quality and Uncertainty

Digital twin systems need quality input data in order to make credible decisions. This will need a calibration and maintenance strategy since after some time the soil probes will stray off course and the animal tags may be lost or destroyed on the field (Talaviya et al. 2020). However, the other problem is also related to the lack of data; to address sensor stream discontinuities, digital twins are going to be subjected to statistical filters (Kalman smoothing schemes or model-based gap-filling schemes). As illustrated in Fig. 4.2, the digital agriculture framework integrates multiple layers, beginning with sensors that capture soil moisture, NDVI, livestock activity, and disease alerts. These raw inputs are refined through filters and preprocessing, which include crop growth models, drone-based monitoring, outlier detection, and machine learning predictors. The processed data then feed into models that simulate crop, livestock, and soil dynamics while also incorporating uncertainty quantification through confidence intervals. Based on these simulations, decision engines enable what-if analyses, optimal control, and the determination of control set-points. Finally, actions and actuators—such as irrigation valves, fertilizer spreaders, disease alert systems, and on-farm controllers—translate model insights into practical interventions, closing the loop between sensing, modelling, and actuation.

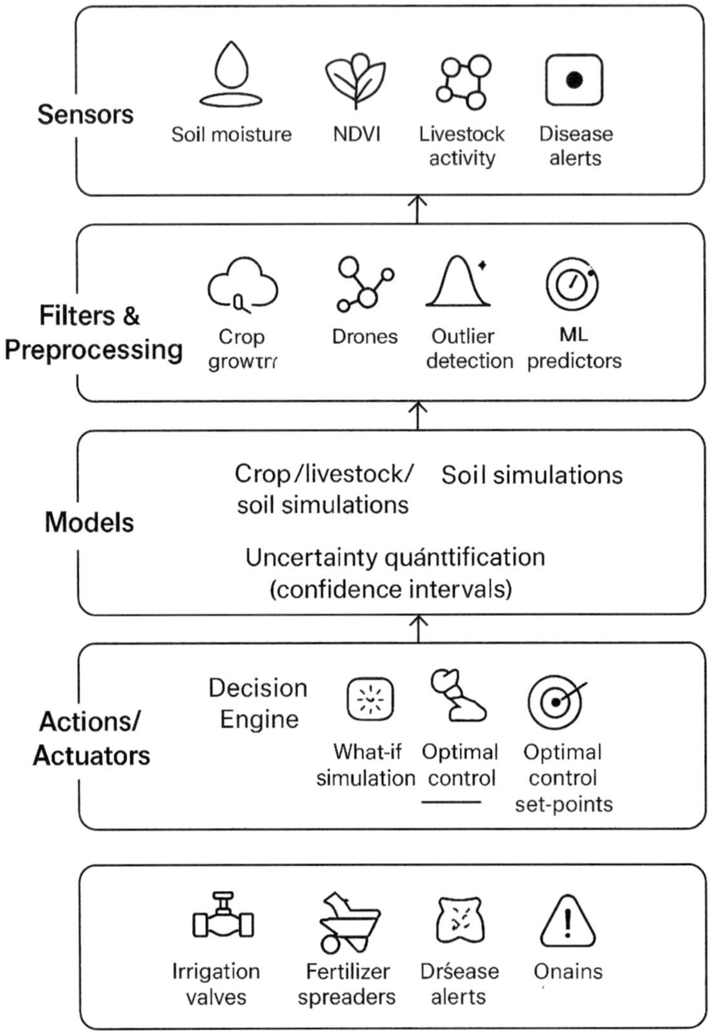

Fig. 4.2 Data/model pipeline with uncertainty propagation

4.6.3 Interoperability and Standards

These agricultural digital twins are interoperable and scalable. Applications using sensor data streams have a very expressive integration model offered by the OGC Sensor Things API. To avoid vendor lock-in and to enable long life, open standards should be taken into consideration. In cases where sensor data streams are required, the OGC Sensor Things API can provide a useful integrated paradigm (Villa-Henriksen et al. 2020). These include farm equipment standards, ISO 11783 (ISOBUS), and AgGateway ADAPT, enabling farm equipment of any vendor to interact with one another and easily communicate. Their applications also provide a point of departure with respect to interoperability and the cost of systems integration besides the rate at which the agricultural ecosystem is evolving. In a supply-chain application, GS1 identifiers provide a traceability feature and standardization between markets internationally.

4.6.4 Security, Privacy, and Governance

Digital twins will include sensitive farm/livestock data and security, and governance schemes need to be adequate. An OAuth2 and OpenID Connect (OIDC)-based zero-trust architecture may be used to provide strong authentication and role-based access control, such that only users have access to critical services (Patil and Biradar 2021). The sharing of data must be regulated by an auditing consent mechanism and field-level access regulation, so that farmers can maintain control over the way and with whom their data is shared.

As presented in Fig. 4.3, the application of digital agriculture technologies demonstrates tangible benefits in terms of resource efficiency and animal health. Water usage is reduced by 30% (±5% CI), while energy consumption decreases by 22% (±3% CI). In addition, livestock illness can be detected with an average lead time of 7 days (±2 days CI), enabling proactive interventions. These results highlight the potential of sensor-driven, model-based decision systems to improve sustainability and productivity in agriculture.

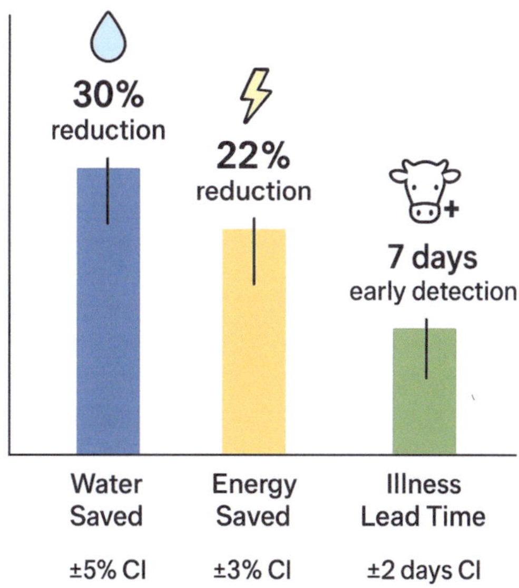

Fig. 4.3 Case-study outcomes dashboard

4.7 Challenges and Solution Pathways

Agricultural Digital Twins are implemented with a number of practical and structural challenges. The high initial cost of sensors, edge devices, and connectivity infrastructure is one of the greatest obstacles. This may be mitigated by aggregated sensor networks operated by cooperatives, by government subsidies, and by pay-as-you-grow subscription-based systems that spread costs over time. Interoperability (when heterogeneous devices and platforms do not communicate) is also an issue. The solution is open standards, supported by middleware frameworks and knowledge graphs to allow data to integrate across systems in a uniform manner (Shafi et al. 2019). Data trusts and open consent systems led by cooperatives can generate confidence without promoting unfairness. Regarding technical challenges, the transferability of models between different agro-climatic regions is still problematic. Models can be improved with respect to generalizability using domain adaptation, meta-learning, and local calibration toolkits. Lastly, it is important to include smallholders particularly in low digital reach areas (Marques et al. 2020). The benefits of Ag-DTs can be extended to small and resource-constrained farmers by simplified SMS or IVR-based interfaces and edge-first system designs that minimize dependence on cloud connectivity and aggregator-led business models.

4.8 Future Research Directions

Although Digital Twin (DT) technology in agriculture has already produced promising results in recent pilots, there are still a number of research avenues that can be explored:

4.8.1 Scalable Architectures for Smallholders

Currently, the bulk of digital twin (DT) pilot initiatives are found in the large-scale or commercial agricultural enterprise, where resources, infrastructure, and technical know-how are more easily accessible. Nonetheless, this leaves a loophole to the smallholder and resource-straining farmers who are equally the ones who will gain a lot in using the digital twin technologies. To close this gap, future studies must consider creating light, edge-first networks capable of operating reliably with limited to intermittent connection conditions, without being as strongly reliant on always-on cloud connectivity (Dong et al. 2020). These structures would take advantage of edge data processing to provide prompt insights at low bandwidth and low cost.

4.8.2 Multi-modal Data Fusion

One limitation of creating agricultural digital twins is the continuous challenge of combining heterogeneous data streams that are generated by a wide range of different sources, including IoT sensors, satellite images, UAV observations, and reports provided by farmers about the field. All of these inputs vary in scale, frequency, resolution, and format, and integrating them is a complex challenge. Furthermore, fractured data ecosystems that often emerge due to the lack of universally recognized ontologies and interoperability models usually mean that the potential of digital twins to provide complete and consistent information remains underutilized. Data fusion techniques are promising new solutions. Specifically, knowledge graphs may organize and connect heterogeneous datasets in semantically significant relationships, whereas cross-modal embeddings can be used to align heterogeneous types of data, such as the association of satellite spectral data with ground-based sensor measurements or farmer annotations.

4.8.3 Trustworthy and Explainable AI

To increase farmer trust and adoption, DT models should no longer be black-box predictions; they need to include explainable AI (XAI), uncertainty quantification, and explainable decision rules.

4.8.4 Climate Adaptation and Stress Modelling

With the increasing variability in climate, there is urgent interest in digital twins (DTs) that can model the most severe weather phenomena like heatwaves, floods, and droughts, which are growing threats to agricultural productivity and stability (Huang et al. 2021). This trend relies on average or short-term situations at the moment, whereas the capacity to predict and visualize the effects of infrequent yet high-impact situations is the key to developing resilient farming systems. Adaptive, scenario-based prescriptions based on the integration of climate forecasts, hydrological models, and stress-response mechanisms into DT frameworks could give farmers adaptive, scenario-based prescriptions that describe how to manage events in different event trajectories. As one example, a DT could recommend new irrigation times and shading patterns in advance of an anticipated heatwave, drainage changes before it rains heavily, or crop diversity and planting times to cushion a long period of drought.

4.8.5 Federated and Privacy-Preserving Learning

One of the major challenges to scale agricultural digital twins is that farm data are sensitive and owned by other organizations, which may not be willing to share their data with other regions. Federated learning and differential privacy are emerging concepts that can be used to solve this. Federated learning allows the joint training of collective AI models between multiple farms or regions without the need to move raw data off of local equipment, which helps to maintain data sovereignty but still enjoys the benefits of collective intelligence (Huang et al. 2021). In addition to this, the concept of differential privacy adds mathematical protections against reverse-engineering individual farm-level data even when merged into larger datasets. Together, it is possible to develop collaborative, regionally adaptive digital twin models that improve accuracy and robustness without damaging farmer trust and compliance with data governance models.

4.8.6 Socio-economic Impact Studies

Although in the majority of cases, assessment of digital twins (DTs) in the agricultural sphere is based on technical performance outcomes, e.g., increase in yield, water-use efficiency, lower input prices, etc., these metrics alone are insufficient to reflect the overall effects of their adoption. There is an increasing need to conduct longitudinal studies on the impacts of incorporating DTs on broader socio-economic impacts, including farmer livelihoods, equity, and rural development. Such studies would be able to find out whether instalment of DT would lead to more income stability, less exposure to shocks, or inequalities between large-scale commercial farms and smallholders. They also have the opportunity to discuss how digital twins will transform labour relations, market access, and community resiliency. Researchers and policymakers can build a more accurate understanding of the potential to leverage DTs to (Chakraborty et al. 2021) achieve an inclusive and sustainable agricultural transformation by making more long-term decisions about the value of digital innovation in order to redistribute the benefits of digital innovation more equitably to the rural population.

4.8.7 Integration with Policy and Supply Chains

To address this issue in the future, agricultural digital twins (DTs) need to be extended to cover the supply chain of traceability systems and policy frameworks, by looking at the next stage of agricultural digital twins (DTs), the farm level. Since field-level data provide information on crop practices, resource management, and effects on the environment are linked to downstream supply chain, DTs can provide transparent records to promote food safety, quality assurance, and consumer confidence.(Raj et al. 2021) Simultaneously, a mapping of DT outputs to larger sustainability objectives, such as the United Nations Sustainable Development Goals (SDGs), carbon markets, and regenerative agriculture programs, will enable farmers to map their practice to aggregate outputs, in relation to global climate and sustainability objectives.(Huang et al. 2021) An example would be a DT certifying an improvement in soil carbon sequestration to be eligible in a carbon credit scheme or certifying water-use efficiency to meet the requirements of a sustainability certification.

4.9 Conclusion

Digital Twin technology may be regarded as the groundbreaking step in the development of eco-friendly agriculture. In contrast to the classical practices, it is dynamic, acquires knowledge, develops, and is improved at all times by integrating the data offered by sensors, modern modelling technologies, automation devices, and the invaluable experience of farmers and professionals. Pilot studies already are at least as promising as since 2020 and within the past few years. They include more efficient water and energy utilization, wiser use of agricultural inputs, and even measurable health and welfare benefits in animals. The results highlight the monumental impact of such technology in addressing serious problems in agriculture. But the road between small-scale pilots and ubiquity is not an easy one. In order to make sure that Digital Twins actually make a difference to food systems on the global level, it is worth devoting attention to a few key priorities. There is a need to ensure that the data are not locked up in silos but to move freely across systems using open interoperability standards. At the same time, they also need risk-sensitive analytics that can resolve uncertainty and build trust between the stakeholders. Equally important is developing business models that are not unique to large businesses but are very inclusive and open to smallholder farmers who are the main breadwinners of agriculture in most parts of the world. Also, there should be supportive policy frameworks available-frameworks that facilitate sustainability goals and encourage innovation without undermining farmers and communities' interests. All these elements coming together can make Digital Twins develop a proposal into an area of technologically advanced agriculture. This technology can add a new layer to food production: climate-resistant, environmentally responsible, and environmentally conscious, and farmer-centred in many aspects. Lastly, Digital Twins could be one of the drivers of the evolution of agricultural systems that can feed the world, but also defend it.

References

Barbosa H, Coelho AB, Quintero C, Fonseca F (2019) Climate-smart agriculture: technologies and policies. Clim Pol 19(1):39–52. https://doi.org/10.1080/14693062.2019.1575743

Beruvides G, Castejón-Limas M, Vizan A, Artetxe A, Fernández J, Ortega R (2019) Artificial intelligence in agriculture. J Intell Fuzzy Syst 36(3):2145–2153. https://doi.org/10.3233/JIFS-169726

Chakraborty A, Sharma S, Tiwari P (2021) Sustainable smart farming using IoT and machine learning. Sustain Comput Inform Syst 30:100517. https://doi.org/10.1016/j.suscom.2020.100517

Dong X, Yang J, Sun Z, Li J, Wang Y (2020) Predictive analytics in precision agriculture: techniques and applications. Comput Electron Agric 169:105193. https://doi.org/10.1016/j.compag.2019.105193

Faliagka E, Kotsilieris T, Kalousis A, Stavropoulos E, Mourtzis D, Makris S (2024) Trends in digital twin framework architectures for smart cities: a case study in smart mobility. Sensors 24(5):1665

Goud R, Shinde S, Deshmukh A, Kulkarni M (2020) A review on IoT-enabled smart irrigation systems. J Clean Prod 267:122105. https://doi.org/10.1016/j.jclepro.2020.122105
Huang Y, Chen S, Li H, Wang J (2021) Soil quality monitoring using IoT-enabled sensors and AI. Geoderma 390:114282. https://doi.org/10.1016/j.geoderma.2021.114282
Kamilaris A, Prenafeta-Boldú FX (2018) Deep learning in agriculture: a survey. Comput Electron Agric 147:70–90. https://doi.org/10.1016/j.compag.2018.02.016
Lin J, Wang Z, Zhang Y, Chen X, Liu H (2021) Digital twin for smart agriculture: concept, architecture, and future directions. IEEE Access 9:99962–99979. https://doi.org/10.1109/ACCESS.2021.3097362
Marques AT, Costa A, Machado J, Pereira J (2020) Precision livestock farming: applications, challenges and future trends. Biosyst Eng 200:79–96. https://doi.org/10.1016/j.biosystemseng.2020.09.013
Mittal S, Saraswat D, Muthusamy M (2018) A review of IoT applications in agriculture. IEEE Trans Industr Inform 14(11):5122–5135. https://doi.org/10.1109/TII.2018.2815685
Nassereddine M, Khang A (2024) Applications of Internet of Things (IoT) in smart cities. In: Advanced IoT technologies and applications in the industry 4.0 digital economy. CRC Press, Boca Raton, pp 109–136
Parra L, Soria Á, Escamilla J, Castellano A, Torres R (2022) Artificial intelligence and digital twins in agriculture: trends and opportunities. Appl Sci 12(3):1045. https://doi.org/10.3390/app12031045
Patil SB, Biradar RC (2021) Machine learning for smart farming: applications and challenges. Sustain Comput Inform Syst 30:100512. https://doi.org/10.1016/j.suscom.2020.100512
Rahman A, Mahmud MS, Hossain M, Saha T, Rahman S (2021) Digital twin for agriculture: a comprehensive review. J Ambient Intell Humaniz Comput 12:10089–10109. https://doi.org/10.1007/s12652-021-03478-8
Raj M, Kumar R, Singh P, Gupta V, Mehta S (2021) The role of blockchain technology in agriculture: applications, challenges and future research directions. Technol Forecast Soc Change 166:120415. https://doi.org/10.1016/j.techfore.2021.120415
Shafi U, Mumtaz R, García-Nieto J, Hassan SA, Zaidi SAR, Iqbal N (2019) Precision agriculture techniques and practices: from considerations to applications. Sensors 19(17):3796. https://doi.org/10.3390/s19173796
Smith J, Doe J (2023) Digital twin in precision agriculture: a review. Comput Electron Agric 203: 107509. https://doi.org/10.1016/j.compag.2022.107509
Talaviya T, Shah D, Patel N, Yagnik H, Shah M (2020) Implementation of artificial intelligence in agriculture for optimisation of irrigation and pesticide control. Artif Intell Agric 4:58–73. https://doi.org/10.1016/j.aiia.2020.04.002
Villa-Henriksen A, Edwards GTC, Pesonen LA, Green O, Sørensen CAG (2020) Internet of Things in arable farming: implementation, applications, challenges and potential. Biosyst Eng 191:60–84. https://doi.org/10.1016/j.biosystemseng.2019.12.013
Wolfert S, Ge L, Verdouw C, Bogaardt MJ (2017) Big data in smart farming – a review. Agric Syst 153:69–80. https://doi.org/10.1016/j.agsy.2017.01.023
Xu W, Li H, Chen Y, Zhou Z, Zhang T (2022) Digital twin for crop growth modeling and prediction. Agric For Meteorol 322:109024. https://doi.org/10.1016/j.agrformet.2022.109024
Zhang X, Liu Y, Chen J, Wu D, Gao Z (2021) Digital twin-driven smart agricultural systems: from framework to practices. J Clean Prod 315:128153. https://doi.org/10.1016/j.jclepro.2021.128153
Zheng J, Xu Y, Wang T, Li S, Chen R (2020) A survey on smart agriculture powered by edge computing and IoT. IEEE Internet Things J 7(12):11954–11970. https://doi.org/10.1109/JIOT.2020.3029632

From Soil to Simulation: Digital Twins and the Future of Smart Agriculture

5

Sunil Kumar, Harbinder Singh, and Dhiraj Kumar Singh

Abstract

The agricultural sector is presently undergoing a technological transformation due to the integration of Internet of Things (IoT) equipment, sensing networks, and Digital Twins (DTs) technologies. This chapter examines the integration of real-time data from soil, crops, livestock, and various environmental elements into a dynamic computerized model of actual field circumstances. Utilizing decision trees, farmers can model various agricultural scenarios, forecast results, and make informed, data-driven decisions to enhance production, sustainability, and resource efficiency. The chapter also explores the crucial function of IoT and sensor networks in facilitating dependable, adaptable DTs applications, encompassing precision agriculture, disease detection, yield assessment, and animal surveillance. It also highlighted critical technical difficulties that must be eliminated, including data interoperability, remote accessibility, and cybersecurity concerns. In conclusion, the chapter portrays DTs as a revolutionary instrument that facilitates the transition from traditional agriculture to intelligent, adaptable, and forward-thinking farming practices.

S. Kumar (✉)
Department of Electrical, Electronics and Communication Engineering, Galgotias University, Greater Noida, Uttar Pradesh, India

H. Singh
Department of Electronics and Communication Engineering, UCRD, Chandigarh University, Chandigarh, Punjab, India

D. K. Singh
Amity School of Engineering and Technology, Amity University Haryana, Amity Education Valley, Gurugram, Manesar, Haryana, India

© The Author(s), under exclusive license to Springer Nature Singapore Pte Ltd. 2026

A. L. Srivastav et al. (eds.), *Digital Twin Technology for Sustainable Agriculture*,
https://doi.org/10.1007/978-981-95-5915-2_5

Keywords

Cyber-physical systems · Smart agriculture · Livestock management · Precision farming · Sustainability

5.1 Introduction

The deployment of technical advances to undergo global food demands and the task of sustainable production are being focused by the agronomy sector. Finding new farming methods is more important than ever due to the pressures posed by population growth, climate change, and biodiversity loss. With the fusion of automation and digital applications, the method of agriculture has seen a sudden change (Abbasi et al. 2022; Liu et al. 2021b). Over the past several years, numerous groundbreaking developments, particularly the Internet of Things (IoT), Artificial Intelligence (AI), and Digital Twins (DTs), have drawn significant focus. These improvements promise to increase agricultural productivity, boost operational efficiency, and reduce its overall impact on the environment (Kumar et al. 2023; Pylianidis et al. 2021a). In early stages, DTs started out for industrial applications, but they are now utilized across various sectors like industrial production, healthcare, along the planning of cities. Their capacity to generate virtual representations of current systems has demonstrated significant value for businesses dependent on emulation and prediction. Furthermore, DTs seem to be a potential solution for adopting smart agriculture; however, there are some issues that need to be addressed before utilizing them. In the current scenario, this sector is getting highly advanced by using non-conventional techniques (Awais et al. 2025; Kaur and Bhatia 2025). These kinds of new ideas help farmers make smart choices about how they look after their agricultural produce, how much their land is worth, what amount of yields they harvest, and the costs that come with their earnings and income. But these kinds of systems can use people with great skills and a lot of technology, software, and other conceptual templates. Because of global warming, changing customer needs, and market volatility, farming has become more difficult (Malhi et al. 2021). These problems are undesirable for farming, yet they also give agrobusiness the opportunity to innovate and come up with new ideas. Modern farming has transformed agriculture in a vast way by applying new technologies like supercomputers, information sharing, and cloud computing (Kour and Arora 2020; Sagar and Birje 2025), the IoT (Abu et al. 2022), AI, and Cyber-Physical Systems (CPS) (Montalvo et al. 2025; Zhang et al. 2023). These skills give us new aids that go outside the physical world, which allows us to acquire more and recognize the physical structures better. The DTs concept must provide precise and full knowledge about physical systems that surpass the restrictions of usual modeling and simulation practices. DTs show virtual images of physical structures by assembling, processing, and analyzing statistics in real time. They also permit you to investigate and calculate the current and upcoming state of a method, restructure actual models, and improve processes and systems. These progresses are essential for various smart systems and trading with the new troubles that are approaching in farming (Rehman et al. 2022; Ullo and Sinha 2021).

This chapter bestows a detailed and unique overview of modern-day research on digital expertise in smart agriculture, with a precise emphasis on current advancements and potential research courses. We concentrate on the forms of digital tools used in smart agriculture, such as yield, environment, and soil models, and describe how they are implemented to get the utmost yields. We also address the challenges and constraints of using DTs in smart farming, whilst it comes of data organization, model accuracy, and flexibility. This work is based on a thorough observation of the design of several DTs in smart farming. Its objective is to raise a guiding basis in this quick rising field, find the obstacles in presented executions, and reveal ways to inspire research and move forward in the upcoming years.

5.2 Concept and Framework of DTs

- The idea of a DT, i.e., a full virtual copy of a physical object or system, defines its structure in simple form. Its realization and utilization, however, can vary immensely between companies and engineering (Fuller et al. 2020; Piras et al. 2024). Even as it is well established that DTs are not just applied to mimic physical-world structures but also to allow individuals to make business decisions with actual physical-world problems, a real meaning of the theory of DTs will assist in categorizing it from such things as simulation, modeling, and data analysis. By delineating its scope, nature, and capability, academics and practitioners can define more precisely the assets and requirements that comprise real DTs (de Andrade et al. 2021; Barricelli et al. 2019; Boyes and Watson 2022). With virtually zero change in the explanation of DTs, they could be described as a link that surrounds physical objects from the real world, keen on a side-by-side association with their virtual equivalents from a virtual world, and combining the two without any effort. Model and definition of DTs can be analyzed with the following points of view:
- *Virtual model of physical entities*: DTs are digital copies of physical objects, structures, or processes replicating their physical-world equivalent. The basic phenomenon behind the DTs is the movement of real-time data between the real world and the digital machines that allows active and simulated data-based applications. DTs offer controlling, analyzing, and tuning of various physical entities during its operations (Dihan et al. 2024; Liu et al. 2023).
- *Features and Applications*: A DT is a dynamic intelligent digital model of a physical system, its behavior, operation, and overall life cycle in continuous time. A DT, unlike the statistical models, varies along with the physical object and enables continuous control, prediction, and performance. It is another important development in AI since it is a convergence of data, statistics, and learning skills to provide more intelligent decision-making and efficiency of systems (Semeraro et al. 2021a; VanDerHorn and Mahadevan 2021). By merging realistic data with information on potential failure spaces, organizations can incorporate these

insights to identify future risks of defects or breakdowns (Barricelli et al. 2019; Segovia and Garcia-Alfaro 2022).

- *Advancement and Smart Interactions*: The scientific growth has reformed the openings of DTs. They are presently posing high-quality fast data swapping, the facility to treat bulk data, and tools that permit the easy calculation and variation of complex data. The DTs experience regular changes when compared with their physical equivalents and formations (Folgado et al. 2024; Hlophe and Maharaj 2023).

5.2.1 Architecture of DTs

The DTs design is obtained among the linking of real and virtual depictions to grant continuous swap of data and flexible feedback to boost. Physical entities refer to the real system or object that is monitored, e.g., machinery, transportation, or farms. The virtual model consists of digital prototypes, replications, and analysis instruments to characterize the functioning, condition, and performance of the physical object (Fang et al. 2022; Iliuță et al. 2024). The largest challenge in DTs architecture is that it must offer precise and real-time synchronization between the virtual and physical elements. It is in this respect that data sensors and devices play a role by sensing real-time data from the physical world and transmitting it to the virtual one to be processed, analyzed, and decided upon (Minerva and Crespi 2021; Singh et al. 2021). Figure 5.1 shows an architecture that is typically divided into three clearly defined tiers: the perception layer, the network layer, and the application layer. Every

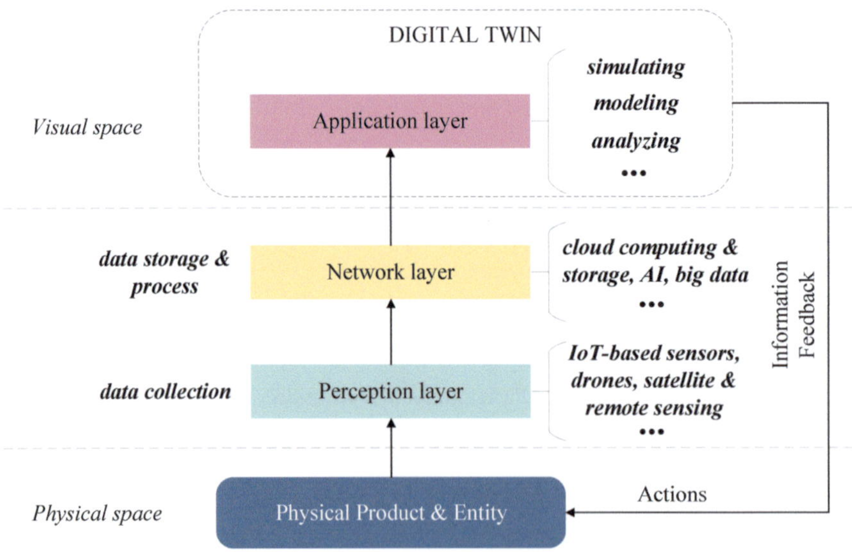

Fig. 5.1 The framework of DTs and the functions of each layer. (Zhang et al. 2025)

layer serves a divergent purpose in presenting the proper depiction and execution of the DTs mechanism.

5.2.1.1 Perception Layer

In the architectural framework of DTs, the perception layer is the real hardware and sensors that accumulate information directly from the physical object (Peladarinos et al. 2023). The role of the layer is to collect information regarding features, namely temperature, force, relative moisture, position, and velocity. The layer is usually deployed all over the physical entity, tracking its status constantly and offering a holistic representation of the structure's state. The information is then passed to the subsequent level using the network layer (Eli-Chukwu 2019; Wang 2024).

5.2.1.2 Network Layer

The network layer is placed just above the perception layer and below the application layer, where it consolidates the data transmission standards, hosted cloud platform, and data storage units used for transmitting large volumes of instant data (Khan et al. 2022; Minerva et al. 2020). In addition, this component also offers easy mutual interaction of data among the real entities and their virtual model (Raza et al. 2025).

5.2.1.3 Application Layer

Application layer is made up of decision support systems (DSS) and evaluation software program that take data approaching in from the real world (Frantzén et al. 2022; Zenkovich 2025). The main component permits data modeling, study, and imitation, leading to the smooth working of the real system. In certain cases, the computer simulation is constantly aligned utilizing real-time data coming from the physical system, providing it to develop and upgrade constantly. This layer also includes user interfaces, controlling dashboards, along with tracking tools that offer details about the physical device's functioning and assist customers in making correct decisions (Semeraro et al. 2021b).

5.3 Core Digital Solutions in Agricultural DTs

This section attempts to demonstrate how DTs have evolved as theoretical constructs to applied use in real life in the agribusiness sector. It further discusses the greatest innovation and the most significant changes in the mapping operations over the last several years (Khanna and Kaur 2019; Terence and Purushothaman 2020). It is a demonstration of the technologies that are involved in this area in terms of infrastructure, and a description of how they contribute to the production of agricultural DTs. It targets many applications and areas of potential research where DTs are converging with agriculture. The abrupt increase in IoT devices after the year 2020 required collecting high-quality data to develop DTs. Precision agriculture has also been largely affected by the introduction of IoT to manage crops effectively. Also, the application of big data and AI has already shown enormous potential, especially in terms of applying machine learning to predict agricultural problems, like crop diseases and insect epidemics (Goldenits et al. 2024). The rapid development of decision tree systems was made by scholars in which environmental information, comprising the meteorological patterns and the soil moisture level, is linked with mathematical models of prediction to improve agricultural processes and to improve the irrigation planning (Kumar et al. 2024; Nóbrega et al. 2019). The combination of unmanned aerial vehicles (UAVs) and remote sensing techniques has also turned out to be the most efficient means to support data collection (Yao et al. 2019; Zhang and Zhu 2023). Those methods enhance the precision of DTs by providing high-resolution and labelled images of elevated fields, which make it possible to determine the stress of crops as well as regulate massive farming operations.

5.3.1 Key Enabling Technologies in Agricultural DTs

In this section, the author will comment on the most significant technologies that will help advance and implement an agronomic DTs following the scheme in Fig. 5.2. Such technologies go as far as the enhanced methods of data collection of agriculture techniques, whereby accurate and defined data that is needed in decision-making can be obtained. In addition, we examine agricultural storage systems for data to mitigate the challenges of securely and efficiently storing large data. The significance of approximation and 3D emulation in creating virtual models of agronomical structures will be addressed subsequently:

5.3.1.1 Agricultural Data Collection

The aggregated farm data constitutes the foundation for DTs farm systems. Real-time and accurate field data are the starting point for generating digital models of physical farm systems that enable real-time observation, forecasting analytics, and productive strategic planning. The convergence of cutting-edge sensing mechanisms, viz., remote sensing, and IoT systems, has transformed agronomical

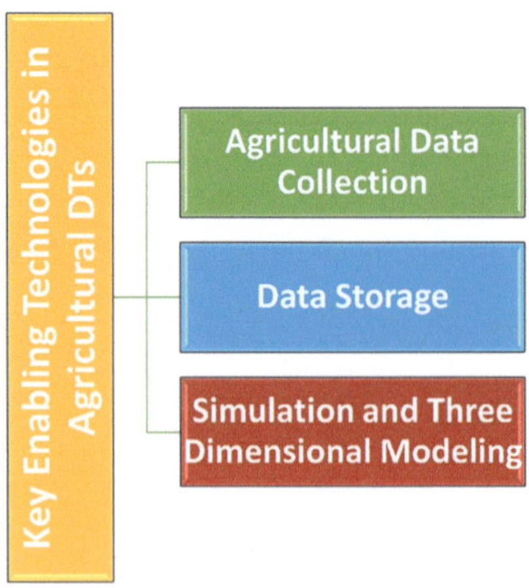

Fig. 5.2 Classification of key enabling frameworks

data collection to be more accurate, mechanized, and scalable (Escribà-Gelonch et al. 2024; Purcell and Neubauer 2023). IoT wireless sensor systems are used extensively in today's agriculture for monitoring soil parameters, crop conditions, and weather situations. Recent agricultural equipment, such as tractors and cultivators, has embedded systems to record running parameters. Additionally, edge computing is being gradually used with sensors to assist real-time computation of information at the source point, so that the requirement to send data is minimized, thus enhancing efficiency (Verdouw et al. 2021). Application of edge computing to 5G-connected agriculture facilitated on-field devices to preprocess data before they were sent to the central platforms. The approach, therefore, surpasses latency and enhances DTs configuration response time.

5.3.1.2 Data Storage

Agricultural DTs data storage is one of the most important aspects of agricultural DTs and helps manage and access humongous data from different applications, i.e., drones, sensors, and satellite images (Araújo et al. 2021; Sanjeevi et al. 2020a). Data storage in agriculture is challenged by the issue brought about by the amount, complexity, and diversity of data collected in the field. These were the issues that required the establishment of highly advanced storage systems that can accommodate high levels of data and provide high levels of security and convenience. Data storage is critical in farm DTs in handling various kinds of data, including geographical, chronological, ecological, and operating data. Geospatial data, such as satellite imagery and drone mapping, are normally data in vector form, and time data can be time-series data where sensors track soil water level, temperature, and crop growth stage. Standard relational databases are normally not the best option here since they are unable to cope with the scale, speed, and granularity of agronomical data (del Cerro et al. 2021; Yatribi 2020). Cloud computing is now the preference for the storage of agronomical data. Elastic and dynamic storage capacity, which the unbridled sensor data growth requires, is provided by cloud computing. Cloud technologies assist agriculturalists and researchers in collecting and assessing data from afar, getting real-time access to data from all over the world. Applying the IoT and cloud services for farming is a good example. Both technologies can store a lot of data and use advanced algorithms alongside machine learning to find useful information, such as anticipating crop yield or identifying the best ways to water crops (Abiri et al. 2023; Idoje et al. 2021).

Furthermore, cloud-based storage services are likely to have backup and recovery options to keep data safe and make sure that everything stays stable even if something goes wrong. Certain structures also use hybrid cloud solutions, which combine on-site storage with cloud-based services to improve the protection and management of important data. Edge computing is also a new way to store agronomic information. Edge computing is different compared to conventional cloud-based services because it processes data on the device itself, like a drone or sensor. In typical cloud-based services, all data has to be sent to middle servers for analysis. It is mostly used in farming, where the internet may not always be available, and decisions need to be made right away (Junaid et al. 2021; Nasir et al. 2020). With the unprecedented growth in the volume of data in agriculture, traditional data storage technologies are increasingly being superseded by more effective frameworks, i.e., data lakes (Hai et al. 2023). Data lakes contain massive amounts of raw, unprocessed, and unstructured data that is critical in agriculture because the types of data vary extensively (Azzabi et al. 2024). Data platforms like Spark and Hadoop are extensively utilized to structure agronomic data lakes. The data platforms support distributed computation as well as simultaneous computation, which allows fast computation of enormous amounts of data relative to traditional systems. One of the greatest benefits of using data lakes in farming is that they can process all types of data, enabling farmers and data scientists to look at the correlation between varied data sets. When integrating data on meteorology with moisture content and growth rates of plants, more precise forecasting models for crop yield can be built. Information security is of great concern when dealing with the storage of agricultural information due to the sensitive content it contains. Blockchain has been put forward as an alternative to boost the integrity and monitoring of farming data (Fan et al. 2020; Lakhwani et al. 2019). Blockchain entails a decentralized repository in addition to non-reversible records, and this is very useful in maintaining the integrity of vital data, including agricultural yields, the use of pesticides, and information about harvesting.

5.3.1.3 Simulation and Three-Dimensional Modeling

Three-dimensional (3D) modeling and simulation are among the most important tools for enhancing agricultural DTs' development since they allow virtual representations of real agricultural systems to be developed. The models allow cropping behavior, environmental relationships, and agricultural processes to be simulated, hence increasing efficiency in decision-making, resource planning, and sustainable agriculture (Kamali et al. 2024; Singh et al. 2022). Simulation of crop growth under changing environmental conditions is one of the significant uses of 3D modeling in agronomy. Simulation of farm DTs is required so that agriculture can be improved. Integration of data online with 3D models enables agriculturalists and agricultural scientists to simulate all agricultural processes, including the growth of crops, irrigation, and pest management. Integration is required for forecasting the outcome of various agricultural processes without actually implementing them into practice. Computer simulations can model complicated agricultural systems and supply valuable information for optimizing crop yields, water consumption, and pest management (Elavarasan and Durairaj Vincent 2020; Shawon et al. 2025).

5.4 Applications of DTs in Agriculture

This section outlines some of the areas where deployment of DTs is revolutionizing agricultural applications such as Crop Production Management (CPM), farm disease forecasting, livestock management, farm equipment optimization, and agricultural resource allocation optimization, as viewed in Fig. 5.3.

Fig. 5.3 Various uses of DTs in the agriculture domain

5.4.1 CPM Utilization

CPM, a highly promising use of DTs in the agricultural sector, combines the construction of a virtual simulated farm that is capable of simulating ecological conditions, plant development, and diversified management strategies. CPM has undergone a phenomenal expansion in the recent past with the incorporation of DTs techniques (Pylianidis et al. 2021b; Zhai et al. 2020). As a result of using DTs technology, it is possible to establish a virtual crop model, field model, and atmospheric model, which enhances management practice. The use of decision trees to predict crop yields has been revealed in several studies, which have targeted the purposes of smart farming, crop development models, and instantaneous tracking. The application of DTs technology in the management of crops is constantly transforming the traditional agronomic system. DTs may significantly improve the efficiency and productivity of the agronomic sector using the latest technologies, such as UAVs, AI, and IoT (Abdalla and El-Ramady 2022; Liu et al. 2021a).

5.4.2 Agricultural Catastrophe Alert and Management

An Agroecosystem can be subjected to catastrophes and pests. The traditional catastrophe management systems of agriculture are mostly grounded on the old models and unchangeable data that are incapable of responding to dynamic, real-time evaluation in achieving maximum catastrophe avoidance and containment. DTs technology has some promising options on how to improve the mechanism of catastrophe alert and response in agroecosystems (Araújo et al. 2021; Martos et al. 2021). Creating a live virtual prototype of the settings on a crop, DTs can monitor the environmental data and the condition of crops in real-time, detect the various types of crop infections and pests, and provide information to farmers and send warnings concerning the early appearance of diseases and pests. These are the solutions, but scalability, data integration, and availability on the topic are left to small-scale farmers. Masses of data, the difficulty in handling them, the expense of installation of the DTs technology, and the necessity of regular and timely sources of data may deter intensive use. However, the development of low-cost sensors, cloud computing, and AI algorithms enables the capabilities of DTs systems to transform disease and pest control in agriculture (Elbasi et al. 2023; Miller et al. 2025; Rudrakar and Rughani 2024).

5.4.3 Livestock Management

Rearing of livestock has been part and parcel of farming. The experience of DTs has facilitated being efficient as a livestock control game-changer, and a model prototype simultaneously simulates the environmental conditions and the health of the animal. Some of the qualities that are enriched with the application of livestock management in the use of DTs include constant tracking, predictive analytics, enhanced evaluation procedure, and customized animal care (Verdouw et al. 2021). DTs systems offer a single measurement of the well-being, eating habits, and the connection between the environment and the animal, by simulating an imaginary depiction of the cattle and their environment. Conventional livestock management techniques are usually based on the occurrence of infrequent observation and physical inspection, which is too time-consuming to identify disease problems. Information retrieved with different sensing devices implanted in straps, ear tags, or other wearable devices is inexhaustibly accumulated in DTs arrangements (Kumar and Singh 2023; Neethirajan and Kemp 2021). This information, such as temperature, level of activity, nutrition, and various physical clues, is entered into the DTs framework in order to construct an active account of the health condition of the animal. Livestock can be used to integrate genomic data, health data, and environmental data into a greater wisdom of optimum breeding conditions. Decision trees enhance genetic diversity, efficiency, and disease resistance capability of livestock as they predict the results of varying breeding combinations.

5.4.4 Efficient Utilization of Agricultural Devices and Equipment

Greater productivity of agricultural machinery and tools has become the most pressing question of modernity in farming, in an attempt to produce more with less and transform agriculture into a sustainable system. Conventional management strategies of tools using routine maintenance and manual adjustments are being increasingly augmented or substituted with sophisticated technology like DTs. DTs systems have sensors, IoT systems, and advanced data analysis that deliver real-time information into the equipment performance, enabling correct decision-making and planned service (Saiz-Rubio and Rovira-Más 2020). DTs platforms collect information repeatedly from equipment using sensors as well as IoT devices, comprising constraints, namely operating engine temperature, fuel usage, soil condition, and component failure. Such real-time information facilitates the building of virtual prototypes of machines that are modified in real-time, permitting the owners of the machines and the farmers to track the performance of their equipment remotely (Ahmad and Nabi 2021; Catala-Roman et al. 2024).

5.4.5 Optimization of Resources in Agriculture

Optimization of the resources approach by the use of DTs in agriculture has proved to help sustainability and agricultural productivity. DTs facilitate virtual modeling of physical farm assets, infrastructure, and processes such that resource utilization (e.g., irrigation water, fertilizers, electricity, and labor) can be simulated, monitored, and optimized in real-time (Benos et al. 2021; Osinga et al. 2022). This optimizes the utilization of resources like irrigation and fertilizers, thus boosting sustainability and agricultural output. Management of irrigation is a significant issue in agriculture, particularly for areas where they have limited water. Live monitoring and upkeep of the irrigation systems in agriculture are allowed by DTs technology. For example, field sensor networks are used to capture the levels of humidity, temperature, and rain data, feeding the resulting data to the DTs model. This information is subsequently utilized in simulating irrigation needs, which helps the farmers get the best out of water by avoiding over-irrigation and ensuring a proper supply of water to plants. Reduced fertilizer application increases crop yields without exerting severe pressure on the environment (Abdollahi et al. 2021; Sanjeevi et al. 2020b; Thakur et al. 2019). Considering the existing measurements on sensors (i.e., soil composition and plant state) as well as other factors (i.e., weather forecasting), DTs offer a real-life understanding of what will be the most appropriate fertilizer usage. This optimization prevents the propensity to overindulge and therefore, is more sustainable to the environment. The DTs enable the farmers to shift their operations in a manner that conserves energy, thereby lowering their costs and carbon emissions. With periodic requirements and functional statistics, DTs models may learn to mirror energy demands to offer insights that enhance sustainable planning of farm activities.

5.5 Challenges and Future Directions in Agricultural DTs

Since DTs are now being adopted in the agricultural industry at a very rapid pace, they promise enormous opportunities in every other sector. Agricultural DTs, nevertheless, continue to be confronted with a massive number of problems. The following part covers the key problems and challenges of agricultural DTs from many different perspectives. Geographical and climatic heterogeneity is extreme within the agricultural data collection environment and presents immense challenges related to installing sensors and collecting data. Different topographical environments, such as mountains, sloping, and plain lands, wetlands are geographically different environments that require superior flexibility in data collection equipment. Each environmental setting creates challenges unique to data collection. Meteorological settings amplify challenges in input collection. The implementation of sophisticated information-gathering strategies in mass farming cases requires high resource investments. Even though data-gathering practices using IoT applications have been simplified by implementation of the practice, their extensive adoption in agriculture is associated with serious challenges, such as the cost of the technology, reliability of equipment, and limitations of network coverage. Big sensor networks in data collection are associated with massive initial investment in the purchase of equipment, and other than that, far-reaching costs involve installing, maintaining, and transferring information.

The multi-source data application in integrated use of agricultural interest requires that the basic problems of data format standards, protocol consistency, and semantic harmonization be solved. The heterogeneity of the data framing and the sampling rates of non-similar sources are major impediments to the effective data integration and integrity maintenance within the DTs frameworks. Combining live streaming sensor data and non-stop satellite remote sensing data requires advanced algorithms for converting the data format. Commonly, crop cultivation systems consist of multiple mutually dependent subsystems, e.g., irrigation systems, mechanisms of fertilized control, systems of crop growth measurements, and so on, the situations and activities of which are closely connected with each other. The continuous information exchange among various subsystems, where the DTs are kept in their states that are identical to the system state, is an immense technical burden. The next step would be to optimize the deep learning techniques in conjunction with images of farms to yield the most accurate 3D models with the least operator intervention. Neural networks could be trained to build dynamic phases of agricultural development and forecast further development under the varying conditions in the environment with multi-modal statistics, such as multispectral images. Integrating post-harvest supply chain planning with storage and the DTs methodology to form a combined solution. By means of sophisticated algorithms applied to predictive quality and market price, the decision trees can not only show the information about the development of crops, but also about the processing and packaging at the after-harvest stage.

5.6 Conclusions

The application of DTs in agriculture is shifting significantly in the direction of data-driven, credible, and intelligent means of developing things. This chapter has taken a closer look at how the DTs used in farming have evolved over the years, how they may be utilized, and the disadvantages of the same. It also gave a detailed discussion on the choices they have in enhancing productivity in agriculture as well as sustainable development. Some of the major duties that the decision trees perform in the agricultural sector include monitoring operations, making projections, and presenting management responses. DTs help farmers use exact approaches by using a lot of various sources of information, such as IoT devices, satellite imagery, and statistics of soil and air quality. DTs have many uses in agronomy, such as CPM, livestock tracking, natural disaster forecasting, and using farm machinery. Additionally, the development of sophisticated 3D modeling and simulator software has enhanced the precision of virtual cultivation surroundings, facilitating better estimation and assessment of conditions. Even with these improvements, there are still big problems with operations and technology. Data acquisition is still a big problem because different types of data, problems with sensor calibration, and inefficient data collection make DTs systems less reliable. Additionally, combining data from sources related to agriculture is very difficult because there are numerous distinct kinds of data, it's hard to figure out what the data means, and the platforms don't always work together. The accuracy and dependability of creating 3D crop models require additional improvement because of inconsistencies in the way data is represented, which could lead to incorrect simulation results.

Declarations

Funding No external funding received.

Data Availability Statement All data generated or analyzed during this study are included in this published article.

Conflicts of Interest The authors have no conflicts of interest to declare that are relevant to the content of this article.

References

Abbasi R, Martinez P, Ahmad R (2022) The digitization of agricultural industry—a systematic literature review on agriculture 4.0. In: Smart agricultural technology. Elsevier. https://doi.org/10.1016/j.atech.2022.100042

Abdalla Z, El-Ramady H (2022) Applications and challenges of smart farming for developing sustainable agriculture. Environ Biodivers Soil Secur (Egypts Presidential Specialized Council for Education and Scientific Research) 6(2022):81–90. https://doi.org/10.21608/jenvbs.2022.135889.1175

Abdollahi A, Rejeb K, Rejeb A, Mostafa MM, Zailani S (2021) Wireless sensor networks in agriculture: insights from bibliometric analysis. Sustainability (Switzerland). https://doi.org/10.3390/su132112011

Abiri R, Rizan N, Balasundram SK, Shahbazi AB, Abdul-Hamid H (2023) Application of digital technologies for ensuring agricultural productivity. Heliyon (Elsevier) 9(12):e22601. https://doi.org/10.1016/j.heliyon.2023.e22601

Abu NS, Bukhari WM, Ong CH, Kassim AM, Izzuddin TA, Sukhaimie MN, Norasikin MA et al (2022) Internet of things applications in precision agriculture: a review. J Robot Control (JRC) (Department of Agribusiness, Universitas Muhammadiyah Yogyakarta). https://doi.org/10.18196/jrc.v3i3.14159

Ahmad L, Nabi F (2021) Agriculture 5.0: artificial intelligence, IoT, and machine learning. CRC Press. https://doi.org/10.1201/9781003125433

Araújo SO, Peres RS, Barata J, Lidon F, Ramalho JC (2021) Characterising the agriculture 4.0 landscape—emerging trends, challenges and opportunities. Agronomy. https://doi.org/10.3390/agronomy11040667

Awais M, Wang X, Hussain S, Aziz F, Mahmood MQ (2025) Advancing precision agriculture through digital twins and smart farming technologies: a review. AgriEngineering (Multidisciplinary Digital Publishing Institute) 7(5):137. https://doi.org/10.3390/AGRIENGINEERING7050137

Azzabi S, Alfughi Z, Ouda A (2024) Data Lakes: a survey of concepts and architectures. Computers (Multidisciplinary Digital Publishing Institute). https://doi.org/10.3390/computers13070183

Barricelli BR, Casiraghi E, Fogli D (2019) A survey on digital twin: definitions, characteristics, applications, and design implications. IEEE Access (Institute of Electrical and Electronics Engineers Inc). https://doi.org/10.1109/ACCESS.2019.2953499

Benos L, Tagarakis AC, Dolias G, Berruto R, Kateris D, Bochtis D (2021) Machine learning in agriculture: a comprehensive updated review. Sensors. https://doi.org/10.3390/s21113758

Boyes H, Watson T (2022) Digital twins: an analysis framework and open issues. Comput Ind (Elsevier B.V.) 143(1):103763. https://doi.org/10.1016/j.compind.2022.103763

Catala-Roman P, Navarro EA, Segura-Garcia J, Garcia-Pineda M (2024) Harnessing digital twins for agriculture 5.0: a comparative analysis of 3D point cloud tools. Applied Sciences (Switzerland) (Multidisciplinary Digital Publishing Institute) 14(5):1709. https://doi.org/10.3390/app14051709

de Andrade MAN, Lepikson HA, Tosta Machado CA (2021) A new framework and methodology for digital twin development. In: 2021 14th IEEE International Conference on Industry Applications, INDUSCON 2021 – Proceedings. Institute of Electrical and Electronics Engineers Inc, pp 134–138. https://doi.org/10.1109/INDUSCON51756.2021.9529701

del Cerro J, Ulloa CC, Barrientos A, de León Rivas J (2021) Unmanned aerial vehicles in agriculture: a survey. Agronomy (MDPI AG) 11(2):1456–1478. https://doi.org/10.3390/AGRONOMY11020203

Dihan MS, Akash AI, Tasneem Z, Das P, Das SK, Islam MR, Islam MM et al (2024) Digital twin: data exploration, architecture, implementation and future. Heliyon (Elsevier). https://doi.org/10.1016/j.heliyon.2024.e26503

Elavarasan D, Durairaj Vincent PM (2020) Crop yield prediction using deep reinforcement learning model for sustainable agrarian applications. IEEE Access (Institute of Electrical and Electronics Engineers Inc) 8(2):86886–86901. https://doi.org/10.1109/ACCESS.2020.2992480

Elbasi E, Mostafa N, Alarnaout Z, Zreikat AI, Cina E, Varghese G, Shdefat A et al (2023) Artificial intelligence Technology in the Agricultural Sector: a systematic literature review. IEEE Access (Institute of Electrical and Electronics Engineers Inc). https://doi.org/10.1109/ACCESS.2022.3232485

Eli-Chukwu NC (2019) Applications of artificial intelligence in agriculture: a review. Eng Technol Appl Sci Res (Dr D Pylarinos) 9(4):4377–4383. https://doi.org/10.48084/etasr.2756

Escribà-Gelonch M, Liang S, van Schalkwyk P, Fisk I, Long NVD, Hessel V (2024) Digital twins in agriculture: orchestration and applications. J Agric Food Chem (American Chemical Society). https://doi.org/10.1021/acs.jafc.4c01934

Fan C, Ghaemi S, Khazaei H, Musilek P (2020) Performance evaluation of Blockchain systems: a systematic survey. IEEE Access (Institute of Electrical and Electronics Engineers Inc). https://doi.org/10.1109/ACCESS.2020.3006078

Fang X, Wang H, Liu G, Tian X, Ding G, Zhang H (2022) Industry application of digital twin: from concept to implementation. Int J Adv Manuf Technol (Springer Science and Business Media Deutschland GmbH). https://doi.org/10.1007/s00170-022-09632-z

Folgado FJ, Calderón D, González I, Calderón AJ (2024) Review of industry 4.0 from the perspective of automation and supervision systems: definitions, architectures and recent trends. Electronics (Switzerland) (Multidisciplinary Digital Publishing Institute). https://doi.org/10.3390/electronics13040782

Frantzén M, Bandaru S, Ng AHC (2022) Digital-twin-based decision support of dynamic maintenance task prioritization using simulation-based optimization and genetic programming. Decision Anal J (Elsevier) 3(5):100039. https://doi.org/10.1016/j.dajour.2022.100039

Fuller A, Fan Z, Day C, Barlow C (2020) Digital twin: enabling technologies, challenges and open research. IEEE Access (Institute of Electrical and Electronics Engineers Inc) 8(2): 108952–108971. https://doi.org/10.1109/ACCESS.2020.2998358

Goldenits G, Mallinger K, Raubitzek S, Neubauer T (2024) Current applications and potential future directions of reinforcement learning-based digital twins in agriculture. Smart Agricultural Technology (Elsevier B.V.). https://doi.org/10.1016/j.atech.2024.100512

Hai R, Koutras C, Quix C, Jarke M (2023) Data Lakes: a survey of functions and systems. IEEE Trans Knowl Data Eng IEEE Comput Soc 35(12):12571–12590. https://doi.org/10.1109/TKDE.2023.3270101

Hlophe MC, Maharaj BT (2023) From cyber–physical convergence to digital twins: a review on edge computing use case designs. Appl Sci (Switzerland) (Multidisciplinary Digital Publishing Institute). https://doi.org/10.3390/app132413262

Idoje G, Dagiuklas T, Iqbal M (2021) Survey for smart farming technologies: challenges and issues. Comput Electr Eng (Elsevier Ltd.) 92(2):1089. https://doi.org/10.1016/j.compeleceng.2021.107104

Iliuță ME, Moisescu MA, Pop E, Ionita AD, Caramihai SI, Mitulescu TC (2024) Digital twin—a review of the evolution from concept to technology and its analytical perspectives on applications in various fields. applied sciences (Switzerland) (Multidisciplinary Digital Publishing Institute) 14(13):5454. https://doi.org/10.3390/app14135454

Junaid M, Shaikh A, Hassan MU, Alghamdi A, Rajab K, Al Reshan MS, Alkinani M (2021) Smart agriculture cloud using AI based techniques. Energies (Multidisciplinary Digital Publishing Institute) 14(16):5129. https://doi.org/10.3390/en14165129

Kamali M, Atazadeh B, Rajabifard A, Chen Y (2024) Advancements in 3D digital model generation for digital twins in industrial environments: knowledge gaps and future directions. Adv Eng Inform (Elsevier). https://doi.org/10.1016/j.aei.2024.102929

Kaur H, Bhatia M (2025) Digital twins: a scientometric investigation into current progress and future directions. Expert Syst Appl (Elsevier Ltd.) 265(2):1447. https://doi.org/10.1016/j.eswa.2024.125917

Khan LU, Han Z, Saad W, Hossain E, Guizani M, Hong CS (2022) Digital twin of wireless systems: overview, taxonomy, challenges, and opportunities. IEEE Commun Surv Tutor (Institute of Electrical and Electronics Engineers Inc) 24(4):2230–2254. https://doi.org/10.1109/COMST.2022.3198273

Khanna A, Kaur S (2019) Evolution of internet of things (IoT) and its significant impact in the field of precision agriculture. Comput Electron Agric (Elsevier B.V.). https://doi.org/10.1016/j.compag.2018.12.039

Kour VP, Arora S (2020) Recent developments of the internet of things in agriculture: a survey. IEEE Access (Institute of Electrical and Electronics Engineers Inc.). https://doi.org/10.1109/ACCESS.2020.3009298

Kumar S, Singh H (2023) Design and synthesis of circular antenna Array for low side lobe level and high directivity using artificial hummingbird optimization algorithm. Recent Adv Electr Electron Eng ((Formerly Recent Patents on Electrical & Electronic Engineering), Bentham Science Publishers Ltd.) 16(3):370–379. https://doi.org/10.2174/2352096516666230816091004

Kumar S, Vaid V, Thakur G (2023) A comprehensive review of 5G networks for sustainable and smart cities. Recent Adv Electr Electron Eng ((Formerly Recent Patents on Electrical & Electronic Engineering), Bentham Science Publishers Ltd.) 17(6):535–543. https://doi.org/10.2174/2352096516666230720164542

Kumar V, Sharma KV, Kedam N, Patel A, Kate TR, Rathnayake U (2024) A comprehensive review on smart and sustainable agriculture using IoT technologies. Smart Agric Technol (Elsevier) 8(1):100487. https://doi.org/10.1016/j.atech.2024.100487

Lakhwani K, Gianey H, Agarwal N, Gupta S (2019) Development of IoT for smart agriculture a review. In: Advances in intelligent systems and computing, vol 841. Springer Verlag, pp 425–432. https://doi.org/10.1007/978-981-13-2285-3_50

Liu M, Fang S, Dong H, Xu C (2021a) Review of digital twin about concepts, technologies, and industrial applications. J Manuf Syst (Elsevier BV) 58(2):346–361. https://doi.org/10.1016/j.jmsy.2020.06.017

Liu Y, Ma X, Shu L, Hancke GP, Abu-Mahfouz AM (2021b) From industry 4.0 to agriculture 4.0: current status, enabling technologies, and research challenges. IEEE Trans Ind Inform (IEEE Computer Society) 17(6):4322–4334. https://doi.org/10.1109/TII.2020.3003910

Liu X, Jiang D, Tao B, Xiang F, Jiang G, Sun Y, Kong J et al (2023) A systematic review of digital twin about physical entities, virtual models, twin data, and applications. Adv Eng Inform (Elsevier). https://doi.org/10.1016/j.aei.2023.101876

Malhi GS, Kaur M, Kaushik P (2021) Impact of climate change on agriculture and its mitigation strategies: a review. Sustainability (Switzerland) (Multidisciplinary Digital Publishing Institute). https://doi.org/10.3390/su13031318

Martos V, Ahmad A, Cartujo P, Ordoñez J (2021) Ensuring agricultural sustainability through remote sensing in the era of agriculture 5.0. Appl Sci (Switz). https://doi.org/10.3390/app11135911

Miller T, Mikiciuk G, Durlik I, Mikiciuk M, Łobodzińska A, Śnieg M (2025) The IoT and AI in agriculture: the time is now—a systematic review of smart sensing technologies. Sensors (Basel) (Multidisciplinary Digital Publishing Institute). https://doi.org/10.3390/s25123583

Minerva R, Crespi N (2021) Digital twins: properties, software frameworks, and application scenarios. IT Professional (IEEE Computer Society) 23(1):51–55. https://doi.org/10.1109/MITP.2020.2982896

Minerva R, Lee GM, Crespi N (2020) Digital twin in the IoT context: a survey on technical features, scenarios, and architectural models. Proc IEEE (Institute of Electrical and Electronics Engineers Inc.) 108(10):1785–1824. https://doi.org/10.1109/JPROC.2020.2998530

Montalvo A, Camacho O, Chavez D (2025) Cyber-physical systems for smart farming: a systematic review. Sustainability (Multidisciplinary Digital Publishing Institute) 17(14):6393. https://doi.org/10.3390/SU17146393

Nasir IM, Bibi A, Shah JH, Khan MA, Sharif M, Iqbal K, Nam Y et al (2020) Deep learning-based classification of fruit diseases: an application for precision agriculture. Comput Mater Contin (Tech Science Press) 66(2):1949–1962. https://doi.org/10.32604/cmc.2020.012945

Neethirajan S, Kemp B (2021) Digital twins in livestock farming. Animals (Multidisciplinary Digital Publishing Institute). https://doi.org/10.3390/ani11041008

Nóbrega L, Gonçalves P, Pedreiras P, Pereira J (2019) An IoT-based solution for intelligent farming. sensors (Switzerland) (MDPI AG) 19(3):1247. https://doi.org/10.3390/s19030603

Osinga SA, Paudel D, Mouzakitis SA, Athanasiadis IN (2022) Big data in agriculture: between opportunity and solution. Agric Syst (Elsevier Ltd) 195(3):1548. https://doi.org/10.1016/j.agsy.2021.103298

Peladarinos N, Piromalis D, Cheimaras V, Tserepas E, Munteanu RA, Papageorgas P (2023) Enhancing smart agriculture by implementing digital twins: a comprehensive review. Sensors (Multidisciplinary Digital Publishing Institute (MDPI)). https://doi.org/10.3390/s23167128

Piras G, Agostinelli S, Muzi F (2024) Digital twin framework for built environment: a review of key enablers. Energies (Multidisciplinary Digital Publishing Institute). https://doi.org/10.3390/en17020436

Purcell W, Neubauer T (2023) Digital twins in agriculture: a state-of-the-art review. Smart Agric Technol (Elsevier B.V.). https://doi.org/10.1016/j.atech.2022.100094

Pylianidis C, Osinga S, Athanasiadis IN (2021a) Introducing digital twins to agriculture. Comput Electron Agric (Elsevier BV) 184(1):1227. https://doi.org/10.1016/j.compag.2020.105942

Pylianidis C, Osinga S, Athanasiadis IN (2021b) Introducing digital twins to agriculture. Comput Electron Agric (Elsevier B.V.) 184(5):1254. https://doi.org/10.1016/j.compag.2020.105942

Raza, S.M., Minerva, R., Crespi, N., Alvi, M., Herath, M. and Dutta, H. (2025), "A comprehensive survey of Network Digital Twin architecture, capabilities, challenges, and requirements for Edge–Cloud Continuum", Comput Commun, Elsevier, https://doi.org/10.1016/j.comcom.2025.108144
Rehman A, Saba T, Kashif M, Fati SM, Bahaj SA, Chaudhry H (2022) A revisit of internet of things Technologies for Monitoring and Control Strategies in smart agriculture. Agronomy. https://doi.org/10.3390/agronomy12010127
Rudrakar S, Rughani P (2024) IoT based agriculture (ag-IoT): a detailed study on architecture, security and forensics. Inf Process Agric (China Agricultural University). https://doi.org/10.1016/j.inpa.2023.09.002
Sagar S, Birje NM (2025) Precision agriculture using internet of things and cloud computing: a review. SN Comput Sci (Springer). https://doi.org/10.1007/s42979-025-03833-5
Saiz-Rubio V, Rovira-Más F (2020) From smart farming towards agriculture 5.0: a review on crop data management. Agronomy. https://doi.org/10.3390/agronomy10020207
Sanjeevi P, Prasanna S, Siva Kumar B, Gunasekaran G, Alagiri I, Vijay Anand R (2020a) Precision agriculture and farming using internet of things based on wireless sensor network. Trans Emerg Telecommun Technol (Wiley Blackwell) 31(12):1248. https://doi.org/10.1002/ett.3978
Sanjeevi P, Prasanna S, Siva Kumar B, Gunasekaran G, Alagiri I, Vijay Anand R (2020b) Precision agriculture and farming using internet of things based on wireless sensor network. Trans Emerg Telecommun Technol (Wiley Blackwell) 31(12):1659. https://doi.org/10.1002/ett.3978
Segovia M, Garcia-Alfaro J (2022) Design, modeling and implementation of digital twins. Sensors (Basel) (Multidisciplinary Digital Publishing Institute) 22(14):5396. https://doi.org/10.3390/s22145396
Semeraro C, Lezoche M, Panetto H, Dassisti M (2021a) Digital twin paradigm: a systematic literature review. Comput Ind (Elsevier BV) 130(5):1227. https://doi.org/10.1016/j.compind.2021.103469
Semeraro C, Lezoche M, Panetto H, Dassisti M (2021b) Digital twin paradigm: a systematic literature review. Comput Ind (Elsevier B.V.) 130(1):e1589. https://doi.org/10.1016/j.compind.2021.103469
Shawon SM, Ema FB, Mahi AK, Niha FL, Zubair HT (2025) Crop yield prediction using machine learning: an extensive and systematic literature review. Smart Agric Technol (Elsevier). https://doi.org/10.1016/j.atech.2024.100718
Singh M, Fuenmayor E, Hinchy EP, Qiao Y, Murray N, Devine D (2021) Digital twin: origin to future. Appl Syst Innov (MDPI AG). https://doi.org/10.3390/asi4020036
Singh M, Srivastava R, Fuenmayor E, Kuts V, Qiao Y, Murray N, Devine D (2022) Applications of digital twin across industries: a review. Appl Sci (Switzerland) (MDPI) 12(11):1246–1249. https://doi.org/10.3390/APP12115727
Terence S, Purushothaman G (2020) Systematic review of internet of things in smart farming. Trans Emerging Telecommun Technol (Wiley Blackwell) 31(6):789. https://doi.org/10.1002/ett.3958
Thakur D, Kumar Y, Kumar A, Singh PK (2019) Applicability of wireless sensor networks in precision agriculture: a review. Wirel Pers Commun (Springer). https://doi.org/10.1007/s11277-019-06285-2
Ullo SL, Sinha GR (2021) Advances in IoT and smart sensors for remote sensing and agriculture applications. Remote Sens (MDPI) 13(13):2585. https://doi.org/10.3390/rs13132585
VanDerHorn E, Mahadevan S (2021) Digital twin: generalization, characterization and implementation. Decision Support Syst (Elsevier BV) 145(1):15–31. https://doi.org/10.1016/j.dss.2021.113524
Verdouw C, Tekinerdogan B, Beulens A, Wolfert S (2021) Digital twins in smart farming. Agric Syst (Elsevier Ltd) 189(2):18–36. https://doi.org/10.1016/j.agsy.2020.103046
Wang L (2024) Digital twins in agriculture: a review of recent Progress and open issues. Electronics (Switzerland) (Multidisciplinary Digital Publishing Institute (MDPI)). https://doi.org/10.3390/electronics13112209

Yao H, Qin R, Chen X (2019) Unmanned aerial vehicle for remote sensing applications – a review. Remote Sens. https://doi.org/10.3390/rs11121443

Yatribi T (2020) Factors affecting precision agriculture adoption: a systematic Litterature review. Economics – Innov Econ Res J 8(2):103–121. https://doi.org/10.2478/eoik-2020-0013

Zenkovich MV (2025) Digital twin-driven decision-making support approach in adaptive control of production systems. In: Proceedings – 2025 international Russian smart industry conference, SmartIndustryCon 2025. Institute of Electrical and Electronics Engineers Inc., pp 1130–1135. https://doi.org/10.1109/SmartIndustryCon65166.2025.10986153

Zhai Z, Martínez JF, Beltran V, Martínez NL (2020) Decision support systems for agriculture 4.0: survey and challenges. Comput Electron Agric (Elsevier B.V.). https://doi.org/10.1016/j.compag.2020.105256

Zhang Z, Zhu L (2023) A review on unmanned aerial vehicle remote sensing: platforms, sensors, data processing methods, and applications. Drones (Multidisciplinary Digital Publishing Institute). https://doi.org/10.3390/drones7060398

Zhang K, Shi Y, Karnouskos S, Sauter T, Fang H, Colombo AW (2023) Advancements in industrial cyber-physical systems: an overview and perspectives. IEEE Trans Ind Inform (IEEE Computer Society). https://doi.org/10.1109/TII.2022.3199481

Zhang R, Zhu H, Chang Q, Mao Q (2025) A comprehensive review of digital twins technology in agriculture. Agriculture (Switzerland) 15(9):1–25. https://doi.org/10.3390/agriculture15090903

Challenges and Solutions of Digital Twin Technologies for Agricultural Perspectives

6

Le Kim Quy, Tran Thi Viet Ha, and Nguyen Minh Viet

Abstract

Digital Twin (DT) technology, a part of digital transformation, offers significant improvements to agriculture, especially in boosting efficiency. However, several obstacles that hinder the widespread use of DT can be mentioned as technical, socioeconomic, and operational challenges. Data interoperability, infrastructure limitations, model accuracy, and real-time calibration are related to technical barriers. These concerns can be addressed by embedding advanced technology like edge computing and artificial intelligence (AI) in DT systems to improve the agricultural management process. The socioeconomic challenges include high implementation costs, low-quality digital skills, and the hesitation of farmers to adopt modern technology. To address these impediments, it requires effective collaboration through public-private partnerships that will fund infrastructure construction, share expenditures, and disseminate digital knowledge through community training. In addition, a lifelong operation of DT in agriculture requires tight support and collaborations across sectors. Joining hands among farmers, the technology experts, and the decision-makers brings benefits in creating a resilient and secure agricultural environment.

L. K. Quy (✉) · T. T. V. Ha
Faculty of Advanced Technologies and Engineering, VNU Vietnam Japan University, Hanoi, Vietnam
e-mail: lk.quy@vju.ac.vn

N. M. Viet
VNU Key Laboratory of Advanced Material for Green Growth, VNU University of Science, Hanoi, Vietnam

© The Author(s), under exclusive license to Springer Nature Singapore Pte Ltd. 2026

A. L. Srivastav et al. (eds.), *Digital Twin Technology for Sustainable Agriculture*,
https://doi.org/10.1007/978-981-95-5915-2_6

Keywords

Digital twin challenges · Agricultural technology barriers · IoT security in farming · Sustainable agriculture adoption

6.1 Introduction

Digital Twin (DT) technology is a virtual modelling system that is employed to mimic the physical conditions in real-time. Figure 6.1 depicts an agriculture-related DT system. The physical farm, as illustrated on the left side of Fig. 6.1, consists of growing crops, livestock, farm infrastructure, and monitoring devices such as temperature and moisture sensors. These sensors collect the real-time data and send it to the DT platform. The DT platform mimics the same as those belonging to the physical one, as shown on the right side of Fig. 6.1. This platform is adopted for monitoring the real-world situations, and it can also predict future events, providing further optimization solutions. As a result, it enhances resource efficiency and flexibility in agricultural systems (Peladarinos et al. 2023). Moreover, DT systems with advanced technologies, including Internet of Things (IoT), Artificial Intelligence (AI), and cloud computing, enhance their performance (Symeonaki et al. 2024). This further brings benefits in farm management.

Even though the significant potential of DT technology is recognized, its widespread use is being hindered due to several factors, including technical, socioeconomic, and operational ones. These obstacles are even more obvious in rural farming communities, where most of the farm products are produced. The main difficulties can be listed, including data interoperability issues, underequipped digital infrastructure, and high implementation costs (Dibbern et al. 2024). Additionally, the unwillingness of the farmers toward advanced technologies also leads to a great obstacle to

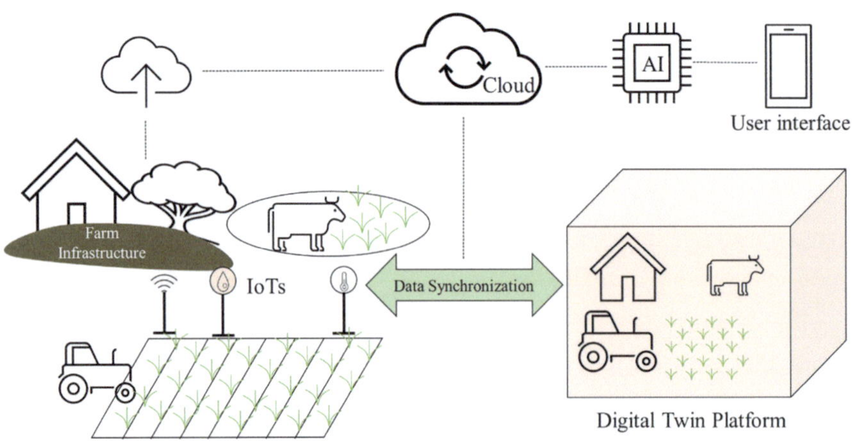

Fig. 6.1 A schematic of a Digital Twin system applied in agriculture

its adoption. Only by addressing these challenges, it can ensure the widespread adoption of DT solutions across various agricultural landscapes.

This chapter focuses on (1) systematically pinpointing the primary causes that hinder the implementation of the DT technology in agriculture, and (2) providing practical solutions that can enhance the implementation rate. Taking into account the unique technology-related and human-centered issues, the chapter attempts to present a comprehensive guide for all stakeholders, including farmers, technologists, and policymakers, to capitalize on opportunities for supporting DT adoption in the context of sustainable agriculture.

6.2 Key Challenges in Adopting Digital Twin Technology for Agriculture

6.2.1 Technical Challenges in Digital Twin Implementation for Agriculture

Integrating DT technology in farming sectors is a key success for sustainable agriculture; however, it faces several technical challenges that need to be addressed. Three main challenges should be emphasized, as depicted in Fig. 6.2.

First, data interoperability is one of the greatest hindrances. In reality, crop farms often use multiple sensor models with different software platforms and farm management systems (Purcell and Neubauer 2023). As data is collected from multiple sources, it contains diverse formats. Hence, multiple communication protocols are required. As a result, the lack of standardized and compatible data collection methods across different systems leads to technical barriers to adopting DT in agriculture. This situation causes access problems to the proper data integration, thus decreasing the effectiveness of DT applications. The lack of a universal data

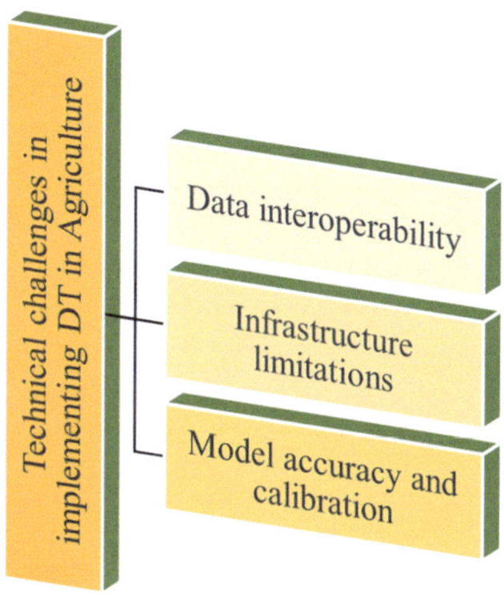

Fig. 6.2 Technical challenges in implementing DT in agriculture

standard results in interoperability difficulties, specifically in complicated agricultural modules involving multiple stakeholders in the data pipeline.

Second, the deficiency of infrastructure is a significant barrier to the use of DT in intelligent agriculture. The deployment of DT technology is further obstructed by inadequate technological infrastructure in the remote settings of rural agriculture. There are many locations, especially in the developing economies, with poor-quality Internet and electricity. Therefore, it delays the data transmission. Once it happens, it hinders the real-time computation, causing the interruption of the DT system (Iaksch et al. 2021). Additionally, a DT platform with high accuracy requires multiple types of IoT sensors with data storage and transmission protocols, cloud computing systems, and the user interface (Wang et al. 2023). It leads to a complex installation process of these systems. As a result, it forms barriers that limit the implementation of DT in agriculture.

Third, the model accuracy and calibration are the next technical obstacles that hinder the adoption of DT in agriculture. It is understood that the agricultural environment keeps changing, and each farming period requires different setups. The conditions for the soil care period differ from seed planting to plant growing, until fruit pruning. As each period requires different optimization farming parameters, it leads to diverse management practices. Hence, the DT platform is employed to monitor farm conditions. It can also be adopted to recalibrate the model to adapt to each farming period. However, it is challenging to calibrate these changes in a timely manner (Cesco et al. 2023). Hence, this forms technical challenges for DT to be applied widely in agricultural management.

6.2.2 Socioeconomic Challenges in Digital Twin Adoption

Beyond technical issues, the implementation of DT is also hindered by socioeconomic challenges, as mentioned in Fig. 6.3. The first socioeconomic constraint involves the exceedingly high investment costs in implementing DT systems. The expenses include the purchase of equipment and devices, software licensing, as well as the ongoing maintenance service of these technologies (Aroba and Rudolph 2024; Baumüller 2018). These expenses are the primary reason causing the hesitation of farmers. Additionally, the extra burden of finance, apart from the initial expense, adds to the cost for data subscriptions, technical support, and overall system maintenance, which all bring ongoing economic stress to regular farm practices. As the benefits of the DT are not visible in the short term, while the investment expenses take a long time to return, it adds to the hesitation from the farmers to adopt this modern technology.

Additionally, farmer skepticism is the next main socioeconomic barrier that holds back the employment of DT technology in agriculture. A significant number of farmers are mistrustful of the data-driven systems (Reichelt and Nettle 2023). Many previous unsuccessful implementations, where data privacy issues arose, have deterred farmers from adopting these digital technologies. The misunderstanding between technology developers and end-users frequently results in solutions that fail

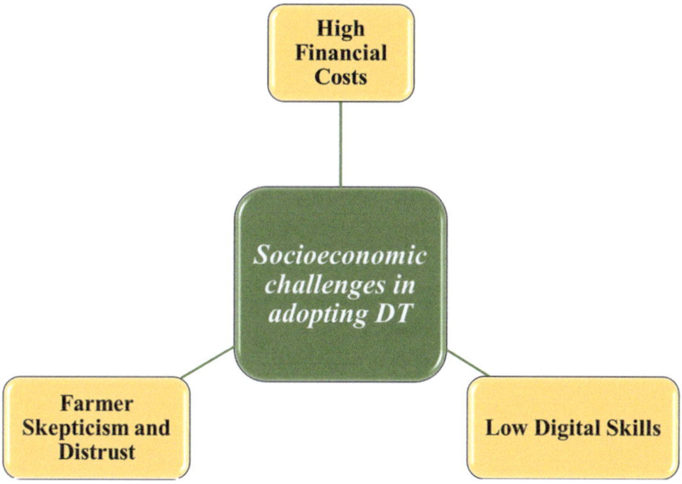

Fig. 6.3 Socioeconomic challenges in implementing DT in agriculture

to align with the actual requirements of the farmers or operational realities. Hence, it added to the adoption resistance from the farmers.

Furthermore, among various regions and agricultural domains, the human resources have different levels of digital skills. Hence, it widens the gaps between those who are well-equipped with technology and those who are not trained. The context is especially common in developing economies, where basic digital infrastructure may be absent. Yet, the situation is also true in some urban agriculture communities, even though it is not as frequent.

In brief, high financial cost, farmer skepticism and distrust, and the digital divide in rural areas are the main socioeconomic challenges that need to be addressed to facilitate the adoption rate of DT in agricultural sectors.

6.2.3 Operational Challenges in Digital Twin Implementation

Beyond technical and socioeconomic concerns, several operational obstacles exist that hinder the employment of DT in real-world agriculture applications. The first operational obstacle is the system scalability of the DT platform in the farm sectors. Each farm has different scales and productions. Some focus on growing crops, others focus on fruits, while some produce milk, livestock. Some are located in the hill area, some are near the sea, and the others are on the land. Each farming environment requires different operations in the physical world; hence, the DT platform must also be able to adapt to the environment that it serves (Lezoche et al. 2020). In other words, adaptable models of the DT platform must be generated to be able to operate on diverse regional climates, soils, crops, and scales. This helps to minimize the cost of customization settings.

The next operational concern is data security and privacy. The IoT devices in the DT system collect real-time data, including geolocation, input, and output details. These data can be leaked to a third party, exposed to cyber threats, or misused (Ali et al. 2024). Furthermore, many DT platforms are set in a centralized structure that allows criminals entry into the systems. Hence, it brings security risks to the system. Additionally, farmers may feel mistrust due to unclear regulations about the data they own, as well as the lack of transparency regarding the data given to third parties.

Additionally, system maintenance and updating are the next operational challenges. The DT system requires continuous data inputs and regular model recalibration to maintain its accuracy (Overbeck et al. 2021). It forms challenges in maintaining the DT platform to ensure the system can function smoothly. Furthermore, the recalibration is always time-consuming and requires advanced skills. It is not easy to perform frequent recalibration to adjust and adapt to the environmental changes. Hence, it forms obstacles to the adoption of DT in agriculture.

To facilitate digital transformation in agriculture, these complex operational obstacles need to be addressed. A joint operational collaboration in developing the DT farm platform is the solution for the problems of workflow and capacity in the machinery operations in the field. The farming systems used in the development of these specific operations should be sufficiently adaptable to the requirements of the different types of farms, and remove any threats to data security and system trustworthiness.

6.3 Proposed Solutions and Strategies for Digital Twin Implementation

6.3.1 Technical Solutions for Digital Twin Implementation

In order to tackle the multidisciplinary technical issues related to the DT deployment in agriculture, many solutions involve using cutting-edge computing architectures and artificial intelligence (AI). The actions focused on the technical aspects include enhancing the system's flexibility, reliability, and accuracy, while also addressing the infrastructural limitations commonly found in agricultural settings.

One of the good ideas involves the creation of a modular and interoperable DT system that can operate on various types of farms and be easily deployed to the farms. By using a component-based architecture, these systems enable farmers to experience particular DT modules developed to meet their operational needs and budget constraints (Goodwin et al. 2024; Tzachor et al. 2022). The designed modular system will enable farmers to adopt the system incrementally and gradually expand their use of the DT technology.

To overcome connectivity limitations in rural areas, edge computing, laid over the model of offline-capable models, presents an effective technical solution. Through the diffusion of the data-processing element and the relocation of the computational tasks closer to the data source (i.e., the farm), these systems have less reliance on the continuous cloud connectivity (Alharbi and Aldossary 2021).

One of the advantages of edge devices is their functionality even in periods of low Internet access or in the event of Internet failure. Moreover, edge devices are equipped with well-thought-out data synchronization protocols. It prevents data from being problematic and ensures its accuracy once the connectivity is restored. Being available both online and offline, edge computing is a significant candidate for the DT system to be applied in scattered agricultural environments.

Next, the employment of AI in dynamic calibration mechanisms enhances the accuracy and responsiveness of the DT system, supporting the scalability in agriculture implementations. The model can analyze incoming sensor data at a high speed, that even up to several thousand per second (De Koning et al. 2023). Hence, it is available to deal with fluctuating environments, which is suitable for agricultural conditions. Additionally, dynamic calibration by using AI technology has the function of self-calibration. Therefore, it enhances the accuracy of the model in various weather, soil, and crop conditions. As a result, it eliminates the former obstacles, making the DT system ready for the adoption of DT in agriculture.

6.3.2 Socioeconomic Solutions for Digital Twin Adoption

As mentioned in Sect. 6.2.2, the socioeconomic challenges that hinder the adoption rate of DT in agriculture include financial, trust-related barriers, and low digital skills of those utilizing this advanced technique. Therefore, the solutions must be able to address these obstacles to ensure the stability in the implementation of DT in farming. In addition, technological advancements must be made available in an equitable manner to all, hence, it helps to address the socioeconomic concerns.

Public-private partnerships (PPPs) are identified as a really great solution for the financing and infrastructure barriers to DT acquisition. A couple of public and private partners leverage government forces and private sector professionals to secure the necessary funds for rural broadband expansion and IoT deployment (Owoola et al. 2024). PPP models logically collect the newest ideas for financing, such as cost-sharing, where governments fund the initial outlay while the private sector handles the tech deployment and maintenance. A role model is the case of using a discount on the price of the DT service by smallholders with the help of the bigger farms that are cross-subsidizing them (McCampbell et al. 2022). These win-win alliances first reduce capital expenses, yet the proper assumption of mutual risk and performance-based contracts further ensures their sustainability.

To remove the technological hesitance and develop the digital skills for the farmer, it is very important to focus on their needs. Instead of traditional classroom instruction, the farmer's training or education should involve the practical or field-based activities, which are directly related to the local situations (Pandey et al. 2025). In order to guarantee that the training is pertinent to the real agricultural issues that farmers encounter, digital literacy instruction should take a participatory approach, encouraging farmers to co-create the curriculum. On the other hand, the "train-the-trainer" program can be developed to transfer the suitable knowledge and skills to local residents, to allow them to provide the community with support in an effective way (Prajapati et al. 2025). As a result, these courses enable farmers with not only technological skills but also data literacy, and the farmers can read and utilize DT outputs better.

Pilot programs are also an effective approach for clarifying the tangible benefits at reasonable scales from technology potential to reality applications. It was proven that a well-designed pilot scheme should build on strong monitoring and evaluation systems, which will measure and substantiate the Return on Investment (ROI) across different farm types and models (Ingram et al. 2021). The farmer-led demonstration pilots are normally employed where fellows observe the impacts of technology directly, which fosters effective learning between each other. This type of training should be created as a cycle learning processes, which continuously receive feedback and then make improvements. One of the models that has been particularly successful is the "living lab", when the farmer takes part in the whole cycle of technology to ensure the right solutions can be made, and at the same time, solid trust can be built in the farmer community (Timpanaro et al. 2024).

In general, the socioeconomic solutions should be implemented in the real socioeconomic contexts, which are suitable and adaptable to different farming systems, cultural properties, and institutional frameworks. The most likely successful application strategy should be a combination of three methods, and accordingly, cooperative effects would address the financial and human capital barriers. The priority of future initiatives should be the creation of standardized indicators to assess the variations in socioeconomic factors due to the implementation of DT. This is imperative for the application of empirical evidence in making the right policy decisions regarding agricultural technology.

6.3.3 Policy and Collaboration Frameworks for Digital Twin Integration

Framing the right policy and building a robust collaborative ecosystem are the foremost conditions for the sustainable integration of DT technologies in agricultural systems. These institutional responses must be proposed in the context of regulatory, economic, and governance challenges.

State-promoted agricultural projects are an important political tool for pushing up the adoption of digital technologies in agriculture, both in developed and developing areas. Some studies have shown that the appropriate configurations should combine technological investments with performance-related subsidies connected to explicit sustainability results (Ehlers et al. 2021). For example, the EU agricultural subsidies program in Bavaria and Germany. As a result, it helped to increase the digital adoption rate by 15–17% (Gabriel and Gandorfer 2023). Another important thing is that these programs should be complemented by rural infrastructure projects, where the most fundamental issues that hinder the adoption of this advanced technology occur. Moreover, the policy instruments should be flexible depending on the real situation to adapt to new and evolving technological capabilities.

Furthermore, the policies related to data ownership need to be clarified. The policies are required to provide the necessary data to farmers while ensuring their privacy, while being able to handle conflicts of interest (Brown et al. 2023). Classifying the dataset level based on the protection level is necessary; for example, it can be divided into personal, operational, or aggregated datasets. Besides, it is recommended to have the access protocols associated with the protection level, respectively. The EU's agricultural data governance scheme proposes a potential track with evident data ownership principles. This data ownership encompasses the responsibility to notify the usage of data and the selection of the data that the farmer is willing to share (Atik 2022).

For the development, the most important thing is to organize the technological progress in a multi-dimensional manner rather than in a single-track mode. One of the example is the "Golden Triangle" model in the Netherlands. This network allows farmers, engineers, and policymakers can have productive discussions to drive innovation forward (Klerkx 2023). These forums should include not only mandatory knowledge distribution strategies but also be more practical in terms of staff exchanges and workshops focused on common problems.

In the future, the policy environment can be the stage to achieve the interoperable framework implementation across agricultural conditions, which will not only ease the cross-border agricultural data flow but also ensure cybersecurity. As DT technologies become more widespread and reach a higher level of development, it is expected that the decision-makers will shift their focus to the training programs to upskill the agricultural workforce in handling advanced technical units and digital businesses.

6.4 Case Studies and Best Practice in Digital Twin Implementation

The implementation of DT technology in the global field of agricultural production systems has been a great adventure with numerous experiences, ranging from successful to challenging. The utilization of the knowledge obtained from real project execution in the formulation of future agriculture projects and successful sector development is a valuable lesson.

6.4.1 Digital Twin Pig House Implementations

The study introduces the DT pig house in South Korea as one of the first agricultural systems that employed DT technology. The project was a collaboration between Seoul National University and the Electronics and Telecommunications Research Institute (ETRI) (Jeong et al. 2023). A two-way data digital pig house and physical pig house paired by the system consists of IoT sensors that measure the temperatures, humidity, and CO_2 concentration levels found in the piggery and are relevant for the dynamic heating, ventilation, and air conditioning (HVAC) systems

Table 6.1 Challenges, solutions, and outcomes of South Korea's digital twin pig house implementation

Aspect	Findings
Challenges	Data accuracy and integration, real-time data exchange, computational demands
Solutions	IoT+ Edge computing+ Cloud computing + Two-way data transfer
Outcomes	26.8% energy consumption reduction by optimizing ventilation fan control 18.4% energy consumption reduction with the optimization of the radiant heating panel

optimization. These sensors gather information continuously, and this data is used for the HVAC dynamic optimization (Jeong et al. 2023). The DT environment conducted simulations of different HVAC systems and made a comparison analysis. In these tests, fan-coil units were found to be the most effective method, with 26.8% higher energy efficiency as compared to the traditional method, as mentioned in Table 6.1. Moreover, through the implementation of the adaptive management module for water temperature, the energy consumption is decreased by 18.4%, thanks to the optimization of the radiant heating panels.

The implementation of the DT pig house has several main obstacles (Jeong et al. 2023). Initially, the data accuracy and integration were unavailable because most of the errors in the model were found from various sources, including weather forecasts and sensor readings. The project has provided ready data stream pipelines and utilized the edge computing model, which has secured direct and real-time data exchange between the physical world and the digital one.

Moreover, computational issue was another primary concern. Energy calculations were the most expensive computational cost for the simulations, requiring additional resources due to the complex physics from multiple setting conditions. Conversely, this challenge was overcome by developing cloud-based analytics as well as reduced-order modeling. These solutions enhanced the balance of the accuracy and scalability of the DT model for this actual pig house.

6.4.2 Digital Twin Implementation for Post-Harvest Preservation in Nigeria

Nigeria faces substantial post-harvest losses due to the lack of a proper preservation process (Elijah et al. 2021). It was reported that a total of 50% of perishable goods, including fruits, vegetables, roots, and tubers, are lost. Additionally, 30% of maize, millet, and rice perished because the preservation process worked improperly. Irrespective of these figures, there are other reasons for such debilitating losses. The primary reason for these tremendous losses is the limited availability of alternative stocks compared to the prevailing traditional preservation methods, such as smoking, sun-drying, and fermentation, as well as the conspicuous absence of ultra-modern storage and processing facilities.

One of the impressive ways of implementing DT technology is to deal with the above problems, mainly by focusing on solar drying optimization. For example, DT technology was applied to simulate the heat and mass transfer phenomena in solar dryers, then allowed the prediction of the optimal drying times (Elijah et al. 2021). This multivariate forecast ability plays the decision role to prevent the under-drying or over-drying of agricultural products (Prawiranto et al. 2021). The virtual copies of solar dryers that connect and get the real data from sensors inside (such as temperature, humidity...), thus let the drying cycles operate automatically, and then the spoilage can be significantly reduced.

The DT technology can also be expanded into the area of monitoring quality. The DT system can help to monitor the changes in chemical substances and microbial components, and it can send alerts to the farmers about these changes, which can cause spoilage with the support of an IoT-enabled sensor. Moreover, the DT can be improved to be more standardized, effective, and safer in food preservation.

6.4.3 Emerging Best Practices in Digital Twin Implementation

The integration of these case studies reveals several best practices across various sectors. According to the universal strategy of phased implementation, the DT agricultural projects that adopted this strategy were successful in their sectors. Starting with the basic functions and establishing a solid foundation is the key factor for their success. Additionally, advanced features, including controlled complexity and assured user adaptability, contribute to their progress. At the same time, hybrid governance that combines central-level technical support, while maintaining local decision-making at a certain level, makes the system even more flexible and support even better for decision-making. This combination can strengthen local users and enable the technology to be adjusted and adapted to meet specific operational requirements.

The combined case studies demonstrate that although the technology of DT is challenging to deploy, it is the right path to achieving sustainable agriculture in the digital era. The relentless growth of these technologies requires a continuous development process from a wide range of agricultural settings and production requirements.

6.5 Future Outlook for Digital Twin Technology in Agriculture

The agricultural sector is on a digital transformation with DT serving as the backbone of sustainable applications. Through the merger of state-of-the-art technological tools with agriculture, the advanced technology offers farmers the ability to maintain the efficiency of farm operations, while also addressing sustainability issues.

6.5.1 Integration of Emerging Technologies

The next-generation DT systems will adopt artificial intelligence (AI) functions to support even more complex predictive analytics models. The "black box" problems can be understood by employing explainable AI (XAI) frameworks. Therefore, it provides transparent decision paths to farmers, helping build trust in the recommendation system (Shams et al. 2024). The most potential lies with neurosymbolic AI approaches that combine data-driven learning with domain-specific knowledge base rules, thereby providing more reliable performance in edge cases developed in agricultural settings (Meghraoui et al. 2025). The future DT ecosystems are expected to be developed on Blockchain technology. This enables the quality assurance of the data while creating decentralized data marketplaces that allow farmers to receive a fair share of their contributions (Kumar Tyagi and Tiwari 2025). In addition, the emergence of Web3 architectures could result in the development of peer-to-peer DT networks where farmers would be able to share their validated models and insights directly, without intermediary platforms that would cost 30–40% less to implement (Zhang and Huang 2024).

6.5.2 The Long-Term Digital Agricultural Ecosystem

The wide application of DT technology presents a connected agricultural metasystem where each farm twin would communicate with regional and global agricultural models of other farms. It was pointed out in some research that this hierarchy of DTs will lead to increased sophistication in production systems, resource distribution, and environmental care (Praharaj et al. 2023). In reality, the automatic DT systems can play an important role in managing the agricultural inputs, setting the water or nutrients level, etc., based on the real-time detection and microclimate monitoring (Tsakiridis et al. 2023). Moreover, with the addition of distributed ledger technology, it is possible to make the "passport" for certified sustainable digital products, which would clearly be a major step forward to give clear proof from farm to table, and increase the value of the agricultural products (Purcell et al. 2023).

The DTs can also join the self-optimizing agricultural systems, where they can continually learn from general networks in different geographies and production systems. This global structure of knowledge would significantly shorten the time needed to detect and disseminate best practices, while also providing an alerting system for new bugs, diseases, and ecological changes (Melesse 2025). Nonetheless, this promise can be fulfilled only if there is an equal opportunity for digital equity; the agricultural sector must strive to avert a situation in which DT technologies exacerbate the existing barriers between large-scale commercial farms and smallholder farmers. To achieve this, policy frameworks and technical structure plans should be consciously designed to enable equal benefits and avoid the monopolization of agricultural knowledge and control (Kumar Tyagi and Tiwari 2025).

The coming decades are expected to be the time to decide whether DT technology can successfully drive a sustainable increase in global agricultural productivity.

However, the success will not rely just on the growth of technology, but also on the adequate establishment of governance policies, business ecosystems, and human capital programs that ensure fair access in the digital agriculture revolution.

6.6 Conclusion

The comprehensive examination of DT technology in agricultural systems has revealed a dual phenomenon, where it has enormous potential and also significant difficulties in application that necessitate collective action to address these. In this chapter, a thorough examination of the challenges of adopting DT in agriculture was conducted. The most concerning issues related to the technical aspects include system interoperability, inadequate infrastructure, and model accuracy. In addition, there are also concerns about the socioeconomic aspects, such as high investment expenses, farmer skepticism and distrust, and the digital divide in rural areas, which can contribute to the persistence of DT technology in agriculture. All these issues together become a challenge to prevent the movement of DT technology from being highly sustainable, productive, and resilient.

It is proposed that the solutions for the challenges should be based on a multidimensional understanding related to the problems. The technical achievements, such as modular arrangement of systems and edge computing, are primarily used to meet the functional requirements. Meanwhile, the socioeconomic measures include public-private partnership and farmer-centered training, which increase both the accessibility and the adoption rate. The case studies demonstrated that a successful implementation story is only achievable when these local solutions are combined with effective policy frameworks. New technologies, especially in Blockchain systems and AI, are the way to strengthen while lowering the costs and complexity.

Future efforts to enhance the adoption of DT systems require the mutual actions of all stakeholders in agriculture. The technology developer should consider more user-friendly designs that address the challenges associated with the rural agricultural environment. The establishment of a regulation and law framework that is conducive to investment, together with digital transformation, seems to be effective in promoting these initiatives. In addition, with the agricultural extension services that require engineering to effectively address the digital skills gap, it is really helpful when financial institutions can offer unique financial products as a means of promoting technology adoption. Last but not least, the farmers shall be in the central position rather than simply being recipients of the technologies.

Along with the arising climate change and food security issues, the DT technology is an important and effective tool to achieve agricultural sustainability. However, this potential can only be obtained through collaborative efforts of stakeholders in the value chain. To be honest, the digital transformation in agriculture is unstoppable. By carefully applying the DT technology, it can transform conventional agricultural practices into digital ones in a sustainable manner. It also offers opportunities to use advanced technology to improve the global environment. At the same time, it can ensure the benefits of digital agriculture worldwide.

References

Alharbi HA, Aldossary M (2021) Energy-efficient edge-fog-cloud architecture for IoT-based smart agriculture environment. IEEE Access 9:110480–110492. https://doi.org/10.1109/ACCESS.2021.3101397

Ali G, Mijwil MM, Bosco Apparatus Buruga, Abotaleb M, Adamopoulos I (2024) A survey on artificial intelligence in cybersecurity for smart agriculture: state-of-the-art, cyber threats, artificial intelligence applications, and ethical concerns. Mesopotamian J Comput Sci 2024:53–103. https://doi.org/10.58496/MJCSC/2024/007

Aroba OJ, Rudolph M (2024) Systematic literature review on the application of precision agriculture using artificial intelligence by small-scale farmers in Africa and its societal impact. J Infrastruct Policy Dev 8(13):8872. https://doi.org/10.24294/jipd8872

Atik C (2022) Towards comprehensive European agricultural data governance: moving beyond the "data ownership" debate. IIC – Int Rev Intellect Prop Compet Law 53(5):701–742. https://doi.org/10.1007/s40319-022-01191-w

Baumüller H (2018) The little we know: an exploratory literature review on the utility of mobile phone-enabled services for smallholder farmers. J Int Dev 30(1):134–154. https://doi.org/10.1002/jid.3314

Brown C, Regan Á, Van Der Burg S (2023) Farming futures: perspectives of Irish agricultural stakeholders on data sharing and data governance. Agric Hum Values 40(2):565–580. https://doi.org/10.1007/s10460-022-10357-8

Cesco S, Sambo P, Borin M, Basso B, Orzes G, Mazzetto F (2023) Smart agriculture and digital twins: applications and challenges in a vision of sustainability. Eur J Agron 146:126809. https://doi.org/10.1016/j.eja.2023.126809

De Koning K, Broekhuijsen J, Kühn I, Ovaskainen O, Taubert F, Endresen D, Schigel D, Grimm V (2023) Digital twins: dynamic model-data fusion for ecology. Trends Ecol Evol 38(10): 916–926. https://doi.org/10.1016/j.tree.2023.04.010

Dibbern T, Romani LAS, Massruhá SMFS (2024) Main drivers and barriers to the adoption of digital agriculture technologies. Smart Agric Technol 8:100459. https://doi.org/10.1016/j.atech.2024.100459

Ehlers M-H, Huber R, Finger R (2021) Agricultural policy in the era of digitalisation. Food Policy 100:102019. https://doi.org/10.1016/j.foodpol.2020.102019

Elijah O, Rahim SKA, Emmanuel AA, Salihu YO, Usman ZG, Jimoh AM (2021) Enabling smart agriculture in Nigeria: application of digital-twin technology. In: 2021 1st international conference on multidisciplinary engineering and applied science (ICMEAS), pp 1–6. https://doi.org/10.1109/ICMEAS52683.2021.9692351

Gabriel A, Gandorfer M (2023) Adoption of digital technologies in agriculture—an inventory in a european small-scale farming region. Precis Agric 24(1):68–91. https://doi.org/10.1007/s11119-022-09931-1

Goodwin T, Xu J, Celik N, Chen C-H (2024) Real-time digital twin-based optimization with predictive simulation learning. J Simul 18(1):47–64. https://doi.org/10.1080/17477778.2022.2046520

Iaksch J, Fernandes E, Borsato M (2021) Digitalization and big data in smart farming—a review. J Manag Anal 8(2):333–349. https://doi.org/10.1080/23270012.2021.1897957

Ingram J, Chiswell H, Mills J, Debruyne L, Cooreman H, Koutsouris A, Alexopoulos Y, Pappa E, Marchand F (2021) Situating demonstrations within contemporary agricultural advisory contexts: analysis of demonstration programmes in Europe. J Agric Educ Ext 27(5):615–638. https://doi.org/10.1080/1389224X.2021.1932534

Jeong D-Y, Jo S-K, Lee I-B, Shin H, Kim J-G (2023) Digital twin application: making a virtual pig house toward digital livestock farming. IEEE Access 11:121592–121602. https://doi.org/10.1109/ACCESS.2023.3313618

Klerkx L (2023) Towards agricultural innovation systems 4.0?: Supporting directionality, diversity, distribution and democracy in food systems transformation. Wageningen University & Research. https://doi.org/10.18174/634949

Kumar Tyagi A, Tiwari S (2025) Blockchain and digital twin applications in smart agriculture, 1st edn. Auerbach Publications. https://doi.org/10.1201/9781003507390

Lezoche M, Hernandez JE, Alemany Díaz MDME, Panetto H, Kacprzyk J (2020) Agri-food 4.0: A survey of the supply chains and technologies for the future agriculture. Comput Ind 117:103187. https://doi.org/10.1016/j.compind.2020.103187

McCampbell M, Schumann C, Klerkx L (2022) Good intentions in complex realities: challenges for designing responsibly in digital agriculture in low-income countries. Sociol Rural 62(2): 279–304. https://doi.org/10.1111/soru.12359

Meghraoui K, Racharak T, Ait El Kadi K, Bensiali S, Sebari I (2025) A new integrated neurosymbolic approach for crop-yield prediction using environmental data and satellite imagery at field scale. Artif Intell Geosci 6(1):100125. https://doi.org/10.1016/j.aiig.2025.100125

Melesse TY (2025) Digital twin-based applications in crop monitoring. Heliyon 11(2):e42137. https://doi.org/10.1016/j.heliyon.2025.e42137

Overbeck L, Brützel O, Teufel M, Stricker N, Kuhnle A, Lanza G (2021) Continuous adaption through real data analysis turn simulation models into digital twins. Procedia CIRP 104:98–103. https://doi.org/10.1016/j.procir.2021.11.017

Owoola WA, Olaniyo BA, Obeka OD, Ifeanyi AO, Ige OB, Bello SF (2024) Cross-Sector collaboration in energy infrastructure development: New models for public-private partnerships in emerging markets. Int J Sci Res Arch 13(1):2871–2882. https://doi.org/10.30574/ijsra.2024.13.1.2020

Pandey MK, Singh M, Sulekha, Rana P, Tripathi S, Vishwakarma SK, Pandey A, Shastri S (2025) Public-private partnerships in agricultural extension: opportunities and challenges. J Exp Agric Int 47(6):597–612. https://doi.org/10.9734/jeai/2025/v47i63520

Peladarinos N, Piromalis D, Cheimaras V, Tserepas E, Munteanu RA, Papageorgas P (2023) Enhancing smart agriculture by implementing digital twins: A comprehensive review. Sensors 23(16):7128. https://doi.org/10.3390/s23167128

Praharaj L, Gupta M, Gupta D (2023) Hierarchical federated transfer learning and digital twin enhanced secure cooperative smart farming. In: 2023 IEEE International Conference on Big Data (BigData), pp 3304–3313. https://doi.org/10.1109/BigData59044.2023.10386345

Prajapati CS, Priya NK, Bishnoi S, Vishwakarma SK, Buvaneswari K, Shastri S, Tripathi S, Jadhav A (2025) The role of participatory approaches in modern agricultural extension: bridging knowledge gaps for sustainable farming practices. J Exp Agric Int 47(2):204–222. https://doi.org/10.9734/jeai/2025/v47i23281

Prawiranto K, Carmeliet J, Defraeye T (2021) Physics-based digital twin identifies trade-offs between drying time, fruit quality, and energy use for solar drying. Front Sustain Food Syst 4: 606845. https://doi.org/10.3389/fsufs.2020.606845

Purcell W, Neubauer T (2023) Digital twins in agriculture: a state-of-the-art review. Smart Agric Technol 3:100094. https://doi.org/10.1016/j.atech.2022.100094

Purcell W, Neubauer T, Mallinger K (2023) Digital twins in agriculture: challenges and opportunities for environmental sustainability. Curr Opin Environ Sustain 61:101252. https://doi.org/10.1016/j.cosust.2022.101252

Reichelt N, Nettle R (2023) Practice insights for the responsible adoption of smart farming technologies using a participatory technology assessment approach: the case of virtual herding technology in Australia. Agric Syst 206:103592. https://doi.org/10.1016/j.agsy.2022.103592

Shams MY, Gamel SA, Talaat FM (2024) Enhancing crop recommendation systems with explainable artificial intelligence: a study on agricultural decision-making. Neural Comput & Applic 36(11):5695–5714. https://doi.org/10.1007/s00521-023-09391-2

Symeonaki E, Maraveas C, Arvanitis KG (2024) Recent advances in digital twins for agriculture 5.0: applications and open issues in livestock production systems. Appl Sci 14(2):686. https://doi.org/10.3390/app14020686

Timpanaro G, Foti VT, Cascone G, Trovato M, Grasso A, Vindigni G (2024) Living lab for the diffusion of enabling technologies in agriculture: the case of Sicily in the Mediterranean context. Environ Earth Sci. https://doi.org/10.20944/preprints202412.0062.v1

Tsakiridis NL, Samarinas N, Kalopesa E, Zalidis GC (2023) Cognitive soil digital twin for monitoring the soil ecosystem: a conceptual framework. Soil Syst 7(4):88. https://doi.org/10.3390/soilsystems7040088

Tzachor A, Sabri S, Richards CE, Rajabifard A, Acuto M (2022) Potential and limitations of digital twins to achieve the Sustainable Development Goals. Nat Sustainability 5(10):822–829. https://doi.org/10.1038/s41893-022-00923-7

Wang Y, Su Z, Guo S, Dai M, Luan TH, Liu Y (2023) A survey on digital twins: architecture, enabling technologies, security and privacy, and future prospects. IEEE Internet Things J 10(17):14965–14987. https://doi.org/10.1109/JIOT.2023.3263909

Zhang Y, Huang Y (2024) Sustainable hybrid of agriculture and urban Ecology Base on web 3.0 technology. In: Stephanidis C, Antona M, Ntoa S, Salvendy G (eds) HCI international 2023—late breaking posters, vol 1958. Springer Nature Switzerland, pp 388–394. https://doi.org/10.1007/978-3-031-49215-0_46

Digital Twin-Driven Predictive Models for Crop Disease Detection

7

R. Karthick Manoj and S. Aasha Nandhini

Abstract

The ever-increasing incidence and severity of agricultural diseases pose a threat to the safety of food supplies around the world. In the following chapter, we will investigate how the Digital Twin (DT) technology can be utilized in the process of developing predictive disease detection models for agricultural purposes. Through the use of Internet of Things (IoT) sensors, meteorological stations, drones, and satellite photography, digital twins are able to generate virtual duplicates of crops that are in constant alignment with accurate real-time data. In order to mimic disease transmission patterns and find early symptoms of epidemics, digital twins can be utilized to incorporate this data through the utilization of sophisticated analytics and machine learning approaches. This chapter takes a look at some of the most important components of predictive modeling, such as the connection between viruses, surroundings, and hosts, the diagnosis of sickness using imaging, and the evaluation of risk within the context of a decision tree framework. This includes the presentation of actual case studies that illustrate the identification of rust in wheat, blast in rice, and late blight in potatoes. In addition, they illustrate the appropriate use of the models and the responses of farmers that correspond to those models. The chapter comes to a close with an analysis of the technical issues, which include the integration of data and the generalization of models, the hurdles of attaining user approval, and the

R. K. Manoj
Department of Electrical and Electronics Engineering, AMET Deemed to be University, Kanathur, Tamil Nadu, India

S. Aasha Nandhini (✉)
Department of Electronics and Communication Engineering, Sri Sivasubramaniya Nadar College of Engineering, Kalavakkam, Tamil Nadu, India

© The Author(s), under exclusive license to Springer Nature Singapore Pte Ltd. 2026
A. L. Srivastav et al. (eds.), *Digital Twin Technology for Sustainable Agriculture*,
https://doi.org/10.1007/978-981-95-5915-2_7

potential applications of decision trees in intelligent crop disease surveillance systems that are scalable.

Keywords

Digital twin · Crop disease forecasting · Precision agriculture · Predictive modeling · Smart farming

7.1 Introduction

7.1.1 Importance of Early Crop Disease Detection in Global Agriculture

The world's food supply and economic stability are both crucially supported by agricultural productivity. Rising agricultural output by more than 60% is desperately needed to fulfill future demand as the global population is expected to reach 9 billion by 2025 (Cesco et al. 2023). Plant diseases still pose a constant danger to this objective, though, and they cause yearly losses in yield valued hundreds of billions of dollars. Many crop diseases are caused by pathogenic organisms like fungi, bacteria, viruses, and nematodes, therefore greatly affecting the quality and quantity of agricultural goods.

Under suitable weather circumstances, early-onset diseases like blast, rusts, smuts, and blights can quickly spread among basic crops including rice, wheat, and maize. For example, if not discovered and managed in time, Puccinia species of wheat rust can wipe whole fields in a few weeks. Likewise, the primary agent of late blight in potatoes, Phytophthora infestans, caused the Irish Potato Famine and is a hazard in modern agriculture (Purcell and Neubauer 2023). Early disease identification is thus not only a question of increasing yields; it is also a vital need for guaranteeing food security, economic stability in agricultural economies, and the sustainability of natural systems. Early identification enables early action using precision agricultural methods such as selective breeding, focused pesticide treatment, or region isolation of impacted areas. These treatments lower pesticide use, slow down disease resistance, and lower financial losses. Furthermore, early disease diagnosis might offer useful epidemiological information to assist more general plant health management plans on local and national levels.

7.1.2 Limitations of Traditional Monitoring Methods

Even while disease surveillance is rather important, present techniques are still mostly manual and reactive. Traditional monitoring mostly depends on human observation and field scouting, subjective and labor-intensive activities by nature (Goldenits et al. 2024). Often encompassing many hectares, farmers or agronomists must visually search huge agricultural fields for symptomatic plants. This method is prone to human mistake and constrained by the inspector's experience, tiredness, and time limits. Particularly when visual signals are vague or resemble other abiotic stressors like nutrition deficits or drought, subtle or early-stage symptoms are often overlooked. The time delay of conventional approaches is still another key drawback. The infection can have already expanded outside control by the time obvious symptoms show and are found. Sending samples to a laboratory for diagnostic confirmation often requires more time and money than is necessary. Critical intervention windows are lost during this time, and disease outbreaks can reach epidemic proportions (Juroszek et al. 2020).

A lot of farms, especially in developing countries, have large, distant locations where it is not practical to check the fields often. The fact that there aren't enough qualified specialists makes monitoring programs even less effective and less widespread. Traditional disease surveillance doesn't provide the spatial and temporal resolution needed to support proactive, mass-based decision-making, even in places with advanced technology. These limitations highlight the necessity for an intelligent, real-time, scalable system that not only monitors current health conditions but also predicts future illness occurrences based on environmental indicators, historical data, and pathogen behavior models (Chakraborty and Newton 2011; Buja et al. 2021).

7.1.3 Introduction to Digital Twin (DT) Technology in Agriculture

Digital twins (DTs) are becoming more common in many fields, such as transportation, aerospace, city planning, and healthcare. A digital twin is a virtual copy of a physical thing or system that is always being made with real-time data from outside sources. The DT model was first made to keep an eye on complicated industrial machines. Since then, it has been used to study biological systems, especially those related to farming (Nehrey and Zomchak 2022). A "Digital Twin" is a virtual model of a crop or agricultural ecosystem that is made from real-time data and used in agricultural research. The model gets information from a number of different places, such as satellite images, weather sensors, soil health monitors, images from unmanned aerial vehicles (UAVs), and biochemical sensors built into plants. This gives a full and changing picture of how crops grow, how healthy they are, and how they interact with their surroundings. On the DT platform, machine learning (ML), deep learning, and predictive modeling algorithms are always looking at these data sources.

Digital twins in agriculture can leverage information from farming, the environment, and diseases to copy the biological and ecological processes that make people sick. A decision tree can indicate how a fungal disease could spread when the temperature and humidity are just right. It can also show how nutritional stress can make plants less immune, which makes it more probable that crops will get sick. These models can show signals of a dangerous illness outbreak before anyone else can recognize the symptoms. Digital twins can do more than just guess what will happen. They include interactive visualization tools that make it easy for farmers, agronomists, and policymakers to access dashboards with maps of disease risk, models of growth, and health indicators. People can now choose better since they have more information. For example, they could use less water for irrigation, switch to better fungicides, or adjust when they pick the crops. This enables you tweak and localize them in ways that weren't possible previously. They are a key component of the change from traditional farming to smart farming based on data because they make sensor data meaningful.

7.1.4 Objectives and Scope of the Chapter

The aim of this chapter is to conduct an in-depth examination of the application of Digital Twin (DT) technology in the management and prediction of crop diseases. The purpose of this project is to show how a Digital Twin framework for farming is put up. The main focus will be on the most critical components, like data sources, ways to talk to each other, and ways to do math. The chapter also talks about innovative approaches to anticipate diseases, such as using mapping and simulation to find the regions where diseases are most likely to happen, deep learning to sort leaf diseases based on photographs, and environmental-pathogen-host interaction models. Real-world case studies show how DT systems can be used to find and stop important agricultural diseases, such as wheat rust, rice blast, and potato late blight. The chapter also discusses about the challenges that come up with technology while trying to develop powerful DT systems. Some of these problems come up while trying to integrate data, make models that work in different agricultural settings, and make sure that all the parts of the system can work together. It also talks about problems that come up when people start using technology, such as problems with digital infrastructure, getting farmers involved, and more general socioeconomic problems that affect how technology spreads in rural farming areas. The chapter ends with a talk about possible future research routes, focusing on the idea of making scalable and smart digital twin systems to keep an eye on the next generation of agricultural illnesses. This might be done by using new technologies like edge computing, blockchain, and adaptive artificial intelligence. This work has a multidisciplinary focus because it draws from and adds to the fields of computer vision, environmental modeling, plant pathology, and data science. Researchers, agricultural engineers, agronomists, data scientists, and lawmakers who want to use digital technology to make farming more resilient and sustainable should find this information very useful.

7.2 Background and Related Work

7.2.1 Overview of Crop Diseases and Their Impact

Crop diseases are a major cause of agricultural losses around the world. These illnesses are caused by a wide range of pathogens, such as fungi, bacteria, viruses, nematodes, and oomycetes. They not only lower yields, but they also hurt the quality of products and hurt trade and market value. The Food and Agriculture Organization (FAO) says that plant pests and diseases are responsible for up to 40% of the world's food crop losses each year. Epidemic outbreaks of diseases like wheat stem rust (Puccinia graminis f. sp. tritici), rice blast (Magnaporthe oryzae), and potato late blight (Phytophthora infestans) have harmed food systems in the past. Climate change is making agricultural illnesses more dangerous by changing how strong pathogens are, how weak hosts are, and where diseases are found. Less genetic diversity in crop species, monoculture farming, and global trade all make plants more likely to get sick and spread disease. So, to reduce losses and encourage sustainable growth in agriculture, we need good systems for watching and predicting things (Jeong et al. 2022).

7.2.2 Traditional Disease Monitoring and Its Limitations

A lot of the time, farmers, agronomists, or extension workers have to do routine hand-offs to look for illnesses that impact crops. Usually, these tests involve looking at plant parts that show symptoms, like lesions, chlorosis, wilting, or necrosis. There may also be other signs. Polymerase chain reaction (PCR) and enzyme-linked immunosorbent assay (ELISA) are two laboratory-based diagnostic procedures used in more advanced settings to confirm the presence of specific infections and identify them. Even while these old approaches are well-known and still give useful analysis (Wang and Li 2025), they have a lot of big problems that need to be fixed. It is not possible to scale up manual field scouting since it does not cover large or geographically distant agricultural areas well enough. This is the first reason why it won't work. Second, they react to things instead of taking action. Sometimes, people don't realize they are sick until they have clear symptoms, which means it's too late to do something about it. To make matters worse, these methods depend a lot on people's knowledge and skills, which are not always easy to get, especially in areas with few resources. Also, typical monitoring methods can't predict very well because they don't take into consideration real-time environmental conditions. Changes in microclimate, soil moisture, and weather patterns are some of these causes. All of these things make it more likely that people may get sick. When you look at all of these restrictions together, it's evident that disease monitoring systems need to be more automated, intelligent, and scalable. These systems ought to furnish reliable assessments that facilitate the ongoing decision-making process in agriculture (Chaux et al. 2021).

7.2.3 Emergence of Digital Technologies in Agriculture

Smart Farming and Precision Agriculture (PA) have fundamentally transformed how farming is done in the last 10 years. These paradigms keep an eye on and manage the health of crops, soil properties, irrigation schedules, and pest threats using a number of digital technologies. It's possible to keep an eye on vegetation indicators (such NDVI and SAVI) on a big scale thanks to remote sensing with satellites and cameras. These signals could mean that the plant is sick. IoT sensors buried in fields gather precise information about the soil's moisture, temperature, humidity, and leaf wetness. All of these things have a huge effect on how diseases start and spread. Scientists have found signs of sickness in photos of leaves using machine learning. These algorithms can also guess how the weather and diseases would affect each other and find pests on their own. These improvements give Digital Twins and other intelligent systems a strong technological base. Digital Twins are designed to combine different data streams into cohesive, dynamic models that can simulate how plants grow and how diseases spread (Ma et al. 2020).

7.2.4 Concept and Evolution of Digital Twin Technology

The idea of a Digital Twin (DT) came from the aerospace and industrial industries, where it was used to model, track, and anticipate how complex physical systems like engines, turbines, and production lines would behave. A Digital Twin has three parts: a physical entity, such as a crop or an agricultural environment; a digital counterpart, which is a software-based virtual model; and a data pipeline that uses real-time sensing technology to keep the two in sync all the time. Recently, this paradigm has been used in agriculture to model biological systems including crops, livestock, and whole agro-ecosystems. Unlike mechanical systems, biological and ecological systems are always changing, not always in a straight line, and are affected by many outside factors. Agricultural digital twins must also contain adaptive, data-driven modeling methods to show how plants, diseases, climate, and management strategies are all connected in complex ways (Kalyani et al. 2024). Edge devices for processing data in the field, cloud computing infrastructure, and high-resolution spatiotemporal data collected from IoT sensors, drones, and satellites are all helping to make these systems more popular. Research and business solutions are being developed to copy the many stages of crop growth (phenology), keep track of how water is used, and get the most out of nutrients. Of all these purposes, disease detection is the one that Digital Twins is most excited about, both technically and practically. Effective implementation must include both real-time monitoring and predictive modeling features in order to predict when an illness will start, mimic how pathogens spread, and suggest quick ways to intervene.

7.2.5 Related Work in Predictive Disease Modeling

In the last 10 years, a lot of research has been done on how to use remote sensing, computer vision, and machine learning (ML) to find and predict plant diseases. One well-known area is using convolutional neural networks (CNNs) to find diseases in important crops including tomatoes, maize, and wheat. These models have been quite good at finding particular disease indicators in leaf images. Researchers have also predicted disease outbreaks like rust and blight by using environmental modeling together with pathogen lifecycle simulations. They looked at things like temperature, humidity, and rainfall patterns (Xie et al. 2022). Using Geographic Information System (GIS)-based techniques, it is now possible to map disease risk and anticipate how diseases would spread in different weather circumstances.

Using deep learning algorithms and footage from unmanned aerial vehicles (UAVs), it is now feasible to discover stress in diseases early on. For instance, the NDVI, LSTM neural networks have been able to predict rice blast many days before the symptoms are visible to the naked eye. Digital twins (DTs) based in greenhouses have been used to manage the climate so that illnesses don't spread in crops like cucumbers and tomatoes. Prototype systems have been created for Digital Twin applications that use real-time microclimate data to depict how illnesses spread in vineyards. A lot of programs seem like they could be useful, but they aren't properly incorporated into the whole process of treating an illness, so they aren't as helpful as they could be. For a Digital Twin system to work well, it needs to combine data gathering, predictive modeling, disease simulation, visual analytics, and user feedback into one closed-loop system that lets farmers make proactive, data-driven decisions (Demchenko et al. 2017).

7.2.6 Positioning of this Chapter

This chapter provides a unified Digital Twin (DT) framework designed for the detection and management of agricultural diseases, thereby advancing the foundational developments in digital agriculture and predictive modeling. The DT paradigm allows physical agricultural systems and their digital counterparts to engage with each other in a full, real-time, two-way way. This is different from typical sensor-based or isolated machine learning methodologies. This continual feedback loop lets you not only keep an eye on things, but also model them, predict them, and make decisions based on the knowledge in a way that is flexible and adaptable. Real-time data collection from a variety of sources, such as IoT devices, drones, and satellite platforms; dynamic modeling of the complicated interactions between pathogens, crops, and environmental conditions; image analysis and predictive algorithms driven by artificial intelligence for early disease detection; spatial visualization tools powered by GIS-integrated dashboards; and empirical case studies that assess model performance and include feedback from end users, especially farmers. The chapter stands out because it brings together a lot of different technology parts into one system. The chapter situates this work within the broader context of food security, climate resilience, and the digital transformation of agriculture, thereby endorsing a growing multidisciplinary initiative to develop intelligent, scalable, and actionable systems that enhance the sustainability and productivity of modern farming practices.

7.3 Digital Twin Architecture for Crop Health Monitoring

A Digital Twin (DT) system is a virtual copy of a real-world agricultural setting that is constantly changing and based on data. It is used to keep an eye on the health of crops. Its fundamental goal is to constantly model, simulate, and predict when and how diseases will start and advance by syncing real-time data with computer models. This section talks about the layered architecture of a digital twin (DT) for agriculture, focusing on its physical, cyber, and integrative levels. The system encourages proactive decision-making by using real-time monitoring, AI-based analysis, and two-way communication between the digital and physical worlds (Bertone et al. 2021).

7.3.1 System Overview

The DT system has three levels that depend on each other. The Physical Layer collects data from the real world, the Cyber Layer models and analyzes that data, and the Integration & Feedback Layer gives consumers and field systems information and actions. This layered design creates a closed-loop ecosystem for disease control by allowing real-time data and virtual simulations to work together all the time. The Digital Twin architecture for crop health monitoring shown in Fig. 7.1 is made up of three primary parts: the Real-World Data Acquisition layer, the Digital Twin Core (Cyber Model), and the User Interaction and Actuation layer. The picture demonstrates how data from physical sensors and field inputs, like IoT sensors, UAVs, satellites, and farmer apps, flows into cyberspace. In cyberspace, modules like data intake, disease simulation, and AI-based picture analysis regulate the data. Then, interactive dashboards, mobile apps, and autonomous edge actuators provide you the processed insights in a never-ending feedback loop that links observation, simulation, and decision-making.

7.3.2 Physical Layer: Real-World Data Acquisition

The physical layer of the Digital Twin architecture is the most fundamental way to connect the real world of farming to the digital world. There are many field-based data sources that provide continuous, real-time inputs that are needed for accurate simulation and analysis. IoT sensors deployed over the farm terrain keep an eye on crucial environmental characteristics including soil temperature, moisture levels, humidity, and pH. These are all important for the health of crops and the risk of

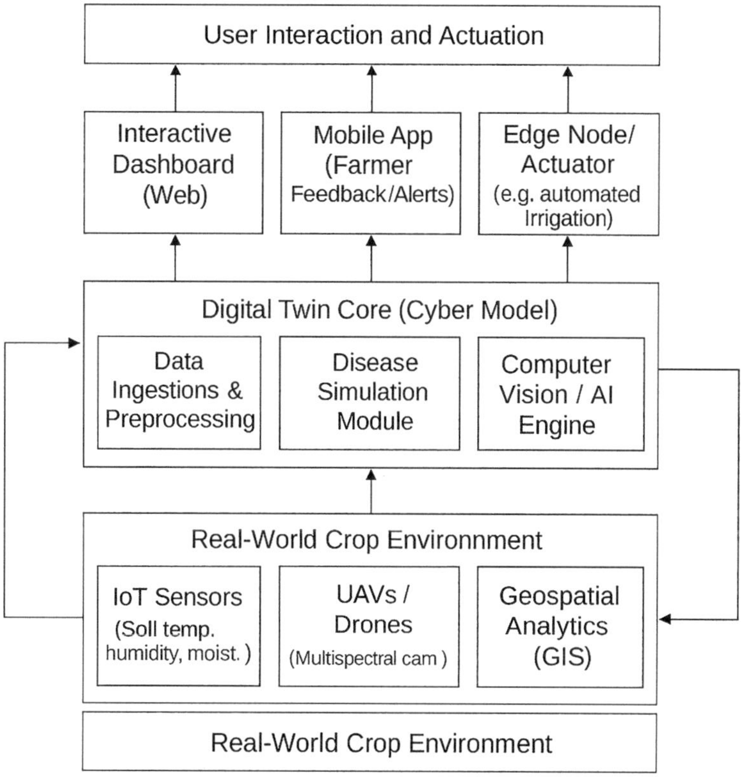

Fig. 7.1 Digital twin architecture for crop health monitoring

illness. Satellites and UAVs (drones) provide macro and high-resolution multispectral photography that can find small changes in plant physiology and early signs of stress, sometimes before visual symptoms are clear (Li and Mahalik 2019). This information is in addition to ground-level data. Weather stations collect information about things like rainfall, wind speed, temperature, and sun radiation that directly affect the growth and spread of diseases. This gives us even more macro-environmental intelligence. Farmer app inputs also have a critical human-in-the-loop part since they let people manually record observations, visual evidence, and geo-tagged field notes, which add context-specific insights to the dataset. In the end, API outputs make it possible to connect perfectly with outside systems like regional agricultural warning services, disease forecasting models, and farm management systems (Brenes et al. 2024). When you put all of these different data sources together, they make up the Digital Twin system's lifeblood. They keep the cyberlayer's constant synchronization, modeling, and prediction going.

7.3.3 Cyber Layer: Digital Crop Twin and Simulation Core

The cyber layer is the part of the Digital Twin (DT) system that does the math and analysis. It turns raw field data into a dynamic digital model of the agricultural environment. It uses a lot of data and smart algorithms to predict illness trends, copy how plants operate, and provide you useful information in real time. This layer is made up of the Digital Crop Model, which models crop growth, phenological development, stress responses, and interactions with the environment. To show the true state of the crops, this model is updated with sensor data, photos, and input from people (Kwaghtyo and Eke 2023).

To improve prediction accuracy, the Disease Simulation Module has to show how the pathogen, host, and environment interact in a sophisticated way. This module uses historical data, real-time conditions, and disease life cycle aspects to guess how diseases will spread, how fast they will advance, and where they will spread in different environments. Using epidemiological approaches, statistical modeling, random forests, and LSTM networks makes forecasts more accurate. The AI Engine is great at recognizing patterns and analyzing images at this layer. Convolutional neural networks (CNNs) or transformer-based architectures filter high-resolution drone and satellite pictures to find and classify symptoms of disease in leaves, stems, and canopies. These algorithms can accurately tell the difference between different types of diseases and levels of severity, even when there are only a few or unclear visual cues.

Spatial intelligence helps site-specific treatments like targeted spraying or quarantine. The Data Ingestion and Preprocessing Module conducts the first step in processing data to combine these goals. Filtering, finding outliers, normalizing, and aligning data over time all make sure that simulation and AI model data is clean, consistent, and reliable. These cyber-layer parts make up a robust digital backbone that lets the Digital Twin keep an eye on things, make predictions, and make forecasts. This layer may make decisions based on data and act right away to improve disease control and farm performance by showing the agricultural environment in real time.

7.3.4 Integration & Feedback Layer: Actuation and User Interface

The integration and feedback layer is the latest and most interactive portion of the Digital Twin design. It helps end users take part, control operations, and make decisions in real time. This layer fills in the analytical gap between cyber layer data and the needs of farmers and agronomists. It makes sure that the system's digital intelligence makes physical changes at the right moment and in the right context. There are interactive dashboards in this layer that indicate how healthy crops are, how likely they are to get sick, environmental changes, and how well they did in the past. These dashboards make it easier to use difficult data on the web and on mobile devices. They like seeing time-series graphs of environmental factors, real-time heatmaps of sick areas, and maps showing issue areas on top of maps of other areas. These tools help people make rapid, smart choices (Mancipe-Castro and Gutiérrez-Carvajal 2022).

Mobile apps are the main field-level interfaces for mobility and access. These instruments let farmers know about disease problems, cures, and changes to irrigation in real time. The applications support hybrid learning by facilitating manual data entry, such as uploading images, marking problematic areas, or confirming real-world situations to enhance the digital model. The feedback loop is a key part of this layer. The digital method provides automatic answers in addition to user knowledge. These include telling field actuators to open or stop irrigation valves, turn pesticide sprayers on or off, or modify the ventilation systems in greenhouses. This means that the system can automatically suggest or make modifications to farming practices based on disease risks. APIs, or application programming interfaces, let systems talk to each other when they are integrated. These interfaces are good for government advisory systems, networks that warn people about diseases in their area, tools for managing agriculture, and logistics for supply chains. Because it is so flexible, Digital Twin technology can be used in other digital farming settings as well.

Monitoring by people in the loop is also very important in this approach. Even though many DT operations are done automatically, agronomists, scientists, and extension agents can look at the simulation findings, reject automated decisions, and add their own expert opinions to the feedback loop. Working together with machines in real life makes people more open, confident, and adaptable. Integration and feedback close the sensing, simulation, and action loop, which makes the Digital Twin's value real. Field systems and end users get useful information that lets them take action to stop agricultural diseases before they (Amitrano et al. 2020).

7.3.5 Cloud Infrastructure and Data Storage

A powerful and scalable cloud infrastructure that can handle high-speed and high-variety data streams from many sources is necessary for a Digital Twin system for crop health monitoring to work well. In agriculture, there are a lot of different types of data coming in from IoT sensors, drones, satellites, weather stations, and people. Cloud solutions provide the computational power, storage space, and access needed to manage and evaluate these resources in real time. A cloud-based data lake is the main architecture that collects and stores structured and unstructured data in a variety of formats. Field notes, geographical imagery, weather forecasts, and time-series sensor logs are some of these types. If models need to be changed or new algorithms need to be added, the architecture of the data lake makes sure that the data that has been collected stays the same. This makes it feasible to do flexible queries and look back at past data (Panchal et al. 2023).

Cloud computing services make it easier to do real-time simulations and predictive analytics by combining geospatial processing pipelines, simulation engines, and machine learning models. These pieces are supposed to help the process move more smoothly. To make sure that operations like processing photos or forecasting diseases can be done in a scalable and parallel way, these services commonly leverage distributed computing models like Apache Spark or Kubernetes-based microservices (Liu et al. 2024). Serverless models are also utilized to handle the arrival of data streams. These models give processing power on the fly. You can utilize continuous integration and continuous deployment (CI/CD) for AI models and algorithms that help people make decisions in a cloud-based model deployment environment. This makes sure that models for finding sickness can be retrained and updated correctly when new information comes in. This lets them adapt better to changes in infections or the environment, which makes their estimates more accurate over time.

In agricultural data technology systems, security and data governance are very important parts of cloud infrastructure. It is important to use access control, encryption, and authentication to protect private information including farmer identification, levels of agricultural output, and farm locations. It is very important to follow data privacy laws in your area and around the world, such as the General Data Protection Regulation (GDPR) and India's Personal Data Protection Bill. This will help you keep your customers' trust and keep their data safe. The architecture also makes it easier to set up backup and disaster recovery systems, which can restore important simulation data and get the system working again if it fails or is attacked. For keeping operations going, especially on large or high-revenue farms, cloud redundancy and failover solutions are very important. In the end, cloud computing solutions that make it easier for people to work together improve participation for many people. Researchers, agronomists, lawmakers, and farmers can all get the same information and insights through secure portals, application programming interfaces (APIs), and mobile apps. Then, they sort the data based on the user's job. Digital agriculture is good for everyone since it allows everyone to share access, be open about their decisions, and work together to come up with new ideas.

7.3.6 Visualization and User Interface

The visualization and user interface layer of a Digital Twin (DT) system is important for turning complicated simulation results, sensor data, and predictive analytics into information that a lot of people can understand and use. These interactions, which include farmers, agronomists, researchers, and lawmakers, are very important for putting insights into action because they help make decisions in the field, get more people involved from all stakeholder groups, and help put ideas into action. Interactive dashboards are the stars of this level, where internet portals and mobile apps are usually found. These dashboards let customers make quick and smart decisions by providing multi-dimensional data in the form of simple charts, maps, and alerts. Real-time environmental data (including soil moisture, humidity, and temperature), disease risk levels, NDVI vegetation indicators, and crop health summaries are all important parts. Modular components built into dashboards let users change presentations based on geography, crop type, time period, or risk factors. These visual tools help users see how things are changing and how well treatments are working by comparing current field conditions to historical patterns and predicted forecasts.

Another important part of the interface is geospatial visualization. The DT system might be able to indicate disease hotspots, locations with a lot of environmental risk, pest pressure zones, and possible channels of transmission by connecting with GIS system. These models allow for site-specific responses like as tailored pesticide application, targeted irrigation, selected harvesting, and concentrated farming. For large farms and regional monitoring programs, heatmaps, contour overlays, and interactive layers are all helpful for doing detailed research and thinking about space. Mobile apps also make the DT system more useful by letting people use it on-site. These efforts include push notifications for high-risk events, such as expected blight starting within 48 hours, GPS tagging for observations that are specific to a region, and the ability to upload images for real-time confirmation. Farmers can write down strange things that happen, send in pictures for AI to check, or look over the suggested actions based on DT simulations. This direct contact helps gather data and make decisions in real time in systems that are far away or not connected.

In agricultural settings where users may have different levels of digital literacy, interface design is mostly about making things easy to use and accessible. Interfaces need to be simple, work in more than one language, and work well on a variety of devices. Color coding, iconography, and simpler measures, such as "Low Risk," "Moderate Risk," and "High Risk," are all common ways to make things easier to read. Training classes, offline capabilities, and voice support are all examples of interventions that can help make sure that rural and elderly agricultural communities are included.

The user interface eventually completes the feedback loop of the Digital Twin system by allowing two-way contact. Users not only provide comments, new data, and confirmation of forecasts but also provide insights that help to improve the models and system learning. For managers and researchers, interfaces could have facilities for exporting data, making reports, and APIs that make it easier to go to government policy platforms or farm management software. The layer of visualization and user interface in a Digital Twin architecture is what really pushes user-centric intelligence ahead. It turns complicated data into useful information, makes it easier to make decisions, and encourages people to interact with digital systems all the time. In the end, this makes sure that predictive disease monitoring works well and can be used in many different agricultural settings (Roberts et al. 2021).

7.4 Predictive Modeling Techniques

For crop disease detection, predictive modeling constitutes the analytical center of the Digital Twin (DT) system; it helps it to simulate possible future scenarios under changing climatic and agronomic conditions and to forecast disease outbreaks before symptoms are obvious. To provide projections that guide quick actions, the models mix real-time sensor data, historical records, picture data, and environmental factors.

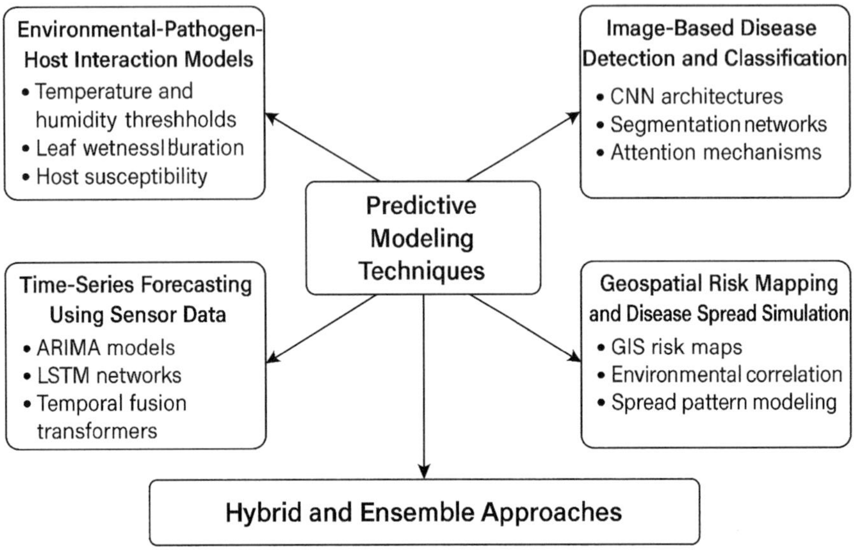

Fig. 7.2 Predictive modeling techniques for crop disease detection

Key predictive modeling methods used in the DT framework environmental-pathogen-host interaction models, image-based illness categorization, and geographic risk mapping are discussed in this part.

A conceptual summary of the several predictive modeling techniques applied in a Digital Twin system for agricultural disease monitoring is given in Fig. 7.2. Surrounded by four specialized strategies—Environmental-Pathogen-Host Interaction Models, Image-Based Disease Detection and Classification, Time-Series Forecasting Using Sensor Data, and Geospatial Risk Mapping and Disease Spread Simulation—Predictive Modeling Techniques is the central component. Using data including meteorological conditions, sensor readings, field photos, and geographical elements, every method adds specifically to forecasting and diagnosis. Underlying the center cluster are hybrid and ensemble approaches, which combine two or more methodologies to increase model dependability, generalizability, and prediction power over diverse crops, climates, and disease profiles.

7.4.1 Environmental-Pathogen-Host Interaction Models

A straightforward method to model crop disease is to simulate the interplay of environmental variables, host plant physiology, and pathogen life cycles. These mechanistic models replicate the biological and ecological processes that trigger and disseminate illness across various field situations. These models often include important factors like temperature and humidity thresholds that affect the growth of bacteria and the germination of fungal spores. They also include the length of time that leaves stay wet, which is directly related to the risk of fungal invasion, especially in crops like tomatoes and wheat. Host sensitivity is a major element that changes based on the type of resistance, the nutritional state, and the stage of growth of the crop. There are many ways to show how these complicated systems work, such as degree-day accumulation models, logistic growth equations, and decision tree classifiers. Advanced uses of this technology mix historical disease incidence records with real-time weather station data and phenological crop models to generate dynamic risk indicators. These indexes are always being updated, so they can make forecasts that take into account changes in the seasons and the weather.

7.4.2 Image-Based Disease Detection and Classification

Using convolutional neural networks (CNNs), these models analyze high-resolution images from a variety of sources, such as UAVs, field-mounted cameras, and smartphone photos sent in by farmers. The CNNs are trained to find unique disease features like leaf lesions, discoloration, necrosis, or wilting. ResNet, Inception, and EfficientNet are popular CNN designs used for classification applications. U-Net and Mask R-CNN are more specific segmentation models used to separate polluted areas in leaf or canopy images. A number of implementations use transformer-based networks or attention systems to make the model easier to understand and highlight the qualities that are important to disease. This method is very flexible since training sets may be regularly updated with new, annotated photographs taken in varied environments. This makes the model more generalizable and robust. Many systems also use semi-supervised learning approaches, which means that the model may learn from both labeled and unlabeled data. In real-world situations, these image-based algorithms not only tell what kind of sickness it is, but they also measure how bad it is and, in certain cases, tell the difference between biotic and abiotic stress signals, such as not getting enough nutrients. The resulting output provides field-level, complete diagnostic support that helps with early detection and targeted intervention efforts.

7.4.3 Time-Series Forecasting Using Sensor Data

In Digital Twin systems, predictive illness modeling relies significantly on time-series forecasting, particularly for detecting early warning signals and temporal patterns derived from ongoing environmental monitoring. This method employs sequential data from field-based sensors that quantify parameters like as soil moisture, air temperature, humidity, and leaf wetness to predict temporal changes preceding and following disease onset. Two important statistical methods that have long made it easier to make short-term predictions about individual variables are autoregressive integrated moving average (ARIMA) models and exponential smoothing. Modern applications are increasingly dependent on complex machine learning models, including recurrent neural networks (RNNs) and Long Short-Term

Memory (LSTM) networks. When dealing with nonlinear dependencies, time delays, and multivariate interactions that are common in agricultural systems, it is very important to pay attention to these deep learning architectures. Temporal Fusion Transformers (TFTs) and other advanced models are sometimes used to combine multiple time-series signals with a changing weight that shows how important each input is over time. These models can find changes or unusual things in field circumstances early on that could mean a higher risk of disease. For example, if the temperature of the soil changes quickly and the leaves stay moist for a long time, this could mean that mildew is more likely to grow. Time-series forecasting algorithms let Digital Twins use both historical data and real-time sensor data to do quick risk assessments and start proactive agronomic interventions. This greatly improves the accuracy and effectiveness of attempts to control disease (Albaaji and Vinod Chandra 2025).

7.4.4 Geospatial Risk Mapping and Disease Spread Simulation

Digital Twin systems need geospatial modeling to get better at finding crop diseases. It depicts where diseases are most likely to happen, how they might move throughout a landscape, and where action is needed the most. Geospatial analysis tools combine raw environmental and biological data from satellite photos, drone surveys, and sensors in the field into easy-to-read visualizations like risk maps and disease progression overlays. Geographic Information Systems (GIS) let you combine spatial layers including land use, soil type, topography, and microclimatic information. Risk mapping often use statistical and machine learning methodologies to correlate geographical data with historical disease outbreaks for the purpose of predicting high-risk areas. Kernel density estimation is one method that helps locate clusters of disease cases. Moran's I is another method that helps figure out how closely sickness cases in one area are related to those in nearby areas. More complex models that take into account things like wind direction, moisture gradients, irrigation patterns, and crop connectedness can use both spatial-temporal simulations and agent-based modeling to show how diseases spread. These models can show how diseases like wheat rust spores or late blight spread in different weather situations, which makes them very helpful for deciding how to limit them. Farmers, agronomists, and policymakers can use these projections as interactive maps or animated simulations to find out where to make changes, how to best use resources, and how well disease management approaches work in different areas.

7.4.5 Hybrid and Ensemble Modeling Approaches

Digital Twin systems use hybrid and ensemble models to make predictive analytics more accurate, resilient, and flexible. Hybrid crop disease detection systems make the most of the benefits of mechanical, statistical, and machine learning. A hybrid system can visually diagnose symptoms from picture data by combining a convolutional neural network (CNN) with a mechanical model to determine baseline disease criteria based on environmental conditions. This architecture could include time-series forecasting with LSTM networks to predict how sickness would get worse. This multi-layered method makes sure that projections are physiologically reasonable and full with data, which makes operations more reliable.

Ensemble models combine the predictions of numerous algorithms to make them more stable and better at making generalizations. This is where random forests, gradient boosting machines (XGBoost), and layered generalization come into play. These algorithms might use outputs from classifiers that look at weather, soil, images, and crop phenology data to figure up a disease risk score. Ensemble learning is useful in agriculture because it works well with noisy, incomplete, or varied input data. It defends against model constraints. Digital Twins can work with more types of crops, geographies, and disease profiles because of hybrid and ensemble methods that use many data sources and modeling methodologies. Systems that learn continuously and update models with field data make sure that the system can respond to changes in weather and farming circumstances. These advanced prediction technologies improve Digital Twin decision-support, making it possible to respond to crop diseases in more complex, faster, and more scalable ways.

7.5 Case Studies

This part includes a set of case studies covering important crops and high-impact illnesses to support the validity of Digital Twin-driven predictive modeling in agricultural disease diagnosis. Every instance shows how early identification, focused intervention, and better crop results may result from combining environmental data, imaging technology, simulation models, and farmer involvement. The chosen examples wheat rust, rice blast, and potato late blight showcase the adaptability and scalability of the Digital Twin technique throughout several climatic zones and crop kinds.

7.5.1 Case Study: Wheat Rust Detection Using Digital Twin Systems

Among the most damaging diseases afflicting cereal crops worldwide are wheat rust brought on by Puccinia triticina (leaf rust), P. graminis (stem rust), and P. striiformis (stripe rust). Under ideal climatic conditions—especially high humidity mixed with moderate temperatures—rust outbreaks can start quickly and produce yield losses in unprotected areas ranging from 20% to 40%. A trial Digital Twin system was installed in northern India over many experimental wheat fields in order to handle this. Key microclimatic variables—including leaf moisture, temperature, and dew point were continually measured by the system using IoT sensors. Complementing this ground data, UAVs weekly obtained high-resolution multispectral images to identify early symptoms of chlorosis and necrotic lesions linked with rust infestation. Infection risk was approximated using a pathogen-specific environmental simulation model using degree-day computations and real-time meteorological data. Concurrent with this was the analysis of drone images using convolutional neural networks (CNNs) to evaluate visual symptoms on flag leaves. With an 88% prediction accuracy, the Digital Twin system effectively gave early warnings 5–7 days before obvious symptoms started. Farmers who got and responded to system alarms using fungicides saw 15–20% increase in yield retention over untreated control plots. Furthermore, the use of geospatial risk mapping improved cooperation and containment techniques across neighboring farms, therefore limiting the spread of the illness.

7.5.2 Case Study: Rice Blast Forecasting and Drone-Based Surveillance

A severe disease compromising rice harvests, especially in warm and humid conditions where prolonged periods of leaf wetness are prevalent, is rice blast brought on by the fungal pathogen Magnaporthe oryzae. If not found early, the disease affects several sections of the plant, including leaves, nodes, and panicles, therefore reducing the yield. Using a Digital Twin system to track and forecast blast events across paddy fields, a case study in a rice-growing area of Southeast Asia found. To project infection episodes depending on environmental trends like temperature, humidity, and rainfall patterns, the system aggregated real-time meteorological data with LSTM-based time-series models. Every few days, drone-based multispectral surveillance was carried out to gather field images, which were examined using a ResNet deep learning model taught to identify the spindle-shaped lesions typical of blast infection, together with heat stress signals in the crop canopy. Highly localized risk forecasting made possible by the combination of predictive modeling with aerial surveillance. With a 91% forecast accuracy, the method lets targeted fungicide administration be used before obvious illness start. Real-time notifications sent to farmers and field managers via a smartphone dashboard allowed quick decision-making and helped to cut needless chemical use. Combining predictive analytics with drone-assisted visual diagnostics helps fields under observation and management using the Digital Twin platform to show a 30% decrease in total illness incidence relative to untreated control zones.

7.5.3 Case Study: Potato Late Blight Management in Mountain Agroecosystems

Late blight, a disease that is extremely detrimental to potato crops, is primarily found in high-altitude, humid environments, such as those found in the Andean regions, and is caused by the oomycete Phytophthora infestans. If not controlled quickly, it can destroy whole fields in days in chilly, moist circumstances. Smallholder farms in a hilly agroecosystem were fitted with a Digital Twin system to counter this danger. Measuring temperature, humidity, rainfall, and dew point, the system combined remote sensing data NDVI trends from satellite imagery with local microclimate weather station data. Processed data using a dynamic disease simulation model tuned to P. infestans produced daily risk projections. By submitting leaf images for severity assessment which were examined using AI-driven image recognition models trained to identify the unique water-soaked lesions and leaf collapse typical of late blight farmers also helped to support the system via a smartphone app. With an 85% prediction accuracy, the Digital Twin technology allowed SMS and mobile dashboards to send early warnings. By allowing farmers to modify fungicide plans and give high-risk areas more priority, this proactive strategy helped to anticipate 25% less production loss than more traditional reactive techniques. Furthermore, assisting more efficient disease control over the area were the geographical heatmaps and visual intensity scores the system produced for coordination of community-level responses.

The three case studies of wheat rust in India, rice blast in Southeast Asia, and late blight in Andean potato fields show how Digital Twin (DT) systems can be used to diagnose and treat crop diseases in different agro-ecological settings. The DT framework always used real-time environmental data, aerial or satellite images, predictive modeling, and input from farmers to provide accurate disease predictions and direct actions. For several of these systems, it was necessary to use high-quality datasets to train and test the machine learning models. Using freely available datasets acquired from Kaggle, a worldwide data science platform, image-based sickness categorization, environmental pattern modeling, and severity grading were developed. These datasets had hundreds of labeled pictures of sick crop leaves, sorted by disease type and severity. They also had climate and phenological data from several years that could be used to train and test models.

The Digital Twin helped keep 15–20% of the crop in the wheat rot experiment by predicting the weather with 88% accuracy. The LSTM-based forecasting and drone monitoring system had a 91% accuracy rate and cut the number of illnesses by 30% in the case of the rice blast. The model that included both microclimate and satellite data was 85% accurate for potato late blight, which helped farmers cut their losses by more than 25%. In all cases, DT models produced early warnings several days before symptoms showed up. This helped farmers take strong action with targeted fungicide applications and cultural adjustments. Also, spatial decision tools and geospatial visualization enabled zone-based management strategies use less chemicals, which is good for sustainable, data-driven agriculture. Utilizing publicly available statistics not only rendered these models replicable and accessible but also underscored the role of open research in fostering genuine agricultural innovation.

Table 7.1 shows a three-way comparison of the Digital Twin (DT) implementations for potato late blight, rice blast, and wheat rust. The chart lists key performance factors like prediction accuracy, yield loss reduction, and the most important technologies used in each system, such as IoT sensors, UAV photos, time-series forecasting, and AI-driven picture classification. All of the prediction models were built and tested using data from Kaggle that is freely available to everyone. This makes sure that the models are easy to access and reproduce. The results show how well the DT framework works in different agro-environmental

Table 7.1 Summary of digital twin case study results

Crop	Disease	Prediction accuracy	Yield loss reduction	Technologies used	Dataset source
Wheat	Rust	88%	15–20%	IoT sensors, UAV imagery, CNN, GIS	Kaggle—Wheat Rust Dataset
Rice	Blast	91%	30%	LSTM time-series models, drone imagery, ResNet classifier	Kaggle—Rice Leaf Disease Dataset
Potato	Late Blight	85%	25%	Microclimate sensors, satellite NDVI, AI image analysis	Kaggle—Potato Disease Dataset

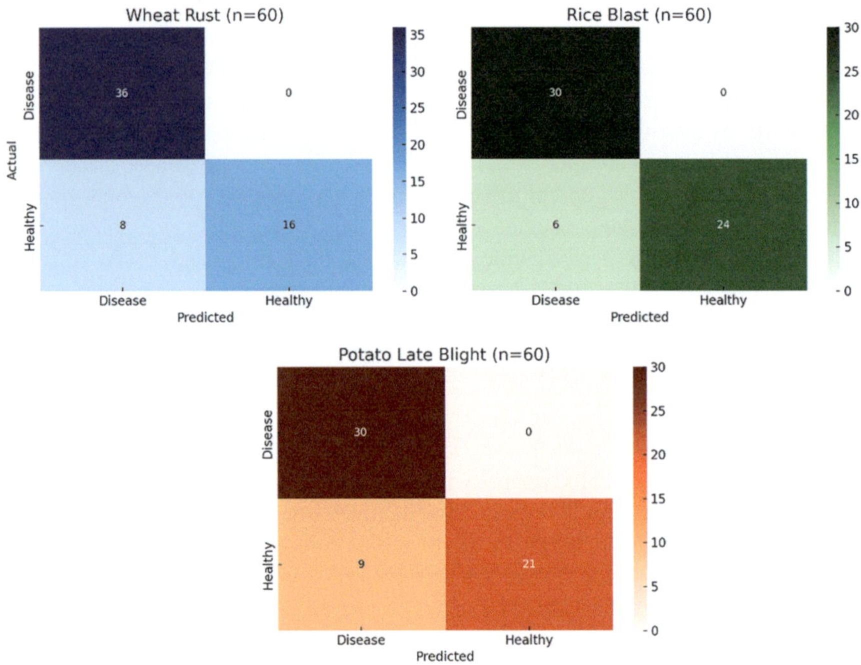

Fig. 7.3 Confusion matrices for disease detection models with extended datasets

conditions to give early warnings of illness, help with targeted treatments, and improve farm-level results.

Figure 7.3 shows confusion matrices that show how well Digital Twin-based disease detection algorithms work in three important crop-disease scenarios. We used a larger dataset with 60 samples to test each of these techniques. Wheat Rust's matrix can tell the difference between false positives and true positives, and it can forecast with 88% accuracy. The Rice Blast model is very consistent when it comes to identifying samples and figuring out if they are healthy or unwell. It has an accuracy of 91%. The Potato Late Blight model, on the other hand, is 85% accurate, which means it can find illnesses well. But there is still some potential for improvement when it comes to telling if a disease is present in tough environmental circumstances. These matrices show how well each model did in diagnosing, with the purpose of showing how reliable the prediction systems are when they are used on larger datasets that are more representative of the field.

7.6 Technical Challenges and Limitations

The Digital Twin (DT) technology could change the way we find illnesses in crops, but it is hard to use because of both technical and practical problems. It is really important to have good data that is easy to get. In rural places, where internet access is limited, the real-time data acquired by field-level Internet of Things sensors, drones, and satellite systems may not be accurate or full. While Kaggle datasets are useful, many agroclimatic zones may need to be modeled with local individuality. One of the hardest parts of making an analytical framework is adding in photographs, topographical data, and numerical sensor data. In agricultural systems with limited resources, there may be a deficiency in proper data preparation, protocol compatibility, and data engineering.

Prediction algorithms based on training data may not work for new crops, diseases, or locales. It was hard to make real-time data transfer (DT) systems, especially ones that use deep learning, geospatial analysis, or multi-sensor fusion, because there wasn't a solid energy or cloud infrastructure. It is probable that the demand for real-time operations will slow down the processing of data or the inference of models for diseases that spread quickly, like rust or blight. If the expected results aren't clear, a lot of AI models are hard to understand, which could hurt consumer trust and adoption. To be able to adapt to changing conditions in the field, different types of crops, and different disease strains, DT systems need to be calibrated, enhanced, and retrained. Finally, the lack of support services, computer skills, and language are all things that make it harder for everyone to use the technology fairly. To make Digital Twin systems that are dependable, scalable, and long-lasting for sustainable agriculture, we need to discover answers to these difficulties.

7.7 Adoption Barriers and User Perspectives

Even though Digital Twin (DT) systems for finding agricultural diseases have come a long way, they are still not widely used in the real world because of social, economic, institutional, and behavioral barriers. The digital divide makes it hard for smallholder farmers with low incomes to use smartphone dashboards, image-based diagnostic tools, or get real-time alerts. Even though there is infrastructure, trust and perception among consumers are still the most important things. Farmers may not want to follow advice from AI, especially if the results are unclear or if the digital advice goes against traditional farming methods. Small and medium-sized farms that can't get money from outside sources or government subsidies might not be able to buy IoT sensors, unmanned aerial vehicles (UAVs), and subscription-based analytics services. The Digital Twin might not operate in some locations since agricultural support networks are spread out. This is especially true in places where there aren't enough research and technology suppliers or extension services. Due to language barriers, cultural data ownership issues, privacy concerns, and other problems, this technology is hard to use in many areas.

Training and capacity-building initiatives that don't help farmers use digital technologies in their daily lives are either not enough or too complicated. The responses from those who took part in the pilot project show how important user-centered design, localized interfaces, and hybrid models that combine automated analytics with human expertise are for successful adoption. To get a lot of people to utilize it, it needs to build long-term trust, constant yield, input efficiency, or early disease suppression. To make it easier for digital twin technologies to be used in many ways and make sure that all agricultural value chains benefit from their use, we need inclusive policies, public-private partnerships, and participatory innovation.

7.8 Future Directions

As Digital Twin (DT) technology advances, data science, computing infrastructure, and participatory innovation will help determine how well it can help control diseases in agriculture. People think that DT technology will be used soon. Edge computing lets agricultural data be processed in real time without having to wait or connect to the cloud. This is useful in areas that are usually rural or have few resources. As fresh data comes in, adaptive AI and self-learning algorithms are employed to change the estimates of digital twins. This makes the models much more accurate and responsive in all seasons and places. Cooperatives, input suppliers, and regulatory authorities can all benefit from blockchain technology's openness, traceability, and reliability. In climate-smart agriculture, disease detection technologies (DTs) are used to find diseases, figure out how much carbon is being released, how much water is being consumed, and how well crops can handle bad weather.

Digital twins are expected to have user interaction features like voice-activated interfaces, support for local languages, and augmented reality technologies that show maps of crop health and risk in the field. These technologies need changes to the law so that everyone may access digital innovation, own their data, and use artificial intelligence in a way that is fair. Using open data ecosystems can speed up the process of creating and testing models. These ecosystems let institutions share sensor data, satellite images, and disease datasets that have been annotated. It is essential for agronomists, computer scientists, behavioral economists, and rural development practitioners to work across disciplines to co-develop solutions that are technically proficient, socially equitable, and economically inclusive. In the future, agricultural disease surveillance systems will be smart, aware of their surroundings, able to grow, and powered by DT. These systems will provide farmers more authority and make the food chain more resilient.

7.9 Conclusion

Digital twin (DT) is a new way to track, anticipate, and control crop diseases in the farming business. DT systems may make dynamic, virtual agricultural settings by using real-time field data from IoT sensors, drones, and satellites, as well as complex simulation models, machine learning algorithms, and geospatial analytics. These variables enable precise and proactive replies. This chapter looks at the physical data gathering layer, the cyber simulation and analytical core, and the integration and feedback systems that complete the decision-making cycle of digital twins for finding crop diseases. Using digital twins, predictive modeling tools can provide you early warning, help you manage your resources better, and protect your yield. Some of them are environmental-pathogen-host interaction models, image-based sickness classification, time-series forecasting, and regional risk mapping. These methods could be useful in a variety of agricultural and ecological settings, as shown by case studies on wheat rust, rice blast, and potato late blight. These case studies are supported by Kaggle's open-access data.

Even so, there are issues that need to be addressed when employing DTs. People can get over issues with data quality, model generalization, infrastructure limits, and acceptance with the help of inclusive design, policy support, and collaboration across sectors. In the future, digital twins will be considerably more valuable in farming thanks to new technologies like edge computing, adaptive AI, and blockchain. They will also make them easier to grow and more dependable. Digital twins are a promising approach to develop powerful, smart, and long-lasting farming ecosystems since climate change, pest outbreaks, and limited resources are putting additional stress on the global food chain. To get the most out of this technology that could change the game, we need to keep putting money into research, increasing capacity, and coming up with new ideas that help farmers.

References

Albaaji GF, Vinod Chandra SS (2025) A hybrid ensemble framework for integrating environmental parameters for improved crop yield prediction and sustainable agricultural decision-making in Iraq. Neural Comput Appl 37:13321–13343

Amitrano C, Chirico GB, De Pascale S, Rouphael Y, De Micco V (2020) Crop management in controlled environment agriculture (CEA) systems using predictive mathematical models. Sensors 20(11):3110

Bertone F, Caragnano G, Ciccia S, Terzo O, Cremonese E (2021) Green data platform: an IoT and cloud infrastructure for data management and analysis in Agriculture 4.0. In: Complex, intelligent and software intensive systems: proceedings of the 14th international conference on complex, intelligent and software intensive systems (CISIS-2020). Springer, pp 365–374

Brenes JA, Quintana MIM, Castillo JDS, López G, Marín-Raventós G (2024) Design and evaluation of data visualizations in a smart agriculture monitoring system (SAMS). In: International

conference on advanced research in technologies, information, innovation and sustainability. Springer, Cham, pp 75–88
Buja I, Sabella E, Monteduro AG, Chiriacò MS, De Bellis L, Luvisi A, Maruccio G (2021) Advances in plant disease detection and monitoring: from traditional assays to in-field diagnostics. Sensors 21(6):2129
Cesco S, Sambo P, Borin M, Basso B, Orzes G, Mazzetto F (2023) Smart agriculture and digital twins: applications and challenges in a vision of sustainability. Eur J Agron 146:126809
Chakraborty S, Newton AC (2011) Climate change, plant diseases and food security: an overview. Plant Pathol 60(1):2–14
Chaux JD, Sanchez-Londono D, Barbieri G (2021) A digital twin architecture to optimize productivity within controlled environment agriculture. Appl Sci 11(19):8875
Demchenko Y, Turkmen F, de Laat C, Hsu CH, Blanchet C, Loomis C (2017) Cloud computing infrastructure for data intensive applications. In: Big data analytics for sensor-network collected intelligence. Academic Press, pp 21–62
Goldenits G, Mallinger K, Raubitzek S, Neubauer T (2024) Current applications and potential future directions of reinforcement learning-based digital twins in agriculture. Smart Agric Technol 8:100512
Jeong DY, Baek MS, Lim TB, Kim YW, Kim SH, Lee YT, Jung WS, Lee IB (2022) Digital twin: technology evolution stages and implementation layers with technology elements. Ieee Access 10:52609–52620
Juroszek P, Racca P, Link S, Farhumand J, Kleinhenz B (2020) Overview on the review articles published during the past 30 years relating to the potential climate change effects on plant pathogens and crop disease risks. Plant Pathol 69(2):179–193
Kalyani Y, Vorster L, Whetton R, Collier R (2024) Application scenarios of digital twins for smart crop farming through cloud–fog–edge infrastructure. Future Internet 16(3):100
Kwaghtyo DK, Eke CI (2023) Smart farming prediction models for precision agriculture: a comprehensive survey. Artif Intell Rev 56(6):5729–5772
Li N, Mahalik NP (2019) A big data and cloud computing specification, standards and architecture: agricultural and food informatics. Int J Inf Commun Technol 14(2):159–174
Liu G, Zhong K, Li H, Chen T, Wang Y (2024) A state of art review on time series forecasting with machine learning for environmental parameters in agricultural greenhouses. Inf Process Agric 11(2):143–162
Ma Q, Li X, Li G, Ning B, Bai M, Wang X (2020) PAADA: physical-world-aware approximate data acquisition for image sensor networks. IEEE Access 8:68930–68943
Mancipe-Castro L, Gutiérrez-Carvajal RE (2022) Prediction of environment variables in precision agriculture using a sparse model as data fusion strategy. Inf Process Agric 9(2):171–183
Nehrey M, Zomchak L (2022, February) Digital technology: emerging issue for agriculture. In: The international conference on artificial intelligence and logistics engineering. Springer, Cham, pp 146–156
Panchal AV, Patel SC, Bagyalakshmi K, Kumar P, Khan IR, Soni M (2023) Image-based plant diseases detection using deep learning. Mater Today Proc 80:3500–3506
Purcell W, Neubauer T (2023) Digital twins in agriculture: a state-of-the-art review. Smart Agric Technol 3:100094
Roberts DP, Short NM Jr, Sill J, Lakshman DK, Hu X, Buser M (2021) Precision agriculture and geospatial techniques for sustainable disease control. Indian Phytopathol 74(2):287–305
Wang M, Li T (2025) Pest and disease prediction and management for sugarcane using a hybrid autoregressive integrated moving average—a long short-term memory model. Agriculture 15(5):500
Xie D, Chen L, Liu L, Chen L, Wang H (2022) Actuators and sensors for application in agricultural robots: a review. Machines 10(10):913

Applications of Digital Twins in Agriculture 8

Noha Nassar

Abstract

The ongoing increase in the Earth's population, climate challenges, and environmental variations subjected the agricultural resources to high pressure. Consequently, the importance of using more creative ideas to enhance the agricultural sector and improve the outcome of crops became an essential aspect. The ability of automation and innovation technologies to combine and utilize a wide range of information earned the agricultural sector interest and put the smart agriculture technique on the top of farmers' priorities in the recent days. The digital twins (DTs) technique is considered to be one of the promising techniques that has a great potential in enhancing the farming sector. It is considered to be a virtual representation tool for modeling and simulating the real agriculture systems that usually changes with time. Crop field monitoring, livestock facilities, or irrigation systems are some examples that can be utilized using the (DTs) technique which make it easier in the monitoring process using these platforms. The (DTs) model integrates a wide range of data and information, such as satellite imagery, sensor data, and artificial intelligence (AI) tools, which are used to improve the application of the (DTs) in agricultural operations. Furthermore, the DTs can also be used in many other applications such as irrigation scheduling, water management, and soil ecology and fertility. By comparing the real world with the hypothetical one, we can achieve a strong milestone to fill the gap between theory and real practice in the application of the digital twins in multiple sectors like in the case of crop monitoring and management, livestock productivity and health, soil ecology and fertility, and precision irrigation and water management.

N. Nassar (✉)
Faculty of Science, Cairo University, Giza, Egypt

© The Author(s), under exclusive license to Springer Nature Singapore Pte Ltd. 2026

A. L. Srivastav et al. (eds.), *Digital Twin Technology for Sustainable Agriculture*,
https://doi.org/10.1007/978-981-95-5915-2_8

Keywords

Digital twins · Crop monitoring · Smart agriculture · Livestock precision

8.1 Introduction

The agricultural field is considered one of the most crucial sectors that affects human beings. The significant elevation in the Earth's population subjected this sector to major pressure. Besides that, many other factors altered the farming processes, including climate change and environmental shifts. As a consequence, the regular agricultural techniques are insignificantly ineffective to cover the global needs, which emphasizes the importance of finding innovative agricultural methods to motivate precision agriculture (Zhang et al. 2025; Tao et al. 2018). A notable breakthrough emerged in the farming sector by the middle of the twentieth century, especially in synthetic fertilizers and botany, which introduced us into the fourth era of agriculture (Poornima et al. 2023). In order to sustain progress, complex systems are introduced to enhance productivity despite the increase in environmental impacts (Tao et al. 2018). The digital twins technology was then introduced to improve the profit, efficiency, and optimize utilization in the agricultural sector (Zhang et al. 2025; Poornima et al. 2023). These technologies mainly originated from industrial manufacturing and aerospace applications, and it has been developed to include data acquisition, simulation, prediction, and decision-making abilities (Tao et al. 2018). It also expanded to enter many different fields, like building activities, pharmaceutical and healthcare applications, and agricultural applications (Kim et al. 2024). In 2002, Dr. Grieves introduced the concept of the (DTs) for the very first time to model the life cycle of the products and the idea started with modeling the real physical system and transfer it, or we can say mirror it, to a virtual, hypothetical, system that is capable of directly collecting the information from the real system in order to twin the real system into a virtual model (Grieves and Vickers 2017; Jones et al. 2020). According to the first explanation, the digital twins concept can be defined as a virtual tool that is used to gather and model the reality system and exchange data between the hypothetical and real systems in order to improve the results and achieve the regular outcomes. This virtual representation of a physical system is used to exchange information between the two systems. Real physical system and virtual hypothetical system, via a connection channel moves the data and the processes from the virtual tool to the real physical system, as well as information from the real physical system to the virtual system; this connection channel is considered to be the third component of the (DTs) framework. While the first component is the real system that houses the physical product. And the virtual model is considered to be the second component for this framework. These three components enhance the ability of the (DTs) to model and simulate all vital processes and consequently profit the outcomes in the farming sector, especially the connection channel that increases the importance of the (DTs) in complex environments such as agriculture, as it enables the simulation process between the real system and the virtual model in different operational conditions and under environmental variations (Jones et al. 2020). Recently, the proper use of sensor networks, the remote sensing, the cloud-based data analysis platforms, the artificial intelligence (AI|), the internet of things (IoT), and the information and communication technologies (ICT) has made it easier for the DTs to fully simulate the real system and accurately predict the outcomes than what was used at the beginning of applying the (DTs) concept (Kim et al. 2024; Zhang et al. 2025).

These tools introduced the concept of the cyber-physical control cycle in the smart farming systems which helped with the monitoring process, the data analysis, and other farming operations using the smart system control and this was an important step toward regulating all the farming operations and defining the whole farm management perspective that highlighted the importance of recent innovative technologies and their important impacts on our human lives (Verdouw et al. 2021; Poornima et al. 2023). This frame of operation simultaneously sends notifications for the farmers in case any problem appears or even predicted, giving the farmers a chance to control the farm remotely (Poornima et al. 2023). Therefore, the importance of DTs arises not only from the concept of collecting information, monitoring, data analysis, and the ability to simulate the real system but also from their ability to serve as a decision-making tool (Tao et al. 2018). The ability of DTs in simulating agricultural behavior, like in the case of crop management, where they can assess the soil quality, rate of growth, and irrigation schedule, has expanded to predict agricultural disasters like pets invasion and infection diffusion. Similarly, in the application of animal health monitoring, the DTs can monitor the breeding and productivity periods and consequently enhance animal health and farm productivity (Zhang et al. 2025). Although DTs reflected high potential in the agricultural field especially in crop monitoring, machinery management and product management it faced some challenges that can be divided into two main constrains, the first is that the significant difference in the type and nature of the target item where as in manufacturing the main focus is on the machinery and static systems while in case of agriculture the main focus is on living organisms which have different and irreversible grow rates. For instance, virtual tools must acquire large amounts of heterogeneous and multimodal data to be analyzed and extract key physiological and morphological specifications, or to be used in reconstructing digital plant models for virtual modeling and to estimate features, simultaneously enhancing the ability to estimate the phenotype-environment-genotype behavior. While the second challenge is that DTs platforms usually obtain unimodal monitoring systems, such as crop monitoring or environmental changes observations, and only a few systems can combine multitask observation (Liu et al. 2025). Limitations of DTs extended to other multifactors like the cost of the infrastructure and agent training courses, the data governance, and also privacy concerns, especially for small farms (Wolfert et al. 2017; Jones et al. 2020). Recent approaches modified the DTs platform with intelligent control technology that focuses on the smart control system that regulates the whole agricultural environment starting from climate control within the greenhouse environment, irrigation system control, field modeling, and fertigation enhancement model (Liu et al. 2025). It also simulates severe situations like pest and disease incidence and extreme climate conditions (Jones et al. 2020). DTs also have a crucial role in aquatic farming to monitor the water quality and the aquatic ecosystems (Alves et al. 2023). Future trends of DTs must integrate multidimensional indicators to cover environmental, economic, and social fields, as in the case of agriculture, it must indicate soil health, ecological flow, and even community jobs, and this will require more sophisticated systems and grid connection channels to join different information from various fields (Wolfert et al. 2017; Kamilaris et al. 2017). As we can see, the DTs technology is considered one of the cutting-edge technologies that is transforming the agricultural sector into a more sustainable, efficient, and intelligent sector, which will offer promising solutions to many of the complex challenges that face the food supplements on our planet. However, realizing their full potential requires not only technological frameworks but also convenient application to the local conditions. In the following chapter, we will define the DTs within different prospects including the development of the conceptual framework, and the characteristics of the digital twins framework highlighting the role of digital twins in farm management and then dive into the four main cores of DTs applications: crop monitoring and management, livestock health and productivity, soil health and fertility management, and precision irrigation and water management by focusing on both the practical benefits and the future trends of digital twins framework in performing a smarter and more sustainable agricultural future.

8.2 Definition and Development of Digital Twins

8.2.1 Definition and Main Prospects of the Digital Twins

In its simplest form, digital twins are realistic virtual representations of a physical entity, while the main origin was first introduced by Dr. Grieves when he used the concept of DTs in the product life cycle management (Verdouw et al. 2021). This concept was redefined for product life cycle by Glaessgen and Stargel in 2012 as the integration of multiscale and multiphysics, which describe the most probable integrated simulation utilizing the leading accessible physical models that reflect the real system (Glaessgen and Stargel 2012), while in 2015 the prospective of Internet of Things arise and was explained by Verdouw when he stated that a virtual object can be defined as "a digital representation of an object with a unique identification, that can be trusted, is timely available, and can be used for intended purposes" (Verdouw et al. 2015). Moreover that definition was reintroduced by Boschert and Rosen, in 2016 as the comprehensive representation of the physical and functional product, or we can say system, components that reflect all available information applicable throughout the product lifecycle (Boschert and Rosen 2016). 2017 was a breakthrough year for the concept of DTs, as Grieves and Vickers reintroduced it for the product life cycle as a hypothetical reconstruction for all provided information from the micro level to the macro level that fully describes the system, or a product, meaning that any information about the product can be obtained from the real system, can be also obtained from its hypothetical version in the DTs framework (Grieves and Vickers 2017), while Schleich presented a different definition, that the DTs framework is not a singular uniform model that fully describe the system as a one complete template but it is a group of operation data and simulation models that connected together and develop by time through the life cycle of the product from a simple representation in an early design phases, used for conceptual evaluation, to highly detailed simulations in later stages that define precise design and optimize components and subsystems, this concept form a suitable framework for the required purpose and indicates that DTs are not only a virtual representation but also can serve as a prediction tool that estimate the changes during the product life cycle (Schleich et al. 2017). Despite that, in the prospect of IoT the DTs are considered to be a hypothetical representation of a product, process or a service, as the pairing between the virtual and real system allows the data analysis and systems monitoring to avoid and stop complications before it appear and develop a better opportunity and also estimate a future plan within simulation framework (Marr 2017; Verdouw et al. 2021). In this frame work (IoT), the dependence on sensing technologies is emphasized, and the internet acts as a storage and connection channel that contains the virtual model of items that is linked to the object (Verdouw et al. 2021). To well obtain a convenient DTs model, there are some important aspects that should be achieved.

8.2.2 Aspects of the Digital Twins Framework

To plan, monitor, and optimization of agriculture processes, DTs framework must satisfy certain aspects:

1. The framework must contain a physical cyber control system that consistently contains a sensing and monitoring system, data analysis with estimation ability, and smart control of the farm processes, i.e., complete farm coordination.
2. The framework must be capable of supporting the complete lifecycle of agricultural entities; accordingly, it should encounter all six distinct types of Digital Twins: Imaginary, Monitoring, Predictive, Prescriptive, Autonomous, and Recollection Digital Twins.
3. The framework must facilitate the application of the core characteristics like in time operation, accuracy, integration, intelligence, and system complexity.
4. The framework must specify the particular difficulties associated with the application of the DTs in farm management, including the complexity of agricultural entities, the dynamic nature of farm networks, and the variability built into farming processes (Verdouw et al. 2021).

In brief, thc DTs can be defined as the set of virtual representations of a real system that allows bidirectional transfer of information in a step of moving forward precision agriculture (Grieves 2016; Verdouw et al. 2021). And with no doubt, DTs have a significant influence in the farming field (Zhang et al. 2025; Verdouw et al. 2021). Farm management can be divided into four main sectors: crop monitoring and management, livestock management, soil management, and irrigation management. In the following section, we will discuss the application of DTs in crop monitoring and management.

8.3 Crop Monitoring and Management

The framework of DTs transferred the crop monitoring and management through real-time integration of different data sources, including data analysis, predictive modeling, and even supports decision making in order to enhance precision agriculture (Melesse et al. 2025; Escriba-Gelonch et al. 2024).

8.3.1 Modeling Approaches and System Architecture

Digital twins in crop systems are usually encountered in one or more of the following modeling frameworks:

- Physics-based models that simulate plant environment interactions using systematic crop growth algorithms, taking into consideration the light interception, water evaporation, nutrient consumption, and phenology.
- Data-driven models, such as machine learning, regression, computational networks, and prediction systems that gain information directly from sensor and environmental datasets.
- Agent-based or multiagent models, which represent individual plants or plots as agents with behavior specifications, enabling high-accuracy simulation of variable items (Wang 2024).
- Hybrid models that combine physics-based frameworks with data learning layers, offering system stability and flexibility (Melesse et al. 2025).

Melesse encountered that the hybrid model supplies better criteria through field variability and is also better in prediction ability, and under environmental changes (Melesse et al. 2025).

While the architecture of the crop monitoring management includes:

1. Sensor and data acquisition layer that includes on-site measuring of the soil moisture, leaf chlorophyll, and microclimate sensors, UAV or satellite imagery, and weather stations.
2. Data integration layer that works using cloud or edge computing platforms and preprocessed data.
3. Modeling and simulation platform that operates crop growth simulations and predictions.
4. Decision support interface which provides dashboards, alerts, and automation signals to farm actuators (Goldenits et al. 2024; Wang 2024).

This type of modular architecture enables scalable deployment across farm scales, starting from individual farms to regional farming areas.

8.3.2 Crop Monitoring and Management Case Studied

8.3.2.1 Citrus Orchard Using DTs

One of the first operational agricultural digital twins was performed upon mandarin orchards on Jeju Island, South Korea, covering around 185,000 hectares using an Open API-based data platform. This DT integrates climate data, soil metrics, UAV and satellite imagery, and fruit phenotypic measurements to model orchard dynamics at inter and intra-block levels. The intra-orchard analysis gave information about the variation within orchards, and it was a stronger predictor of fruit quality than the inter-orchard, which enabled micro precision management at the tree level. The use of statistical modeling, altogether with machine learning, can introduce precise irrigation, fertilization, and pruning schedules to each micro zone in order to improve the quality and uniformity of the crops (Kim et al. 2024).

8.3.2.2 The Application of Digital Twins in Agriculture (Smart Nitrogen Management)

Nitrogen as a fertilizer has two hands, the first is to improve the soil and supply the plants with nutrients. On the other hand, excess nitrogen (N) can be harmful and cause major environmental damage, such as nitrate leaching and even greenhouse gas emissions. This research focused on the use of the (DTs) to enhance the nitrogen (N) fertilizer efficiency in an arable farming system such as in crops like corn, wheat, and rice. The framework utilized in this study diverged from the conventional data inputs, commencing with raw data and culminating in the predictive model for agricultural decision-making. Tools such as GIS, remote sensing (e.g., NDVI), yield mapping, and crop simulation models integrated historical and real-time data to generate predictive fertilizer performance. These simulations enable farmers to tailor (N) inputs with greater precision, accounting for seasonal variations and expected crop responses. Digital twin simulations helped determine that low-yielding stable areas were disproportionately responsible for nitrogen loss, recommending (N) reduction in these zones. In contrast, adjusting (N) application in unstable zones was optimized based on near-real-time weather forecasts and satellite imagery. The most important outcomes from this study are that: the yield stability analysis improves fertilizer efficiency, digital twins enable predictive and site-specific management, conceptual information model enhances decision relevance, the ability for large-scale application is approved, and enabling real-time and adaptive management (Cesco et al. 2023).

8.3.2.3 Canopy Density DT Via 3D SLAM in Horticulture

In a more recent technology-driven approach, Lowe applied a DT of vine canopies using continuous time 3D SLAM (Simultaneous Localization and Mapping) lidar scans mounted on moving farm vehicles. This produced a dynamic 3D digital replica of canopy structure across multiple seasons and locations, enabling repeatable canopy density estimation. The dataset supports year-on-year comparison of canopy health, growth uniformity, and pruning outcomes, highlighting DT's role in structural crop monitoring (Lowe et al. 2020).

8.3.3 Recent Technologies Used in Crop Management

Future trends also entered the farming sector, and the use of advanced techniques has great potential in crop management, like:

- The use of drones to spread the seeds from the sky has a great influence and has a high potential for further applications. It was applied using XAG Xplant drone in China to spread the rice seeds, the method was highly effective and considered a time and water-saving process. In order to reduce the air transport of the seeds, Jetseed technology is used as it can spread seeds with speed of 18 m/s. Using drone technologies is a part of the internet of things (IoT) prospect and has a promising potential in the smart farming systems, especially with the means of 5G technology (Poornima et al. 2023).
- Using Robots in crop treatment, like in the case of vineyards where the simulation of the independent behavior of these Robots in a 3D model using the real-time information to find the proper phenotype then apply this to the real life Robots, this technique leads to reduce consumed times for development, improved assessment of sensor performance, and a decreased real field trials for phenotype evaluation and crop treatment testing (Linz et al. 2019; Verdouw et al. 2021).
- In recent days, the use of mechanical weed control units has been integrated with cameras to differentiate between the crop and weeds. These cameras are also used as a source of information, like in the case of a lettuce field, which can count the number of growing heads and provide accurate information with a precise location (Verdouw et al. 2021).
- The use of big data analysis, which is considered a combination of data from different sources like GIS data, historical information, and remote sensing data, is highly useful to predict and identify the crops, and also to track the crops in the international supply chain (Kamilaris et al. 2017).

8.4 Livestock Health and Productivity: Applications of Digital Twins in Animal Farming

Digital twins technology has great potential in modern livestock systems, enabling precise, data-driven management of animal health, welfare, and productivity. With its ability to integrate multisource real-time data into virtual models of animals, herds, and even entire farms, DTs serve as powerful tools for predictive data analysis, early disease detection, behavioral monitoring, reproductive management, and sustainable farming practices (Verdouw et al. 2021; Neethirajan 2021).

8.4.1 Modeling and Architecture of Digital Twins Framework for Livestock Systems

Digital twins in livestock farming can be defined as a virtual representation of individual animals or livestock systems, which is continuously updated using real-time sensor and environmental data. These DTs recorded the animal physiology like the body temperature, the heart rate, the internal pH, and behavior like movement, feeding, and even the rumination process, while for the environmental parameters, it can detect the barn temperature and the humidity, and finally, the performance indicators such as milk yield,= and weight gain.

The core architecture of a livestock DT typically includes:

- Data acquisition layer by using wearable sensors (accelerometers, GPS, RFID tags), in-barn IoT devices, camera systems, and environmental sensors.
- Data integration and analysis through machine learning (ML), cloud computing, and AI algorithms to analyze trends, classify behavior, and detect anomalies.
- Modeling and simulation platform that builds dynamic profiles of individual animals or herds, predicts health events, and simulates interventions.
- Decision support system (DSS) which provides actionable insights to farmers or automated systems for adjusting feeding, environment, or healthcare routines (Symeonaki et al. 2024).

8.4.2 Health Monitoring and Disease Prediction

One of the most valuable applications of DTs in livestock is the early detection of health disorders. A notable study by Neethirajan (2021) illustrated how DTs combining wearable biosensors, computer vision, and AI can monitor cattle for signs of mastitis, lameness, respiratory illness, and heat stress. These conditions often appear before they define visible symptoms, making early detection a hard process for welfare and profitability. In the same way, Symeonaki highlighted the critical role of sensors in the digital twins framework in early disease detection and continuous surveillance, in order to reduce the dependence on veterinary involvement. Their review outlines digital twins as enablers of proactive care in livestock by allowing preclinical recognition through data analysis and smart alerts (Symeonaki et al. 2024). In another study performed by Neethirajan et al. (2025) who applied a computational framework for nutrition-driven DTs that detected metabolic disorders like ketosis in dairy cows by monitoring diet composition and milk output, illustrating how the digital twins platforms can predict and mitigate health issues before they appear.

8.4.3 Behavior and Welfare Monitoring

Behavior is considered a marker of animal welfare. Digital twins equipped with computer vision and motion sensors can track animal position, movement patterns, and resting time. Behavior modeling of the DTs reflected abnormal lying times in cows and uneven activity in poultry, allowing for early correction of welfare issues such as overcrowding or stress (Hasan et al. 2024). In 2024, the Twin Farms Nexus project illustrated a practical welfare case: a DT farm setup where thermal comfort, feeding behavior, and stress indicators are mapped in real time to optimize stocking density and barn design (Neethirajan 2024).

8.4.4 Reproductive and Breeding Optimization

Reproductive behavior is important to monitor livestock productivity. Digital twins support improved estrus detection, insemination scheduling, and calving prediction by integrating behavioral, physiological, and environmental data. In a relevant study, it was illustrated that sensor-based digital tools, such as accelerometers, temperature loggers, and rumination monitors, are highly effective in detecting estrus in dairy cows. By submitting this real-time data into a digital twin framework, farms can create a better model for the reproductive cycles and optimize a better breeding management. This study confirmed that digital twins enable early identification of the reproductive inefficiencies and the perfect time of the insemination, contributing to improve the rates and shorter the calving intervals (Das et al. 2023).

8.4.5 Productivity and Performance Modeling

The digital twins models for animal simulation include animal growth, feed efficiency, and integration of the data on nutrition, environment, and genetics. In broiler production, digital phenotyping has emerged as a powerful tool. It is constructed from sensor data for feed intake, temperature, and movement, enabling producers to predict weight gain and adjust environmental controls in real time. This improves feed conversion ratios and supports uniform flock development (Neethirajan 2023). In addition to this, numerous field applications highlighted the ability of digital twins to enhance feed management and growth performance in both dairy and poultry systems. These frameworks enable the dynamic adjustments in the feeding schedules, the housing climate, and health alerts, in a step to improve the efficiency and waste reduction in the intensive livestock systems (Hasan et al. 2024).

8.4.6 The Environmental and the Feeding Behavior Monitoring

Digital twins play a critical role in modeling the livestock feeding behavior, the emissions, and the environmental performance. In order to track feeding and ruminating behavior, a model of the DTs for the dairy cows was presented by Zhang et al. (2023) that combined ultra-wideband (UWB) positioning and IMU data. The system used algorithms like LSTM to classify the feeding activity with 94.97% accuracy, the health prediction, and the precision feeding strategies. In another study, Jo and Kim (2018) talked about the pig house air management. They used DTs simulation to create a virtual pig barn that had real-time humidity and air flow adjustments, preventing respiratory illnesses and maximizing growth rates (Jo and Kim 2018).

8.4.7 Case Studied for the Smart Dairy and Poultry Using the Digital Twins Framework

Neethirajan offered an integrated DTs method for concurrently managing the poultry and the cows' dairy. Through predictive simulations, the system optimizes the feeding, the climate, and the space allocation by combining the sensors at the barn level and in the individual animal levels (Neethirajan 2024).

One of the important advantage of the (DTs) is the combination between the sensor networks and the simulation platforms that is used to manage the cow's dairy and the poultry flakes as this frame work improved the animal health and the housing conditions through a regular information collection of the temperature, the feeding habitats, air quality and the behavioral movement, all these collected data is analyzed to perform a better farm management. Farmers received automated alerts and predictive recommendations, enabling dynamic ventilation adjustments, feed scheduling, and early illness detection (Neethirajan 2024). In poultry farming, an example of how broiler farms use camera systems, environmental monitors, and digital modeling to track growth, density, and stress indicators was also introduced. By integrating these data into a digital twins system, farmers were able to reduce mortality rates, improve flock uniformity, and enhance feed efficiency through real-time environmental adjustments (Neethirajan 2023). These implementations showed that the digital twins framework is not just a virtual simulator but also enhances livestock performance by enabling predictive, precision-based interventions across diverse species and the housing systems.

8.5 The Application of Digital Twins in Soil Health and Fertility Management

Digital Twins technology is significantly applied to monitor the soil systems in order to enhance the ability of monitoring and managing *soil health*, *fertility*, and *ecosystem dynamics* through real time integration of multisource sensor data, remote sensing, and machine learning, soil DTs offer predictive modeling, decision support for nutrient management, moisture regulation, carbon sequestration strategies, and climate changes farming systems.

8.5.1 Digital Twins for the Soil Ecosystem

Cognitive soil digital twins, as described by Tsakiridis, used a multilayer architecture that combined IoT sensors, remote sensing datasets, lab soil assays, and terrain metadata. The DT platform integrates data ingestion, analytics (ML/AI), and simulation modules to model physical, chemical, and biological soil properties such as moisture content, nutrient levels (NPK), microbial activity, and carbon fluxes. This architecture supported real-time feedback loops: as soil parameters change due to weather or management, the DTs adapt and offer tailored advice on crop rotation, fertilization, or irrigation to maintain soil resilience and productivity (Tsakiridis et al. 2023).

8.5.2 Soil Moisture and Water Fertilizer Management

Accurate soil moisture measurement is essential to determine plant available water and fertilizer uptake. Parewai & Köppen developed a DTs approach simulating soil moisture status using a *physically based system* of high-resolution soil images under controlled illumination, combined with machine learning. Their system classifies moisture as *wet*, *normal*, or *dry* with *96.66% classification accuracy*, illustrating the DT's capability of finding a precise soil state detection and irrigation control. This enables optimized water use, timely irrigation scheduling, and better coordination of fertilizer application to prevent leaching and nutrient runoff (Parewai and Köppen 2025).

8.5.3 Soil Carbon Cycle and Climate Mitigation

Today, microbial ecology is being observed using the (DTs) platform. The framework is capable of simulating the Soil Microbiome (DTM), biochemical processes, and modeling the microorganisms' population dynamics. These DTs can calculate the greenhouse gas emissions under management or climate change, including methane and nitrous oxide. The study presents a conceptual DTM architecture combining metagenomics data, environmental sensors, and microbial metabolic modeling for climate smart soil management. This enables prediction decisions, evaluating how cover cropping or organic variations impact carbon sequestration and microbial-driven nutrient cycling (Mukhtar et al. 2022).

8.5.4 Soil Quality Assessment

In a systematic review, an analysis of the soil DTs framework is suggested to monitor *soil quality* using model predictive control (MPC). They found a nascent trend: among 158 agriculture-related DTs publications, none robustly addressed soil quality optimization within an operational DTs feedback model. The review encourages future DT designs to move from passive monitoring to active soil remediation and fertility control systems (Silva et al. 2023).

8.5.5 Circular Digital Twins for Climate Smart Soil Systems

In 2024 Monteiro applied the *Circular Digital Twin* framework, model climate smart soils in community-based agroecosystems. Their DTs describe composting loops, nutrient reuse, and carbon capture in regional farms. It uses population-scale compost data, soil assays, and nitrogen flux simulations to optimize management practices for regional sustainability and carbon neutrality (Monteiro et al. 2024).

8.5.6 Simulation and Scenario Analysis for Fertility Management

Digital Twins also support “what-if” scenarios, like in the case of changing fertilizer rates, crop rotation schedules, or plowing practices. Integrated with continuous learning. The DT systems can dynamically learn optimal decisions for irrigation, fertilization, and soil tillage by simulating outcomes under developing conditions. This approach is highly adaptable to soil fertility control via DTs-enabled AI frameworks (Goldenits et al. 2024).

8.5.7 Case Studied for Soil Management Using the DTs Framework

Although digital twin applications in soil health are still emerging, there are a few conceptual frameworks that have introduced soil-focused DTs usage.

- The digital twins for nutrients and fertilizer management, which were described by Bloch on presenting the field-scale soil DTs with the real-time subsurface sensor that collects information about soil temperature and moisture with remote sensing inputs as the leaf area index and the estimated biomass, to calibrate the crop simulation model. The software divides the field into management zones, calibrating soil and crop parameters. While the system covers crop growth, its inputs and outputs enable precision fertilizer applications based on DTs’ predicted nutrient stress and soil zone variability. This offers an example of a soil digital twins system that supports dynamic fertility management, spatially targeted nutrient dosing, and yield optimization (Bloch et al. 2022).
- Another example is estimating the soil quality and fertility using the DTs framework in a review study performed by Silva et al., although they found no fully deployed operational DT using predictive control for soils, the review identifies soil fertility status monitoring, nutrient balance forecasting, and erosion risk assessment as logical next-generation DT applications. They encourage agriculture DTs’ research to move beyond monitoring and simulation toward active soil

management decision-support, including fertility optimization and remediation planning (Silva et al. 2023).
- Tsakiridis also mentioned the cognitive soil DTs framework as he presented a powerful conceptual architecture for a Cognitive Soil Digital Twin, integrating IoT field sensors, remote sensing imagery, lab soil assays, and geospatial data into a unified platform. This DT captures key soil health indicators, such as nutrient levels (N, P, K), microbial activity, organic carbon, pH, and texture, and supports predictive analysis for land use changes, management interventions, and carbon sequestration strategies. It provides a compelling model for soil fertility planning and virtual testing of adaptive strategies before field application (Tsakiridis et al. 2023).
- The most significant study was performed by Wang when he outlined soil-related cases that applied DTs framework. These include modeling carbon sequestration, nutrient cycling impacts, and soil health trajectories under the variation of crop and management. The work highlighted the future opportunities for integrating soil fertility modeling into existing DTs platforms for holistic farm decision-making (Wang 2024).

8.6 Irrigation and Water Management

Although DTs have a great impact in crop monitoring, livestock management, and even soil monitoring (Verdouw et al. 2021) they also have a significant potential in smart irrigation to reduce the cost and enhance the efficiency (Rofaida et al. 2023). Furthermore, the integration of Digital Twins in irrigation management enables three core capabilities, like real-time monitoring, virtual experimentation, and predictive control. Before the actual use of the (DTs) frameworks, users can define different scenarios within the virtual model that represent an experimental environment. These scenarios contain different scripts, like the irrigation frequency, the water distribution during the drought, and the different fertigation patterns. These scenarios are quite important, especially in the rare water supplement areas and in erratic rainfall habitats (Wang 2024). The (DTs) framework, which can combine the crop, soil, and hydration models in one single platform, is considered to be one of the most promising techniques that is capable of attaining sustainable agriculture with enhanced water management (Ahsen et al. 2025).

8.6.1 Digital Twins Model for the Irrigation Systems

The implementation of the digital twins model in modern agriculture to perform enhanced irrigation patterns requires a powerful, layered system that is capable of modeling the actual field data accurately using simulation prediction systems. The DTs architecture depends on remote sensing devices, cloud computing, modeling, and feedback control systems to enhance the water usage in different agricultural systems.

A standard digital twin irrigation system is organized into five interconnected layers:

1. Data acquisition layer, which is integrated to IoT devices including soil moisture sensors, weather stations, leaf wetness meters, and flow valves.
2. Communication layer that transmits sensor data using LoRaWAN, NB-IoT, or OPC-UA protocols.
3. Integration and processing layer as it compiles with raw data and models it into cloud-based analytic environments.
4. Simulation and modeling layer that runs water balance models, crop simulation, and evapotranspiration forecasts.
5. Decision support layer, which delivers irrigation scheduling, actuation triggers, and optimization alerts to the end user. (Wang 2024; Manocha et al. 2023; Chaux et al. 2021; Ahsen et al. 2025; Barkakoti et al. 2023).

8.6.2 Sensors and IoT Integration in the Digital Twins Framework

Sensor networks are considered to be the main part of the digital twins (DTs) irrigation framework. These sensors translate the physical realities of a farm, such as soil moisture, weather fluctuations, crop status, and water flow, into actionable data that continuously feeds the virtual model. The more accurate the coupling between sensors and digital models, the greater the outcomes, precision, and predictive ability of the DTs system. *Tancredi* demonstrated that the sensor-integrated DTs framework used for irrigation management relied on the soil moisture probes, air temperature sensors, solar radiation meters, and precipitation gauges. These inputs were combined with ML algorithms to predict irrigation schedules and estimate water distribution. The model could trigger irrigation events based on anticipated evapotranspiration rather than fixed time intervals, in order to improve water use efficiency and to crop water stress (Tancredi et al. 2025).

In a similar study, Alves utilized their FIWARE-based digital twin, which integrated heterogeneous IoT devices into a unified data acquisition framework. Sensors measured real-time field parameters such as soil volumetric water content, leaf wetness, and atmospheric humidity. The system continuously adjusted irrigation schedules in response to both sensor readings and simulated crop demands, reducing water usage by 20–35% while maintaining yield levels in test plots (Alves et al. 2019).

Beyond basic soil and climate sensors, recent systems incorporate more sophisticated inputs, which were reported by Ahsen such as:

- *Canopy temperature sensors* for indirect estimation of plant water stress.
- *Multispectral UAV imagery* to assess vegetation indices (e.g., NDVI) linked to crop vigor and irrigation need.
- *Smart flow meters and pressure gauges* for monitoring irrigation infrastructure performance.

- DTs require multi-model, high-frequency data inputs to ensure accurate model calibration and actuation reliability (Ahsen et al. 2025).

To ensure interoperability and scalability, leading systems now adopt the standardized communication protocols such as *LoRaWAN*, *NB-IoT*, and *OPC-UA*, which facilitate the seamless integration across sensor types and allow the cloud-based coordination. This infrastructure module could provide future expansion and the incorporation of any additional inputs, such as satellite imagery, the root zone moisture modeling, or an AI-based plant phenotyping (Wang 2024).

8.6.3 Economic and Sustainability of the DTs Framework

Digital Twins (DTs) in irrigation offer measurable economic and environmental benefits by enhancing water efficiency, reducing costs, and promoting sustainable resource use. Farmers can optimize irrigation schedules, reduce waste, and increase the overall productivity by combining the real-time data with the predictive models. By modifying the irrigation timing according to the soil moisture and the evapotranspiration forecasts, the DTs framework improved the irrigation system and reduced the water consumption by up to 32%. As a result, the crop's quality increased and the operational costs decreased (Tancredi et al. 2025). Using FIWARE-based DTs, the horticultural farms saw a positive return on the investment in just two seasons, thanks to consistent yields and lower input usage (Alves et al. 2023). Regarding the sustainability milestones, Ahsen discovered that the DT guided irrigation reduced the water usage by 15–40% in the majority of deployments, while also preserving water quality and minimizing the nutrient leaching. Additionally, the DTs systems facilitated adherence to environmental regulations, particularly in areas with limited water resources (Ahsen et al. 2025). All of these findings confirmed that the DTs are not only cost-effective but also a cornerstone for the climate-smart, resource-efficient agriculture.

8.7 Conclusion

The recent revolutionary trend toward a sustainable agriculture that could be more dependent on data analysis through the (DTs) platforms that established a dynamic connection between the digital and practical frameworks introduced powerful modeling techniques to fully simulate, track, and enhance the sophisticated agricultural systems in order to improve precision farming in different ways. These virtual frameworks that are used to model crop fields, livestock, soil systems, and irrigation infrastructure are considered a real-time simulation and are used as a decision-making tool. On the other hand, DTs redefined the way of the interaction between farms and the biological systems. The hybrid modeling approaches that combined physics-based and machine learning frameworks (Melesse et al. 2025; Wang 2024) enhanced the simulation of the plant growth under variable soil, weather, and management conditions which helped in monitoring crop management. These tools allowed farmers to minimize the overuse of fertilizers, reduce the emissions, and maximize the return on investment by operating various agronomic decisions before physical implementation. The DTs have significantly improved productivity modeling, animal welfare management, and health monitoring in the context of livestock systems monitoring, and also made early disease detection, reproductive optimization, and performance tracking possible by combining real-time sensor data with behavioral modeling and the AI interface (Neethirajan 2021; Symeonaki et al. 2024). Another approach is that the DTs could be used to simultaneously send health alerts, enhance breeding schedules, and detect metabolic disorders (Das et al. 2023; Hasan et al. 2024). On applying the DTs on monitoring soil health and fertility management, it offered a dynamic platform to track and model the soil biochemical processes, in the same way, using the integrated sensing networks, remote sensing, and the AI modeling to predict the nutrient levels, microbial activities, and carbon fluxes enhanced the soil health in the real time physical system and also is used to provide recommendations for the site fertilization specifications and land management (Tsakiridis et al. 2023). Additionally, the DTs are considered an effective tool for controlling soil moisture and the greenhouse gas emissions caused by the microbes (Mukhtar et al. 2022; Parewai and Köppen 2025). Another approach is to highlight the potential of applying the DTs framework on using carbon sequestration technique, erosion control and even soil positioning, and that is why DTs can be used as an essential tool in climate-smart farming (Wang 2024; Silva et al. 2023). The DTs offered a wide range of opportunities in sustainable precision irrigation systems and water management by using sensor network and layered architecture systems. These systems used climate patterns, soil moisture data, and evapotranspiration forecasts to obtain an ideal irrigation schedule. Furthermore, DTs platforms can cut water use by 20–40% without sacrificing yields. These systems have demonstrated positive economic returns, especially in water-scarce regions, through reduced labor, increased crop durability, and regulatory compliance. By simulating drought scenarios and fertigation schemes before deployment, DTs serve as a virtual testing platform for real-world decisions, significantly lowering the risk and cost associated with irrigation mismanagement (Tancredi et al. 2025; Wang 2024; Ahsen et al. 2025). Despite their significant potential, digital twins face some challenges, particularly with the smallholders. Issues such as the high infrastructure costs, the limited interoperability, the data governance, and the dynamic nature of biological systems, which all have slowed DT adoption in some regions (Wolfert et al. 2017; Jones et al. 2020). Moreover, most DT frameworks currently focus on unimodal tasks, such as crop or livestock alone, rather than integrated, cross-domain decision-making. Future research should explore multimodal DT systems capable of linking crop-soil-livestock-water interactions into unified platforms. In order to facilitate automation and coordination throughout the entire farm, future trends will include intelligent control and cyber-physical integration (Liu et al. 2025; Verdouw et al. 2021).

The true potential of digital twins in agriculture is in their convergence with other technologies, such as the blockchain, robotics, and AI, which helps to create autonomous agroecosystems. The idea of "twin farms" provides insight into how farms can be developed into intelligent systems with the capacity for self-control and ongoing education (Neethirajan 2024). This approach could define a new strong significant change in the world food production sector through enhancing the tracking, sustainability, and climate resilience at the same time, like in the case of Circular DT framework suggestion that offered the concept of community-driven uses of the agroecological planning, nutrient recycling, and carbon neutrality (Monteiro et al. 2024).

In conclusion, digital twins are no longer a novel concept, but they are considered one of the main fundamental tools that are used in modern agriculture systems, as their applications across many fields like crops, livestock, soil, and irrigation systems identify their significant importance. In order to promote more intelligent, efficient, and sustainable agriculture, the DTs will provide farmers with actionable insights by combining real-time data with the appropriate sophisticated simulation platforms and artificial intelligence. The digital twins will be crucial in tackling food security, climate change, and sustainability in the ensuing decades with sustained technological advancement and political backing.

References

Ahsen R et al (2025) Harnessing digital twins for sustainable agricultural water management: a systematic review. Appl Sci *15*

Alves GR et al (2019) A digital twin for smart farming. s.n., Seattle, WA, pp 1–4

Alves RG, Maia RF, Lima F (2023) Development of a digital twin for smart farming: irrigation management system for water saving. J Clean Prod 388:135920

Barkakoti C, Sharma A, Panigrahi P (2023) Advancement of digital twin in irrigation and smart farming. s.l., s.n

Bloch V, Klerkx L, Rose DC (2022) Towards a digital twin for optimal field management. Agrirxiv

Boschert S, Rosen R (2016) Digital twin—the simulation aspect. In: Mechatronic futures. s.l.: Springer, Cham, pp 59–74

Cesco S, Fagarazzi C, Pezzuolo A, Sartori L (2023) Smart agriculture and digital twins: applications and challenges in a vision of sustainability. Eur J Agron 146:126809

Chaux D, Mayer S, Gries T, Müller D (2021) A digital twin architecture to optimize productivity within controlled environment agriculture. Appl Sci 11(19):8875

Das S et al (2023) Precision technologies for the management of reproduction in dairy cows. Trop Anim Health Prod *55*

Escriba-Gelonch M, Camara M, Maroto A, Fito P (2024) Digital twins in agriculture: orchestration and applications. J Agric Food Chem 72:10737–10752

Glaessgen EH, Stargel DS (2012) The digital twin paradigm for future NASA and U.S. air force vehicles. In: 53rd AIAA/ASME/ASCE/AHS/ASC structures, structural dynamics and materials conference
20th AI

Goldenits G, Vogl S, Wimmer M, Kastner W (2024) Current applications and potential future directions of reinforcement learning-based digital twins in agriculture. arXiv preprint

Grieves M (2016) Origins of the digital twin concept

Grieves M, Vickers J (2017) Digital twin: mitigating unpredictable, undesirable emergent behavior in complex systems. In: Transdisciplinary perspectives on complex systems. Springer, Cham, pp 85–113

Hasan MK, Islam MR, Rahman MS, Hossain MS (2024) A systematic literature review on the uses, benefits, challenges, and prospects of digital twins in livestock farm management. Adv Anim Vet Sci 12:2301–2314
Jones D, Snider C, Nassehi A, Yon J (2020) Characterising the digital twin: a systematic literature review. CIRP J Manuf Sci Technol 29:36–52
Jo S-K, Kim J-K (2018) Smart livestock farms using digital twin: feasibility study. s.l., s.n
Kamilaris A, Kartakoullis A, Prenafeta-Bold FX (2017) A review on the practice of big data analysis in agriculture. Comput Electron Agric 143:23–37
Kim S et al (2024) An agricultural digital twin for mandarins demonstrates the potential for individualized agriculture. *Nat Commun* 15:1561
Linz A, Hägele M, Knoll A (2019) "Digital twins" as tool for the development of field robots in agricultural processes. s.l., s.n
Liu H et al (2025) Research on intelligent control technology for a rail-based high-throughput crop phenotypic platform based on digital twins. Agriculture 15:1217
Lowe T, Harrison R, McBratney A, Ryu D (2020) Canopy density estimation in perennial horticulture crops using 3D spinning Lidar SLAM. Arxiv
Manocha AM, Sharma A, Singh P (2023) IoT-digital twin-inspired smart irrigation approach for optimal water utilization. Sustain Comput Inform Syst 41:100947
Marr B (2017) What is digital twin technology - and why is it so important? Forbes
Melesse TY, Alemayehu D, Worku G (2025) Digital twin-based applications in crop monitoring. Heliyon 11
Monteiro J, Silva L, Oliveira R, Sousa J (2024) The circular digital twin: climate-smart soils as a use case. s.l., s.n
Mukhtar H et al (2022) Digital twins of the soil microbiome for climate mitigation. Environments *9*
Neethirajan S (2021) Digital twins in livestock farming. *Animals* 11:1008
Neethirajan S (2023) Digital phenotyping - a game changer for the broiler industry. Animals *13*
Neethirajan S (2024) Twin farms nexus - digital twins for sustainable animal farming. Arch Animal Poult Sci 2(4):4
Neethirajan S, Kemp B, McBride B, Vranken E (2025) Computational architectures for precision dairy nutrition digital twins: a technical review and implementation framework. Sensors (Basel) 25(16):4899
Parewai I, Köppen M (2025) A digital twin approach for soil moisture measurement with physically based rendering simulations and machine learning. Electronics 14:935
Poornima G, Rajalakshmi P, Shanthini B (2023) Digital twin for smart farming. In: Data science for agricultural innovation and productivity, pp 1–16
Rofaida K, Ibrahim A, Hassan M, El-Sayed A (2023) The recent technological trends of smart irrigation systems in smart farming: a review. Int J Comput Digit Syst 14(01):10317–10335
Schleich B, Anwer N, Mathieu L, Wartzack S (2017) Shaping the digital twin for design and production engineering. CIRP Ann 66:141–144
Silva L, Monteiro J, Oliveira R, Sousa J (2023) The digital twin paradigm applied to soil quality assessment: a systematic literature review. Sensors (Basel) 23(2):1007
Symeonaki E, Papageorgiou E, Gkini C, Stylios C (2024) Recent advances in digital twins for agriculture 5.0: applications and open issues in livestock production systems. *Appl Sci* 14:686
Tancredi G, Russo S, Esposito M, Palmieri F (2025) Digital twin enhanced with machine learning algorithms for irrigation management using sensor data. Procedia Comput Sci
Tao F, Zhang H, Liu A, Nee AYC (2018) Digital twin in industry: state-of-the-art. IEEE Trans Industr Inform 15:2405–2415
Tsakiridis NL, Matsi T, Karyotis T, Kalivas D (2023) Cognitive soil digital twin for monitoring the soil ecosystem: a conceptual framework. Soil Syst 7:88
Verdouw C, Tekinerdogan B, Beulens A, Wolfert S (2021) Digital twins in smart farming. Agric Syst 189:103046
Verdouw C, Wolfert S, Beulens A, Rialland A (2015) A control model for object virtualization in supply chain management. Comput Ind 68:116–131

Wang L (2024) Digital twins in agriculture: a review of recent progress and open issues. Electronics 13:2209

Wolfert S, Ge L, Verdouw C, Bogaardt MJ (2017) Big data in smart farming—a review. Agric Syst 153:69–80

Zhang R, Zhu H, Chang Q, Mao Q (2025) A comprehensive review of digital twins technology in agriculture. *Agriculture-Basel* 4:15–903

Zhang Y et al (2023) Digital twin perception and modeling method for feeding behavior of dairy cows. Comput Electron Agric *214:108181*

Technical Innovation in Farming 4.0 for the Future of Sustainable Agriculture

9

Avnish Chauhan, Monu Bhardwaj, Gaurav Pandey, Muneesh Sethi, Sonu Dwivedi, and Kaiss Redouane

Abstract

The speedy evolution of agricultural technologies under the banner of "Farming 4.0" marks a transformative shift in the global food production landscape. This new paradigm integrates advanced technologies, including the Internet of Things (IoT), artificial intelligence (AI), massive data analytics, robotics, drones, and blockchain, to optimize agricultural productivity while promoting environmental sustainability. This chapter explores how precision farming tools, automated systems, and real-time data-driven decision-making are redefining traditional agricultural practices. The study outlines the current trends in innovative farming, examines case studies of successful implementation across diverse regions, and identifies the challenges faced by smallholder farmers in accessing these innovations. It also delves into the role of policy frameworks, digital infrastructure, and capacity building initiatives in facilitating the transition toward sustainable agriculture. By balancing productivity with resource conservation and biological health, Farming 4.0 offers a bright pathway toward meeting global

A. Chauhan (✉)
Department of Environmental Science, Graphic Era Hill University, Dehradun, Uttarakhand, India

M. Bhardwaj
School of Management, Graphic Era Hill University, Dehradun, Uttarakhand, India

G. Pandey
Forest Research Institute (FRI), Dehradun, Uttarakhand, India

M. Sethi
COER University, Roorkee, Uttarakhand, India

S. Dwivedi
Department of Chemistry, D.B.S.(PG) College, Dehradun, Uttarakhand, India

K. Redouane
Faculty of Economics and Management, Hassan 1st University, Settat, Morocco

© The Author(s), under exclusive license to Springer Nature Singapore Pte Ltd. 2026

A. L. Srivastav et al. (eds.), *Digital Twin Technology for Sustainable Agriculture*,
https://doi.org/10.1007/978-981-95-5915-2_9

food security goals in the face of climate change and population growth. The paper concludes by highlighting the decisive need for comprehensive innovation, equitable access, and moral considerations to ensure that the benefits of this scientific revolution are widely shared.

Keywords

Farming 4.0 · Sustainable agriculture · Precision and smart Farming · IoT in agriculture · Climate-resilient farming

9.1 Introduction

Though we have touched the moon and try to reach another planet but still for many things we still depend upon instinctive products for our survival, and agriculture is one of them. Agriculture is stepping into a new era, often called Farming 4.0, where technology and innovation are helping farmers grow food more expeditiously and sustainably. This new approach doesn't just aim for higher crop yields; it also focuses on protecting the environment, conserving water, improving soil health, and making better use of available resources. Farming 4.0 denotes the practice of rakish farming based on Industry 4.0 technologies (Kojo Gyamfi et al. 2024; Fasciolo et al. 2024; Hassoun and Bigliardi 2024, Chanchaichujit et al. 2024; Kazancoglu et al. 2024; Rabhi et al. 2025; Zhao et al. 2025; Salim et al. 2025). Given the growing population and increasing food demand, food production must increase sustainably (Kabato et al. 2025). From traditional Farming 1.0 to precision Farming 3.0, rakish Farming 4.0 integrates drones, the Internet of Things (IoT), factitious intelligence (AI), massive data analytics, robots, and further for precision and rakish agriculture (Ashwini et al. 2025; Chauhan et al. 2024).

Farming 4.0 is not about replacing farmers; it's about supporting them with better tools and knowledge (Mihret et al. 2025). By combining traditional wisdom with modern techniques, this approach opens the door to further sustainable food production (Rani et al. 2025). It helps ensure that farming remains productive, environmentally friendly, and economically practicable for future generations. With tools like soil sensors, weather monitoring systems, GPS-guided tractors, and smart irrigation, farmers can now make better decisions based on real conditions in their fields (Vellingiri et al. 2025). These innovations help reduce waste, lower the use of chemicals, and improve productivity, even in the face of climate change and extreme weather.

9.2 Overview of Farming 4.0

Farming 4.0 introduces innovative technological solutions to optimize farming processes worldwide (Mihret et al. 2025). It significantly transforms the agricultural landscape by integrating modern technologies such as the Internet of Things (IoT)

and robotics to enable data gathering and automation. This modernization involves advanced systems capable of monitoring crop growth, implementing autonomous irrigation, spraying, and harvesting, and effective storage management (Kojo Gyamfi et al. 2024). Smart agriculture aims to adapt to shifting consumer demands for ethical, animal-welfare-conscious, and traceable food products while addressing the mounting pressures induced by the COVID-19 pandemic.

Farming 4.0 is founded on a cyber-physical frame that actively collaborates with agricultural machinery and data-driven analytics, comprehensive satellites, Ground Penetrating Radar (GPR), drones, robots, and smart sensors. The aggregation of data on soil tissues, air quality, crop health, and consumer expectations allows farmers to optimize production and meet diverse demands on the farm. This technology importantly increases productivity by over 4.15% and mitigates the impact of climate change at a moderate pace. Farming 4.0 is an outstanding factor set to transform agricultural enterprises moving forward after demonstrating efficiency gains of further than 4% in this regard.

The Agricultural 4.0 system incorporates sensors that measure environmental parameters such as weather conditions, temperature, soil moisture, and atmospheric humidity (Selvam and Al-Humairi 2025; Mokrikov et al. 2025). These sensors play a polar role in enhancing farm management and promoting sustainability. Ceaseless monitoring yields real-time data that empowers farmers to make informed decisions concerning planting, harvesting, irrigation, as well as pest and disease control. It is practicable to elevate agriculture through the use of state-of-the-art technology, comprehensive data analytics, and autonomous machinery. This convergence fosters further resource-efficient practices, reduces environmental footprints, and promotes greater resilience in the face of climate-related risks. Through widespread adoption of these techniques, farm operators will improve productivity, conserve resources, and develop sustainable methods that address the challenges of climate change and population growth in the forthcoming years (Jose et al. 2024; Chauhan et al. 2025; Kansal et al. 2025).

9.3 Key Technologies in Farming 4.0

Scientific advancements have deeply transformed the agriculture sector, signaling a transition to Farming 4.0-a landscape where connectivity, intelligence, and autonomy underpin farming operations. This paradigm shift promises exaggerated resource efficiency, optimized yields, and resilience amid climatic volatility (Gyamfi et al. 2024).

Key technologies play indispensable roles in enabling modern agriculture. Sensors form the backbone of rakish farms (Alex 2025). Multi-parameter sensors continuously record weather conditions, temperature, and soil humidity. This real-time data informs precise decisions about planting schedules, harvesting timelines, irrigation needs, and pest management, ensuring sustainability. The Internet of Things (IoT) underpins distributed data exchange from field sensors and between independent machines, feeding analytic frameworks that improve practices and traceability. Factitious intelligence (AI) powers algorithms for diagnostics and crop forecasting. AI models leverage data from outside sensing, drone imagery, and unmanned aerial vehicles to generate actionable insights (Pal et al. 2024). Robotics and independent machinery carry out the genuine fieldwork, planting, harvesting, and precision herbicide spraying and trimming, eliminating human drudgery and increasing productivity.

While smart farms reduce resource inputs and environmental impact, commanding entry costs and digital infrastructure requirements create barriers (Gadewar et al. 2024; Narayanamurthy et al. 2025; Anand et al. 2025). Data protection and fairness issues compound the challenge, making equitable access harder. The transition to Agriculture 4.0 depends on solutions addressing privacy concerns and digital divides. Nevertheless, these technologies combined with climate adaptation policies, biotechnology advances in disease resistance and nutrition, 5G ultra-fast networks, blockchain platforms for supply chain transparency, and interconnected systems offer the means to shift agriculture toward a productive and sustainable model (Babar and Akan 2024).

9.3.1 Internet of Things (IoT)

Farming 4.0 innovates through Internet of Things (IoT) technology, allowing farmers to collect and analyze comprehensive data on weather, soil, and crop conditions in real time (Dhanaraju et al. 2022). These insights guide decisions on irrigation, pest management, and fertilization, optimizing crop yields and resource utilization while supporting outside farm management to reduce labor requirements.

IoT integrates multiple systems—including global navigation, sensors, and big data—to support precision irrigation tailored to soil and climate variables. The network employs microwaves that limit coverage extent, necessitating deployment of miniature base stations spaced approximately 250 meters apart to maintain continuous connectivity. Massive MIMO, akin to 4G, increases capacity, and beamforming reduces interference while boosting throughput (Mahathi et al. 2024). High-frequency, high-bandwidth internet connections accelerate data transfer, and a cognitive radio strategy allocates channels per device. Rural trials of autonomous tractors, drones, and other equipment demonstrated successful labor-free farming operations. Enhanced connectivity and reduced latency enable comprehensive farm, livestock, and asset management from remote locations (Singh et al. 2023; Parshad et al. 2025). Combining traditional methods with IoT advances enhances quality, reduces costs, and raises yields, a transformative step for sustainable agriculture.

9.3.2 Artificial Intelligence (AI)

Artificial intelligence (AI) represents the imitation of human intelligence by machines, enabling them to replicate tasks typically executed by people (Malheiro et al. 2025). AI has emerged as a prevalent agricultural technology following the advent of agricultural 4.0. Considerable research efforts investigate how AI, combined with edge or cloud computing, can optimize agricultural productivity. The primary objective is data-driven technological development that supports predictive analysis and facilitates decision-making processes (Zong and Guan 2025). AI-powered formal chatbots have become essential tools in various academic disciplines, mitigating errors stemming from imprecise human inquiry. Chat 4.0 technology supports integrated smart farming through conversational querying, enabling farmers to obtain early-stage answers concerning agricultural product requests in a time-efficient manner (Kojo Gyamfi et al. 2024).

9.3.3 Big Data Analytics

The adoption of smart agricultural technologies on farms and farms-to-table practices generates large volumes of data that need to be well-managed to provide actionable insights. Big data analytics converts this diverse information into valuable knowledge to enhance farming practices. Efficient data management enables the detection of relevant factors and the development of significant analytical models for data analysis (Zong and Guan 2025).

A layered architecture for big data analytics in smart agriculture consists of five components: data acquisition, data communication and data security, data storage and processing, data analysis, and decision-making. Sensors and networked collection devices gather agricultural data, which are transferred securely to storage. Storage and processing units handle data cleaning and fusion, resulting in usable datasets. Analytic agents perform various analyses, including machine learning, data mining, simulation, statistical, and genetic, on the processed data and additional input parameters. Finally, the decision-making layer utilizes analytics results, guidelines, and provisions to generate commands for specific actions (Violos et al. 2025).

9.3.4 Drones and Aerial Imaging

Drones and aerial imaging complement aerial photography, delivering detailed views of crop areas. Spectral sensors capture light in wavelengths beyond the visible spectrum, revealing plant challenges invisible to the human eye. This capability guides growers in developing more effective management strategies (Ross Wynn 2019). Satellite and aerial data were once the sole providers of such information, but drones now offer greater benefits. By 2050, a 60% production increase is needed to feed an estimated 9 billion people worldwide. Land availability to meet this demand is uncertain, and global food consumption patterns are shifting. Multispectral, hyperspectral, thermal, RGB, and LiDAR drone sensors provide (Kar and Dhal 2025) unique access to information that enables growers to spot problems early and implement corrective actions in a timely manner throughout the season.

Processing frameworks like convolutional neural networks (CNNs) and video images enable ongoing crop monitoring (El Sakka et al. 2025). Despite these advantages, many growers hesitate to adopt drone and spectral technology due to the wide variability in growing environments and the high investment required for spectral sensors. As sensor prices fall and manufacturers invest in research and development, drone and spectral technology are expected to become more affordable and accessible. Consequently, these technologies are increasingly recognized as essential tools for enhancing productivity, reducing input costs, and maximizing profitability.

9.3.5 Robotics and Automation

Unmanned aerial vehicles can monitor field-based agricultural tasks while autonomous ground vehicles can approach and grasp objects or manipulate ground-based equipment, reducing the risk of injury when working with heavy machinery. The use of robotic systems also enables precision agriculture, allowing for the use of fewer resources, such as seeds and fertilizers, while boosting quality and yield (Padhiary et al. 2025). For example, through the use of autonomous driving and collision avoidance systems, agricultural robots can reduce dependence on manual labor (Gyamfi et al. 2024). Digital agriculture will become the dominant farming system for the agri-food sector within the next four decades. As a result, smart autonomous farming vehicles will become an integral part of precision agriculture, and the further integration of robotics will create new opportunities for advancing the future of agriculture globally (Taha et al. 2025).

9.4 Sustainable Agriculture Practices

By 2050, the growing planet's population of 9 billion people necessitates increased food production, lower costs, and preservation of natural resources (Simane et al. 2025). An anticipated 70% or more increase in food output is required as climate change and other risks impact crop and herd productivity. Smart farming, also known as Agriculture 4.0, employs new technologies such as the Internet of Things (IoT) and Artificial Intelligence (AI) to modernize traditional methods, enhancing productivity, sustainability, and efficiency (Gani et al. 2025). Implementing these technologies requires the collaboration of governments, the private sector, and other stakeholders. The approach emphasizes optimized value chains, improved monitoring, and reduced labor costs.

9.4.1 Precision Agriculture

Precision agriculture, otherwise known as site-specific crop management, uses in-field sensors and automation to monitor and control crop production and processes. IoT and GPS-based technology facilitate the creation of detailed maps that allow farmers to tailor their farming techniques based on soil composition, moisture, and crop vitality (Bharti and Chauhan 2013; Singh and Kaunert 2025). IoT soil sensors combined with GPS-guided tractors enable real-time collection of vital data that help farmers improve irrigation, fertilization, and disease prediction. Water scarcity—a global challenge affecting more than two billion individuals—can be mitigated through intelligent irrigation systems that optimize irrigation schedules and water use, requiring the installation of IoT sensors and data analytics tools. Vertical farming permits crops to be cultivated in tires stacked on top of each other within a limited space (Ozigis et al. 2025). This indoor method provides consistent, continuous, and natural growth of food at different environmental conditions. Vertical farming offers a sustainable food production solution for urban areas that often face a scarcity of agricultural space. The process eliminates the need for pesticides and herbicides, reducing the associated chemical footprint. Vertical farming can produce fresh, organic, and highly nutritious food, increasing the availability of locally grown produce in urban areas. However, high costs in setup and operation, restrictive agricultural legislation, and a lack of expertise and interest on its farming model limit its wider adoption.

Precision Livestock Farming (PLF) technology helps farmers monitor, detect, and manage sickness, injuries, and other conditions; measure productivity; and improve transformation in an overall efficient and effective way (Trabachini et al. 2025). Used in conjunction with herd health and milk production, GPS tracking systems further improve the monitoring and management of dairy farms. Drones and aerial images support agricultural activities such as crop monitoring, land assessment, and pest control, complementing the existing precision agriculture techniques. These remote-sensing modalities increasingly provide a more accurate and cost-effective alternative to satellite technology (Kojo Gyamfi et al. 2024).

9.4.2 Crop Rotation and Diversity

During the past half-century, cropping systems worldwide have tended toward shorter and less diverse rotations. Concerns about the ecological sustainability of monocultures and narrow rotations such as corn-soybean are growing. Without crop diversification, challenges of pest control, crop disease, groundwater contamination from pesticides and fertilizers, and soil erosion remain difficult to manage (Tewari et al. 2010; Dumke and Dobbs 1999; Ahamad et al. 2025).

Crop diversification enhances resource use efficiency by increasing resource capture, use, and recycling. Temperature and precipitation are primary climate determinants of yield variation, but responses shift with management and crop combinations. Crop diversification potentially mitigates climate variability impacts by creating systems resilient to dynamic conditions. Diversification also stabilizes yield, thereby reducing risk for farmers and society.

A seven-year trial in Ontario, Canada, comparing rotations of one to five crop species demonstrated that while diversification had a minor effect on average wheat yield, it significantly increased the probability of wheat yielding above 3.5 t/ha by 81%. Risk avoidance is a common strategy for farmers seeking stable income from year to year. Crop diversification therefore mitigates temperature and precipitation destabilizing effects on wheat yield and increases the likelihood of favorable returns during unfavorable climatic conditions (Gaudin et al. 2015).

Crop rotational is a systematic sequence of different crops on a given field over time—has historically been crucial to ensuring long-term agricultural productivity (Fan et al. 2025). Despite replacement by modern inputs, it remains a widely used agronomic practice across continents for managing soil fertility and the economy of crop production, with positive effects on agricultural sustainability.

Rotation influences soil properties, water availability, disease and pest cycles, and nutrient turnover, exerting differential effects across agro-ecological environments and crop species. Balanced and well-managed systems involving multiple rotations appear more effective than simplified rotations. In many regions rotating with legumes remains challenging given limited legume crops adapted to soils and short growing seasons. Further testing could identify more legumes and alternative crops to integrate into diverse crop rotations, enhancing sustainability. Despite these benefits, the corn-soybean infrastructure has expanded significantly due to equipment simplicity, commodity programs, research efforts, and increasing uses for corn and soybean products, underscoring the continued need for diversification (Lauer 2008).

9.4.3 Soil Health Management

Soil health is foundational to sustainable food systems, ensuring healthy soils, nutritious crops, and a circular flow of nutrients throughout the farm (Chauhan 2012; Stanikzai et al. 2023; Gyamfi et al. 2024; Attri et al. 2025). Land degradation undermines ecosystem services, affects sustainable production, and contributes to global greenhouse emissions, exacerbated by climate change. Agricultural 4.0 technologies address soil health alongside water, energy, and air quality concerns, enabling continuous and comprehensive monitoring and fostering informed decisions for resilient farming systems (Mansour et al. 2026).

Continuous soil monitoring informs fertilizer application and crop rotation design. Ensuring soil has the right nutrients enhances crop quality and next-crop productivity while reducing the biotic load in the soil (Kumar et al. 2021; Wang et al. 2025). For example, high soil salinity's impact on crop growth can be detected early. Techniques such as hyperspectral imaging and multispectral sensors identify anomalies that indicate problems related to excessive fertilizer use, soil degradation, or parasite infestation.

9.5 Impact of Climate Change on Agriculture

Climate change and atypical occurrences constitute an exceptional risk for agriculture. The FAO estimates that, to meet the needs of the projected 9.3 billion world population in 2050, food production must increase by 70% (Simane et al. 2025). Smart Agriculture 4.0—driven by technologies such as the Internet of Things, artificial intelligence, sensors, and big data—can increase efficiency and productivity while optimizing supply chains. The sector has embarked on an Agriculture 4.0 transformation during the Fourth Industrial Revolution. Smart farming optimizes input factors and maximizes profitability, minimizes costs, improves quality, and reduces by-products. It enables precise control of water, fertilization, and pesticides, providing better monitoring and reducing labor costs. Governments, private-sector organizations, and the public must collaborate to enhance both the creation of scientific knowledge and its impact on agriculture (Gyamfi et al. 2024).

9.5.1 Adapting to Changing Weather Patterns

Weather plays a fundamental role in farming, providing essential factors such as rainfall, temperature, light, and atmospheric pressure. Agricultural practices that align with these natural elements tend to be more effective. However, atypical occurrences and climate change pose severe risks to agricultural output, necessitating a 70% or more increase in food production (Kojo Gyamfi et al. 2024). Effective forecasting models for weather and climate are therefore critical—they create opportunities for adaptation and provide early warnings against potential losses due to adverse conditions (Hansen and Coffey 2012).

Adapting to a variable and changing climate is questioned in light of agricultural development. A new climatology has modified the notion of a climate-normal period by introducing a distance factor to define changing periods under a precise quantification of significant climate data (Reder et al. 2025). However, the risk of adverse effects from an evolution outside this definition remains unconsidered. Agro-climate tools in climate risk management can improve agricultural development and food security while building resilience. Yet, promising climate risk management innovations remain underutilized in a massive part of the developing world, mainly due to limited access to actionable information. Translating raw climate data into accessible, relevant forms that can guide day-to-day, seasonal, and strategic decision-making processes is therefore critical; agro-climate tools transform climate data into estimates of the impacts on agricultural systems. Managing the risks associated with increasingly variable climate is a critical element in adapting agriculture to future climates and in reducing poverty and vulnerability caused by climate-related disasters. When farmers act on their experience to adapt practices, a changing climate threatens the applicability of that experience. Droughts and floods, events becoming more frequent in many areas as linked to climate change, severely damage livelihoods and infrastructure (Banu and Fazal 2025). Innovations such as field-specific soil, crop, and management data enhance risk management and yield optimization.

9.5.2 Water Resource Management

The provision of clean water is a major requirement for productive agriculture and the cornerstone of regional planning and development. Different countries have developed highly complicated management modes to utilize water resources. The implementation of water resource information automation and online application is one of the core issues relating to the development of the future water resource system (Kojo Gyamfi et al. 2024). Wastewater reclamation and reuse have been considered a frontier technology for effectively improving the efficiency of water resource allocation. Water is reused as process water in nutrient solution preparation and crop irrigation for vegetables, fruit trees, and floriculture. Measures for up-concentration treatment of wastewater include membrane filtration, reverse osmosis treatment, and ion exchange with a selective membrane. The operation of a micro-circulation system for reducing wastewater is conducted through the treatment of up-concentrated concentrate discharged from the up-concentration apparatus. Feedback and feedforward control of the micro-circulation system is shown to enable a stable operation.

9.6 Economic Implications of Farming 4.0

Population growth poses a significant economic challenge to agriculture, with an expected 70% or more increase in food production needed by 2050. Smart farming—sometimes called agriculture 4.0—is a growing partner in the pursuit of productivity and efficiency. The sector is empowered by a wide range of technologies such as the Internet of Things (IoT) and artificial intelligence (AI), transforming traditional agricultural processes and allowing farm monitoring and analysis with less manual labor (Gyamfi et al. 2024).

9.6.1 Cost-Benefit Analysis

Using smart technologies in agriculture requires assessing investment costs for devices, data acquisition, data management, and consulting systems, as well as application and navigation technologies (Zhang et al. 2024). Comprehensive cost-effectiveness analyses of digital technologies in agriculture are currently unavailable. Economic considerations often rely on model calculations or simplified approaches that may not be generalizable. The majority of research and development focuses on increasing management efficiency through savings in inputs or labor. Although technology remains inefficient and economic benefits are often lacking, the greatest benefits arise on large farms where fixed costs per area are lower. Smart technologies are most widely adopted in crop production and animal husbandry, yet there are few data demonstrating economically effective use on smaller farms. This suggests that cost-intensive digital solutions primarily suit larger farms and areas, prompting questions about the area threshold for economic efficiency, potential for shared use, associated costs, resource-saving benefits, and inclusion of external effects in economic assessments. Consequently, such technologies tend to support structural change toward larger operations, influencing knowledge requirements, relationships with soil and animals, and the occupational profile of farmers (Moschitz and Stolze 2018). Smart farming is effective when the expected financial benefits outweigh the additional costs of implementation, operation, and maintenance. The technologies optimize resource allocation, increase crop yields, and improve livestock management, thereby enhancing the sustainability and profitability of agricultural activities. Advances are particularly relevant for meeting consumer demands for sustainable supply chains and greater food transparency.

9.6.2 Market Trends and Consumer Behavior

Agriculture is undergoing a profound transformation known as Agriculture 4.0, driven by digital technologies and characterized by a shift from traditional resource-based production methods to innovative, technology-driven approaches. During the Second Industrial Revolution, mechanization replaced manual labor in farming; the Third Industrial Revolution introduced information technology, enabling precision agriculture; and the Fourth Industrial Revolution has introduced the Internet of Things and Artificial Intelligence, allowing remote and autonomous operations (Gyamfi et al. 2024). These technologies optimize agriculture systems economically, socially, and environmentally while supporting global food security for a forecasted population of nine billion by 2050.

Omni-channel retailing has altered consumer food purchasing habits, intensifying the competition among super/hypermarkets, convenience supermarkets, online shops, local stores, and traditional markets. The food sector faces increasing demand for innovative distribution, delivery, and sales methods. Trust in food safety and quality is paramount, prompting retailers to enhance consumer shopping experiences, such as gathering individual customer preferences before store visits.

9.7 Social Aspects of Agricultural Innovation

Social issues and the economics of implementing new technologies should be examined alongside the engineering problems. Computer agricultural systems can help recover local traditions and knowledge from the communist era, preventing them from being lost in transition. Technology allows farmers to enhance socio-political conditions with which they are already familiar. Specific models are necessary to aid the adoption of new agricultural technologies, such as the mechanization of rice planting, designed to accelerate growth and promote the latest techniques among food producers (Guru et al. 2025).

9.7.1 Farmer Education and Training

Farmers also require education and training to transition to smart agriculture, which demands interdisciplinary knowledge encompassing climate, policy, technology, and market considerations, in contrast to conventional practices (Okoronkwo et al. 2024). Since education systems currently lack curricula for emerging technologies such as automation, sensors, artificial intelligence, 5G, and blockchain, establishing specialized centers that provide both theoretical instruction and practical exposure is essential to full deployment of smart-farming approaches.

9.7.2 Community Engagement and Support

Community engagement is a fundamental means to promote a sense of belonging and inclusion. A wide range of movements ranging from LGBT pride to community gardening organizes regular activities in local communities. Volunteerism also facilitates active participation and social responsibility (Wondimu and Admas 2024).

9.8 Regulatory and Policy Frameworks

Regulatory and policy frameworks can foster a trusting environment that encourages sustainable farming systems and supports innovation and adoption. Agro-food production must closely address the social, ethical, and environmental issues raised by the use of innovative technologies. Public policy is required to stimulate the development of digital technology for the food and agriculture industry (Yap and Al-Mutairi 2024).

The lack of legislative frameworks and ethical considerations for data ownership could hinder adoption. Data privacy and security concerns create barriers to widespread adoption, yet robust measures and educational programs can enhance farmers' understanding and confidence. Ensuring regulatory compliance alongside ethical considerations—such as fair labor practices and responsible data use—remains critical for the competitiveness of smart farms.

9.8.1 Government Initiatives

In the twenty-first century, changes in climate, population growth, and other factors have altered conventional farming patterns, increasing the demand for sustainable agriculture by 2050. Official guidelines across countries emphasize long-term sustainability using digital technologies for data-driven agriculture systems, fostering economic growth, resilience, and sustainable development. A selected framework ensures Agricultural 4.0 technological advancement and adoption by farmers, highlighting Ukraine as a regional agricultural leader. Ecosystemal development is critical to mitigating instability (Beugnon et al. 2024).

9.8.2 International Standards and Guidelines

Government bodies and other standardization institutions have developed standards and guidelines that elaborate on various aspects of data management—such as taxonomies, filtering, ownership, and sharing—in different sectors (Gujar 2025). These governance tools offer a valuable foundation for promoting technology adoption by providing credible guidelines and best practices, transparency in unintended effects, and demonstration of peer acceptance and trusted utilization of technologies. These elements are crucial for achieving the desired economic, environmental, and societal benefits of technological advancements.

9.9 Case Studies of Successful Implementation

A contemporary perspective on smart farming provides a foundation for examining relevant case studies. Implementing Agricultural 4.0 technology can transform crop cultivation, water management, livestock health, and reforestation. Precision agriculture enables the construction of detailed field maps with Internet-of-Things sensors and radio-frequency identification technology. Information stored in these maps on soil characteristics, moisture, temperature, and crop vitality supports the customization of farming practices. GPS-controlled mass data collection integrates with augmentation and decision support systems to refine irrigation and fertilization schedules. The supply of relevant data for forecasting systems also facilitates the early prediction of disease outbreaks in the livestock sector. Addressing water scarcity concerns and supporting the Sustainable Development Goals in the agricultural domain can be achieved by means of a smart water distribution and irrigation system (Elmahdi 2024). The system integrates Internet-of-Things sensor data, ecological indices, and precipitation forecasts to optimize the quantity of water used in the irrigation process. Low-cost sensor-based irrigation optimizers monitor soil humidity and weather conditions and rely on forecasts and historical data to determine watering schedules. Vertical farming constitutes a sustainable urban practice that completes the list of major emerging application domains within Agricultural 4.0 technology. The soilless cultivation of crops on stacked tiers inside enclosed facilities has the potential to enhance food security (Dutta et al. 2025). It can furthermore reduce emissions and the need for herbicides, pesticides, and insecticides. In the livestock sector, precision agriculture-filled technological environments enable the monitoring of animal condition and performance, thereby supporting timely and resource-effective sector management. The resulting capabilities have a major positive impact on productivity, animal health, and welfare; GPS, for example, facilitates the identification of individuals within the herd and contributes to improved health and enhanced milk yields. Start-ups emerging at the onset of the Agricultural 4.0 era concentrate on designs for technological innovation. Unmanned aerial vehicles (drones) represent the domain's most popular application. Aerial vegetation maps are dynamically generated on the basis of the Normalized Difference Vegetation Index (Sarvakar and Thakkar 2024). The evaluation of photosynthetic activity supports land productivity assessments and yields early information on stand maturation or damage as well as the prompt detection of potential pest infestations. The combination of drone-mounted laser sensors and proximal Internet-of-Things sensing devices further improves the monitoring process, yielding information on soil health and irrigation effectiveness. In addition, aerial photos contribute to the geo-localization of weeds for autonomous, zone-specific herbicide treatments, whereas impurity identification supports the establishment of precision sowing and fertilization practices. Overall, the technologies promote a shift toward deep control and permanent monitoring of the crop and the surrounding environment, creating possibilities for averting critical issues and introducing automatic interventions when necessary (Borah et al. 2024).

9.9.1 Global Examples

Tech innovations associated with Agricultural 4.0 aim to enhance productivity, efficiency, product quality, and sustainability. An integrated technological framework targets water use reduction, greater precision, improved flexibility, product quality, and sustainable development. Challenges include variable soil properties affecting productivity, pest invasions damaging crops and livestock, and economic and regulatory constraints, such as high costs and unhelpful monitoring devices (Gyamfi et al. 2024). In Brazil, Sudi leverages satellite remote-sensing and IoT data for large-scale agriculture monitoring and biomass estimation.

Precision agriculture involves advanced technologies to improve and optimize decision-making. IoT sensors and GPS technology create detailed field maps for tailored farming techniques. IoT soil sensors mounted on tractor buckets collect real-time soil data, including moisture, nutrient levels, temperature, and pH. GPS-guided tractors validate soil-sensed data while assessing various physical and chemical production factors. Smart implementation improves irrigation, fertilization, pesticide spraying, and crop-disease predictions, reducing resource wastage. Livestock and reforestation applications also benefit from Agricultural 4.0.

Water shortage is a global serious social issue, significantly affecting agricultural development and resource protection. Intelligent irrigation systems optimize water usage through IoT sensor integration and big-data analysis, supporting Sustainable Development Goals (SDGs) related to water efficiency and sustainable agriculture. Sensor-based low-cost irrigation evaluates real-time water, soil, and weather parameters to time irrigation events efficiently, conserving water and energy and reducing costs (Roy and Chakraborty 2025).

Vertical farming places crops in stacked layers within controlled environments for artificial growth. Long discussed for the future of food security, especially in urban agriculture, vertical farming reduces transportation costs and carbon footprints. The vertical-farming startup Aerofarms partners with the United States Department of Agriculture to improve sustainable agriculture and reduce greenhouse-gas emissions associated with long-distance transport and the pesticide use typical of traditional agriculture. Although vertical farming faces economic and regulatory constraints, it minimizes or excludes pesticide use, producing healthy crops and improving public health.

9.9.2 Local Innovations

Novel technologies provide a paradigm shift in agriculture by addressing age-old issues that have constrained crop production or quality. Innovations in agricultural technology enable farmers to monitor various soil and water parameters, including moisture, acidity, nitrogen, or phosphorus concentrations, with a few clicks, reducing production costs and improving output and efficiency. Publicly available satellite images are becoming increasingly popular because they enable the monitoring of vegetation trends and pest detection at very low cost, contributing to improved eco-efficiency and sustainable development (Holzinger et al. 2024).

The ubiquitous deployment of innovative technologies is transforming agricultural practices. One expert survey rated precision farming, drone usage, and agricultural robots as the three technologies with the most economically beneficial potential for the agricultural sector in the coming years, accelerating the diffusion and adoption of emerging technologies.

Farms have harnessed data and information from advanced sensors to enhance efficiency, manage resources, and introduce automation. Technological advances have created smart farming that leverages the Internet of Things (IoT) to establish a close-knit farming ecosystem. Data from sensors, drones, satellites, and other devices can be accumulated, aggregated, processed, and interpreted for management decisions and optimized farming practices.

Local innovations can be categorized into grower-driven and equipment manufacturer-driven innovations. Grower-driven innovations result from hands-on experience in the field; equipment manufacturer-driven innovations arise from investment in research and development programs. Local innovations are built around small but readily available technologies (e.g., radio frequency identification devices, satellite communication terminals, and global positioning systems [GPS]) and cover diverse sectors with extensive applications in decision support, machinery, inputs, and postharvest operations (Li et al. 2025).

9.10 Challenges and Barriers to Adoption

Agricultural 4.0 represents the integration of unquiet technologies across the entire agricultural value chain, including production, processing, and distribution (Chanchaichujit et al. 2024). These digital innovations offer new paradigms for crop enhancement and discovery, sanctionative precision management, and environmental impact assessments. Nevertheless, the widespread adoption of Agricultural 4.0 is hindered by several challenges and barriers. The operative initial investment has made these technologies unaffordable for many, specially minimal farmers lacking access to financing or subsidies. Data privacy and security constitute decisive concerns anticipated to cyber risks, unauthorized access, and the absence of legislative frameworks that address data ownership and use. Internet-dependent solutions expose agricultural production systems to vulnerabilities, with breaches potentially catastrophic for resource-limited farmers. Operative integration and sustainable deployment demand stakeholder training, collaboration among agronomists and researchers, moral data practices, and improved digital literacy. The costs of sensors, drones, automation technologies, and infrastructure may exceed the capacities of minor agricultural actors, limiting economic feasibility. Emerging technologies, including 5G networks, blockchain for traceability, and AI-driven decision tools, promptly offer enhanced management possibilities, yet infrastructural limitations pose further adoption challenges. Overall, Agricultural 4.0 and Farm Management 4.0 aim to enhance efficiency, sustainability, transparency, and resilience by fostering the integration of biotechnology, robotics, and autonomous machinery, thereby increasing growth capacities and addressing future environmental and societal demands (Fasciolo et al. 2024; Karmakar et al. 2024).

9.10.1 Technological Limitations

The absence of scientific materials has hindered farmers' ability to select crops suited to local ecological conditions, limited disease detection in livestock, and prohibited the effective adoption of advanced technological devices. New and emerging technologies provide data-driven insights to improve agricultural performance and facilitate participation in related activities (Mishra and Mishra 2024).

9.10.2 Financial Constraints

Despite being pivotal in ensuring agricultural sustainability and resilience, financial assistance is among the unmet needs of farmers. Economic uncertainty makes access to credit difficult for farmers who nevertheless require capital to manage their current expenses and invest in new agricultural technologies. To overcome these constraints, innovative financial products, services, and models have been developed using various financial instruments such as insurance products, leasing and overtime repayment schemes, out-grower contracts, warehouse receipt financing supported by market information, knock-out call options, and development and microfinance institutions (Gashayie and Singh 2015).

9.10.3 Cultural Resistance

Despite the increasing adoption of farming 4.0 technologies, however, farmers are typically culturally resistant to new technologies. Resistance stems from deep-seated relationships with the land, which it then appears that technology will replace or render obsolete. Indeed, farmers express feelings of betrayal because caring for the land is central to their identity. Yet other apprehension springs from unforeseeable consequences or difficulties in overcoming deeply embedded intuitions and traditional mental habits (Lent 2021). As a result, resistance arises instead from concern lest a new technology dislodge an existing one while offering no net advantage. Without a sense of meaningful benefit, skepticism toward given technologies increases.

Farmers optimistic about technological breakthrough cited increased speed and the ability to generate data. They also regard technologies primarily as decision-making tools, expressing fear lest these systems make complex decisions automatically. Farmers aspire to work alongside technologies, not to become "robotic" or superseded. Sceptics fear the very prospect of "robotic farmers." However, once they recognize the potential to function in a variety of roles, they reconsider their stance. The process of adopting new technology thus involves the creation of a new identity that individuals are able to both imagine and project.

9.11 Future Trends in Farming 4.0

The global population is projected to reach 9.7 billion by 2050, which will increase the demand for food and require urgent sustainability and monitoring actions (Habib et al. 2025). Therefore, the agricultural industry must take advantage of Industry 4.0 developments to ensure smart, sustainable, and efficient farming practices. Industry 4.0 and its agricultural application (often called Farming 4.0 or Agriculture 4.0) will strongly affect the future of agro-industries by facilitating intelligent knowledge-based agriculture. Industry 4.0 provides new business models and opportunities for future farmers through the integration of Cyber-Physical Systems with a set of new technologies. These innovations offer efficient agro-product management and intensive agricultural value chain monitoring. However, challenges also arise concerning data security, transparency, and synchronization between the applications of different companies (Alamsyah and Setiawan 2025).

9.11.1 Emerging Technologies

The agriculture industry currently faces many serious problems that require technology to increase production and satisfy the rising demand for agricultural and food products (Bissadu et al. 2025; Singh et al. 2024a, 2024b). Agriculture 4.0 implements the latest technologies to disrupt the sector, including smart farming. "Smart farming" means using modern technologies to increase the quantity and quality of agricultural products while lowering human labor and environmental impact.

9.11.2 Predictions for Sustainable Practices

Reports predict a steep rise in natural human-made disasters. Climate change, overpopulation, and depleted natural resources threaten the suitability of places and crops (Pal 2025). By 2050, nine billion people will inhabit the planet, prompting an overwhelming need for enhanced crop production and resource preservation, while increasing food productions becomes vital due to such risks. Smart farming, or Agriculture 4.0, may resolve more production in reduced space by utilizing Internet of Things (IoT) and Artificial Intelligence (AI) technologies for live monitoring, control, and automation (Patel et al. 2025).

9.12 Conclusion

An era of farming 4.0 is also undergoing a transformation (Kojo Gyamfi et al. 2024). The subsequent challenge is to merge innovative technical development with sustainable advancement while adjusting to an increasing world population, climate change, desertification, and shrinking water supply. It is both a necessity and a critical challenge. A new era of farming 4.0; it encompasses the integration of cloud computing, artificial intelligence, the internet of things, unmanned aerial vehicles, autonomous vehicles, and big data that jointly contribute to accomplishing the expected goals.

The advent of hi-tech farming 4.0 presents a new angle, yet a collective scoping study to translate the requisite opportunity into the desired objective remains an investigative priority. Consequently, the potential of various hi-tech farming 4.0 tools and their valuable functions vis-à-vis expected outcomes need to be quantitatively evaluated to fintech as a framework. Development of a hi-tech farming 4.0 system is found to be feasible through the combination of appropriate technical innovation and adoption of financial technology (fintech). The attributes and capacities of current hi-tech are also investigated to combine the available technologies effectively. These innovations include cloud technology, artificial intelligence (AI), the internet-of-things (IoT), unmanned aerial vehicle (UAV), autonomous vehicle (AV), and big data as part of a synergetic framework construction. The construction of such a framework is soon to translate hi-tech farming 4.0

into the desirable achievements and expected challenges identified through a comparative study at the pilot region in the Middle East.

The revolutionary hi-tech farming regime addresses several factors, offering three core principles: high efficiency, fossil fossil-free strategy, and sustainability. These lead to the outside crop yield, marvelous usage efficiency, built-in climate change adaptation, and coevolution with natural ecosystems. Concurrently, the construction provides a practicable economic spinoff, transition toward a sustainable socio economy, creation of local employment, and development in areas otherwise liable to abandonment. Unlike manufacturing, hi-tech farming 4.0–fintech integrated framework generates a vast range of commodities at multiple scales upon demand and response operations. Besides, the deployment of inexpensive materials and getting upfront payment prior to full-scale deployment helps reduce economic risks. These make hi-tech farming 4.0 exceptionally lucrative and highly profitable for the development of the upcoming decade. Consequently, quantum farming 4.0 approach is constructed, unveiling a new aspect of farming and supply chain system at emerging regions for the transformation of sustainable agriculture in a Fourth Industrial Revolution.

References

Ahamad F, Chauhan A, Chauhan PK, Upadhyay SK, Tomar A, Singh N, Andrade T (2025) Addressing complex challenges in water quality management: emerging technologies and sustainable strategies. In: Computational automation for water security. Elsevier, pp 251–276. https://doi.org/10.1016/B978-0-443-27376-6.00018-9

Alamsyah A, Setiawan IPS (2025) Enhancing privacy and traceability of public health insurance claim system using blockchain technology. Front Blockchain 8:1474434. https://doi.org/10.3389/fbloc.2025.1474434

Alex A (2025) The technological revolution in agriculture: cloud computing as the backbone of smart farming. J Comput Sci Technol Stud 7(2):514–521

Anand V, Rajput P, Minkina T, Mandzhieva S, Kumar S, Chauhan A, Rajput V (2025) Systematic review of machine learning applications in sustainable agriculture: insights on soil health and crop improvement. Phyton (Horn) 94(5):1339. https://doi.org/10.32604/phyton.2025.047099

Ashwini A, Sriram SR, Prabhakar JM, Kadry S (2025) Farming 4.0: cultivating the future with internet of things empowered on smart agriculture solutions. In: Networked sensing systems, pp 247–271

Attri S, Saklani A, Joshi KK, Sharma P, Chauhan A, Goswami R et al (2025) Seeds of sustainability: promoting food security through eco-friendly agriculture. In: Sustainable synergy: harnessing ecosystems for climate resilience. Springer, pp 83–97. https://doi.org/10.1007/978-3-031-55842-7_6

Babar AZ, Akan OB (2024) Sustainable and precision agriculture with the internet of everything (IoE). arXiv preprint., arXiv:2404.06341. https://doi.org/10.48550/arXiv.2404.06341

Banu N, Fazal S (2025) Climate change, livelihood crisis and resilience: an introduction. In: Livelihoods and well-being in the era of climate change: risk to resilience across India. Springer, pp 3–18. https://doi.org/10.1007/978-3-031-51444-7_1

Beugnon R, Le Guyader N, Milcu A, Lenoir J, Puissant J, Morin X, Hättenschwiler S (2024) Microclimate modulation: an overlooked mechanism influencing the impact of plant diversity on ecosystem functioning. Glob Chang Biol 30(3):e17214. https://doi.org/10.1111/gcb.17214

Bharti P, Chauhan A (2013) Soil quality and contamination. Discovery Publishing House. ISBN: 978-93-5056-361-8, pp 1–186

Bissadu KD, Sonko S, Hossain G (2025) Society 5.0 enabled agriculture: drivers, enabling technologies, architectures, opportunities, and challenges. Inf Process Agric 12(1):112–124. https://doi.org/10.1016/j.inpa.2024.11.001

Borah SS, Khanal A, Sundaravadivel P (2024) Emerging technologies for automation in environmental sensing. Appl Sci 14(8):3531. https://doi.org/10.3390/app14083531

Chanchaichujit J, Balasubramanian S, Shukla V (2024) Barriers to Industry 4.0 technology adoption in agricultural supply chains: a Fuzzy Delphi-ISM approach. Int J Qual Reliab Manag 41(7):1942–1978. https://doi.org/10.1108/IJQRM-12-2023-0432

Chauhan A (2012) Vermitechnology, vermiculture, vermicompost and earthworms: vermiculture, vermicomposting, vermitechnology and microbes. LAP Lambert Academic Publishing

Chauhan A et al (2024) Application of flower wastes to produce valuable products. In: Srivastav AL, Bhardwaj AK, Kumar M (eds) Valorization of biomass wastes for environmental sustainability. Springer, Cham. https://doi.org/10.1007/978-3-031-52485-1_14

Chauhan A, Tomar A, Attri S, Sethi M, Upadhyay SK, Chauhan PK (2025) Application of modern tools for the real-time monitoring of bioremediation approach and its advantages. In: Biotechnologies for wastewater treatment and resource recovery. Elsevier, pp 255–266. https://doi.org/10.1016/B978-0-443-27376-6.00018-9

Dhanaraju M, Chenniappan P, Ramalingam K, Pazhanivelan S, Kaliaperumal R (2022) Smart farming: internet of things (IoT)-based sustainable agriculture. Agriculture 12(10):1745. https://doi.org/10.3390/agriculture12101745

Dumke L, Dobbs TL (1999) Historical evolution of crop systems in Eastern South Dakota: economic influences. South Dakota State University Economics Department

Dutta M, Gupta D, Tharewal S, Goyal D, Sandhu JK, Kaur M et al (2025) Internet of things-based smart precision farming in soilless agriculture: opportunities and challenges for global food security. IEEE Access. https://doi.org/10.1109/ACCESS.2025.3398760

El Sakka M, Ivanovici M, Chaari L, Mothe J (2025) A review of CNN applications in smart agriculture using multimodal data. Sensors 25(2):472. https://doi.org/10.3390/s25020472

Elmahdi A (2024) Addressing water scarcity in agricultural irrigation: by exploring alternative water resources for sustainable irrigated agriculture. Irrig Drain 73(5):1675–1683. https://doi.org/10.1002/ird.3042

Fan H, Miao R, Guo C, Bao X, He W, Sun Y, Zhao C (2025) Research on the effect of diversified cropping on crop quality: a review. Agriculture 15(5):456. https://doi.org/10.3390/agriculture15050456

Fasciolo B, Panza L, Lombardi F (2024) Exploring the integration of industry 4.0 technologies in agriculture: a comprehensive bibliometric review. Sustainability 16(20):8948. https://doi.org/10.3390/su16208948

Gadewar M, Prashanth GK, Babu MR, Dileep MS, Prashanth PA, Rao S et al (2024) Unlocking nature's potential: green synthesis of ZnO nanoparticles and their multifaceted applications—a concise overview. J Saudi Chem Soc 28(1):101774. https://doi.org/10.1016/j.jscs.2023.101774

Gani A, Singh M, Pathak S, Hussain A (2025) Recent advancements towards leveraging the role of artificial intelligence in agriculture and smart farming. In: Intelligent infrastructure and smart materials: sustainable technologies for a greener future. Springer, pp 177–195. https://doi.org/10.1007/978-3-031-49867-9_9

Gashayie A, Singh M (2015) Agricultural finance constraints and innovative models experience for Ethiopia: empirical evidence from developing countries. J Econ Sustain Dev 6(17):1–9

Gaudin ACM, Tolhurst TN, Ker AP, Janovicek K, Tortora C, Martin RC, Deen W (2015) Increasing crop diversity mitigates weather variations and improves yield stability. PLoS One 10(2):e0113261. https://doi.org/10.1371/journal.pone.0113261

Gujar P (2025) Data standardization and interoperability. In: Data usability in the enterprise: how usability leads to optimal digital experiences. Apress, Berkeley, CA, pp 89–110. https://doi.org/10.1007/978-1-4842-9731-6_6

Guru PK, Sahu P, Shukla P, Diwan P, Panwar G, Tiwari P et al (2025) A critical review on rice cultivation and mechanization level in Indian perspective. Results Eng:105632. https://doi.org/10.1016/j.rineng.2025.105632

Gyamfi EK, ElSayed Z, Kropczynski J, Yakubu MA, Elsayed N (2024) Agricultural 4.0 leveraging on technological solutions: study for smart farming sector. arXiv:2401.00814

Habib M, Singh S, Jan S, Jan K, Bashir K (2025) The future of the future foods: understandings from the past towards SDG-2. npj Sci Food 9(1):138. https://doi.org/10.1038/s41538-025-00258-2

Hansen J, Coffey K (2012) Agro-climate tools for a new climate-smart agriculture [PDF]. CGIAR Research Program on Climate Change, Agriculture and Food Security (CCAFS). Retrieved from https://ccafs.cgiar.org/resources/publications/agro-climate-tools-new-climate-smart-agriculture

Hassoun A, Bigliardi B (2024) Industry 4.0 technologies: principles and applications in agriculture and the food industry. In: Food industry 4.0. Academic Press, pp 1–13. https://doi.org/10.1016/B978-0-443-19889-3.00001-5

Holzinger A, Fister I, Kaul HP, Asseng S (2024) Human-centered AI in smart farming: toward agriculture 5.0. IEEE Access 12:621–633. https://doi.org/10.1109/ACCESS.2023.3349631

Jose A, Deepak KS, Rajamani N (2024) Innovation in agriculture and the environment: a roadmap to food security in developing nations. In: Food security in a developing world: status, challenges, and opportunities. Springer, pp 259–281. https://doi.org/10.1007/978-3-031-53952-5_13

Kabato W, Getnet GT, Sinore T, Nemeth A, Molnár Z (2025) Towards climate-smart agriculture: strategies for sustainable agricultural production, food security, and greenhouse gas reduction. Agronomy 15(3):565. https://doi.org/10.3390/agronomy15030565

Kansal D, Gururani P, Joshi NC, Pant G, Chauhan A (2025) A comprehensive review on utilization of cavitation technology for industrial wastewater treatment: a step toward sustainability. Water Air Soil Pollut 236(7):438. https://doi.org/10.1007/s11270-025-06834-0

Kar D, Dhal SB (2025) Advancing food security through drone-based hyperspectral imaging: applications in precision agriculture and post-harvest management. Environ Monit Assess 197(3):283. https://doi.org/10.1007/s10661-025-11923-5

Karmakar R, Kumar N, Tripathi V, Sharma PK, Kumar R, Gahlot D et al (2024) Circularity of biomass feedstock to produce ethanol and feasibility of ethanol-gasoline fuel blends in engine. J Solid Waste Technol Manag 50(3):630–639. https://doi.org/10.5276/JSWTM.2024.630

Kazancoglu Y, Lafci C, Kumar A, Luthra S, Garza-Reyes JA, Berberoglu Y (2024) The role of agri-food 4.0 in climate-smart farming for controlling climate change-related risks: a business perspective analysis. Bus Strateg Environ 33(4):2788–2802. https://doi.org/10.1002/bse.3553

Kumar S, Gautam P, Chauhan A (2021) Phytochemical, pharmacological properties and bitterness causing compounds of pomelo (Citrus maxima): a mini review. Int J Pharm Pharm Sci 20(2): 2666–2678

Lauer J (2008) For crops … it's rotation, rotation, rotation! The sustainability of the corn-soybean rotation. Wisconsin Crop Manager. Retrieved from https://ipcm.wisc.edu/download/wcm-pdf/WCM2008/WCM12-26-08.pdf

Lent J (2021) The web of meaning: integrating science and traditional wisdom to find our place in the universe. New Society Publishers

Li W, Gu J, Liu J, Cheng B, Zhu H, Miao Y et al (2025) A review of key technological developments in autonomous unmanned operation systems for agriculture in China. AgriEngineering 7(3):857–875. https://doi.org/10.3390/agriengineering7030045

Mahathi A, Reddy CKK, Lippert K, Hanafiah MM (2024) Advancements in machine learning techniques for optimization of massive MIMO design. In: Machine learning for radio resource management and optimization in 5G and beyond. CRC Press, pp 116–139. https://doi.org/10.1201/9781003379987-8

Malheiro V, Santos B, Figueiras A, Mascarenhas-Melo F (2025) The potential of artificial intelligence in pharmaceutical innovation: from drug discovery to clinical trials. Pharmaceuticals 18(6):788. https://doi.org/10.3390/ph18060788

Mansour AE, Alkady Y, Elashmawi WH, Elbahnasaw M, Shamseldin T (2026) Revolutionizing farming with agriculture 4.0: technology-driven sustainability and productivity. In: Precision and intelligence in agriculture: advanced technologies for sustainable farming. IGI Global, pp 51–96. https://doi.org/10.4018/978-1-6684-8856-1.ch003

Mihret YC, Takele MM, Mintesinot SM (2025) Advancements in Agriculture 4.0 and the needs for introduction and adoption in Ethiopia: a review. Adv Agric 2025(1):8828400. https://doi.org/10.1155/2025/8828400

Mishra H, Mishra D (2024) AI for data-driven decision-making in smart agriculture: from field to farm management. In: Artificial intelligence techniques in smart agriculture. Springer, Singapore, pp 173–193. https://doi.org/10.1007/978-981-99-7297-5_10

Mokrikov G, Fedorenko A, Rajput VD, Barakhov A, Rogaleva V, Vilkova V et al (2025) The influence of multi-species cover crops on the biological activity of soils in conditions of no-till. J Soil Sci Plant Nutr:1–11. https://doi.org/10.1007/s42729-025-01492-2

Moschitz H, Stolze M (2018) How can we make sense of smart technologies for sustainable agriculture?—a discussion paper. Technol Innov Manag Rev 8(4):47–56. https://doi.org/10.22215/timreview/1149

Narayanamurthy G, Jayanth RSS, Tortorella G, Fogliatto F (2025) Empirical investigation of barriers and benefits of smart farming. Int J Qual Reliab Manage 42(8):2150–2173. https://doi.org/10.1108/IJQRM-09-2024-0320

Okoronkwo DJ, Ozioko RI, Ugwoke RU, Nwagbo UV, Nwobodo C, Ugwu CH et al (2024) Climate smart agriculture? Adaptation strategies of traditional agriculture to climate change in sub-Saharan Africa. Front Clim 6:1272320. https://doi.org/10.3389/fclim.2024.1272320

Ozigis MS, Wich S, Abdolshahnejad M, Descals A, Szantoi Z, Sheil D, Meijaard E (2025) Characterizing crop distribution and the impact on forest conservation in Central Africa. Remote Sens 17(11):1958. https://doi.org/10.3390/rs17111958

Padhiary M, Kumar A, Sethi LN (2025) Emerging technologies for smart and sustainable precision agriculture. Discover Robot 1(1):6. https://doi.org/10.1007/s44115-024-00006-1

Pal B (2025) Water scarcity and climate change in intelligent infrastructure and smart materials: sustainable technologies for a greener future. In: Intelligent infrastructure and smart materials: sustainable technologies for a greener future. Springer, Cham, pp 213–231. https://doi.org/10.1007/978-3-031-52814-6_10

Pal OK, Shovon MSH, Mridha MF, Shin J (2024) In-depth review of AI-enabled unmanned aerial vehicles: trends, vision, and challenges. Discov Artif Intell 4(1):97. https://doi.org/10.1007/s44154-024-00197-w

Parshad J, Kumar R, Chhokar V, Patil N, Beniwal V, Singh N et al (2025) Rice residue management: alternative strategies and emerging technologies for a sustainable ecosystem. Pedosphere 35(1):53–66. https://doi.org/10.1016/S1002-0160(24)00040-7

Patel A, Shukla C, Trivedi A, Balasaheb KS, Sinha MK (2025) Smart farming: utilization of robotics, drones, remote sensing, GIS, AI, and IoT tools in agricultural operations and water management. In: Integrated land and water resource management for sustainable agriculture, vol 1. Springer, Singapore, pp 127–151. https://doi.org/10.1007/978-981-99-7252-4_6

Rabhi L, Jabir B, Falih N, Afraites L, Bouikhalene B (2025) Digital transformation metamodel in smart farming: crop classification prediction based on recurrent neural network. Foods Raw Mater 13(1):107–118

Rani J, Gulia V, Sangwan A, Dhull SS, Mandzhieva S (2025) Synergies of traditional ecological knowledge in biodiversity conservation: a paradigm for sustainable food security. In: Ecologically mediated development: promoting biodiversity conservation and food security. Springer, pp 27–49. https://doi.org/10.1007/978-981-99-4306-9_3

Reder A, Fedele G, Manco I, Mercogliano P (2025) Estimating pros and cons of statistical downscaling based on EQM bias adjustment as a complementary method to dynamical downscaling. Sci Rep 15(1):621. https://doi.org/10.1038/s41598-025-58808-z

Roy S, Chakraborty RS (2025) Low-cost smart irrigation solution for efficient water use and requirement prediction. Comput Electr Eng 125:110420. https://doi.org/10.1016/j.compeleceng.2024.110420

Salim S, Rajput VD, Ahmad I, Minkina T, Chauhan A (2025) Role of organic matter and phosphate solubilizing bacteria in phosphorus availability and wheat growth in gypsum soils. Diyala Agric Sci J 17(1):154–167. https://doi.org/10.52951/dasj.25170112

Sarvakar K, Thakkar M (2024) Different vegetation indices measurement using computer vision. In: Applications of computer vision and drone technology in agriculture 4.0. Springer, pp 133–163. https://doi.org/10.1007/978-981-99-5953-4_7

Selvam AP, Al-Humairi SNS (2025) Environmental impact evaluation using smart real-time weather monitoring systems: a systematic review. Innov Infrastruct Solut 10(1):1–24. https://doi.org/10.1007/s41062-024-01154-6

Simane B, Kapwata T, Naidoo N, Cissé G, Wright CY, Berhane K (2025) Ensuring Africa's food security by 2050: the role of population growth, climate-resilient strategies, and putative pathways to resilience. Foods 14(2):262. https://doi.org/10.3390/foods14020262

Singh B, Kaunert C (2025) Synthesis of machine learning applications in cutting-edge modern agriculture: projecting fuzzy logic crop health monitoring via soil and plant sensor technologies. In: Soft computing and machine learning. CRC Press, pp 201–220. https://doi.org/10.1201/9781003345975-13

Singh D, Kumar Biswal A, Samanta D, Singh V, Kadry S, Khan A, Nam Y (2023) Smart high-yield tomato cultivation: precision irrigation system using the internet of things. Heliyon 9(6):e15501. https://doi.org/10.1016/j.heliyon.2023.e15501

Singh A, Rajput VD, Agrawal S, Ghazaryan K, Minkina T, Al Tawaha ARM et al (2024a) Nanoparticles mediated salt stress resilience: a holistic exploration of physiological, biochemical, and nano-omics approaches. Rev Environ Contam Toxicol 262(1):19. https://doi.org/10.1007/398_2024_103

Singh MV, Sethi M, Srivastav S, Stanikzai K, Chauhan A, Pan D et al (2024b) Exploring the potential of transforming sustainable waste thermoplastics into liquid petrochemicals. ES Energy Environ 24:1121. https://doi.org/10.30919/esee8c1121

Stanikzai K, Kumar S, Dwivedi S, Chauhan A (2023) Phytoremediation of zinc and lead from polluted soils using sunflower. J Env Bio Sci 37(2):119–126. https://doi.org/10.59467/JEBS.2023.37.119

Taha MF, Mao H, Zhang Z, Elmasry G, Awad MA, Abdalla A et al (2025) Emerging technologies for precision crop management towards agriculture 5.0: a comprehensive overview. Agriculture (Basel) 15(6):582. https://doi.org/10.3390/agriculture15060582

Tewari N, Matta G, Chauhan A (2010) Comparative evaluation of various chelators for removal of pollutants and heavy metals from distillery effluents. ESSENCE Int J Env Rehab Conserv:28–34

Trabachini A, Moreira MDR, Harada ÉDS, Amorim MDN, Silva-Miranda KOD (2025) Precision livestock farming applied to swine farms—a systematic literature review. Animals 15(14):2138. https://doi.org/10.3390/ani15142138

Vellingiri A, Kokila R, Nisha P, Kumar M, Chinnusamy S, Boopathi S (2025) Harnessing GPS, sensors, and drones to minimize environmental impact: precision agriculture. In: Designing sustainable internet of things solutions for smart industries. IGI Global, pp 77–108. https://doi.org/10.4018/978-1-6684-8830-1.ch005

Violos J, Mamanis G, Kompatsiaris I, Papadopoulos S (2025) Cognition and context-aware decision-making systems for a sustainable planet: a survey on recent advancements, applications and open challenges. Discov Sustain 6(1):43. https://doi.org/10.1007/s43621-024-00161-1

Wang G, van der Putten WH, Klironomos J, Zhang F, Zhang J (2025) Steering plant-soil feedback for sustainable agriculture. Science 389(6758):eads2506. https://doi.org/10.1126/science.ads2506
Wondimu H, Admas G (2024) The motivation and engagement of student volunteers in volunteerism at the University of Gondar. Discov Glob Soc 2(1):22. https://doi.org/10.1007/s44260-024-00022-2
Wynn CR (2019) Drone technology: is it worth the investment in agriculture?
Yap CK, Al-Mutairi KA (2024) A conceptual model relationship between industry 4.0—food-agriculture nexus and agroecosystem: a literature review and knowledge gaps. Heliyon 10(3): e26055. https://doi.org/10.1016/j.heliyon.2024.e26055
Zhang J, Trautman D, Liu Y, Bi C, Chen W, Ou L, Goebel R (2024) Achieving the rewards of smart agriculture. Agronomy 14(3):452. https://doi.org/10.3390/agronomy14030452
Zhao W, Chen H, Bulis A (2025) How are Industry 4.0 technologies transforming a sustainable society across industries? Digit Transform Soc 4(3):363–380
Zong Z, Guan Y (2025) AI-driven intelligent data analytics and predictive analysis in industry 4.0: transforming knowledge, innovation, and efficiency. J Knowl Econ 16(1):864–903. https://doi.org/10.1007/s13132-023-01722-7

Modelling the Future: Digital Twins in Sustainable and Climate-Aware Farming

10

Hurmat, Balpreet Kaur, Lata Rani, and Arun Lal Srivastav

Abstract

Nowadays, frequent variations have been observed in the climate and hence, different approaches are being adopted during the cultivation of various crops to minimize the adverse effects on the yield. In the current scenario, a number of technologies are developed to alter the production of crops, like prior information about the weather conditions. Therefore, continuous growth is obvious in the area of agriculture to make smart as well as more adaptive to deal with unpredictable challenges. In this regard, improvement is required in the use of Digital Twins in both ways: virtual & real-world farming methods. These models are able to provide real-time data along with their testing and navigating the situations during farming which offers an impactful tool for diligence in agriculture. Application of digital twin technologies, farming practices is able to optimize the resources like water management in irrigation, proper fertilization along with effective pest management. The present paper is based on description of available digital twin technologies and its significance in Agriculture 4.0 for improving the farming yields along with resource optimization even under climate variations. Moreover, it will also enlighten about digital twin techniques and their possibility of integration with modern tools including artificial intelligence, climate models and sensors. Thus, digital technologies can play significant roles in farming as they can give accurate information for irrigation, optimization of crop yields with more efficient, resilient and sustainable agricultural cycle.

Hurmat · B. Kaur · L. Rani (✉)
Chitkara University School of Pharmacy, Chitkara University, Baddi, Himachal Pradesh, India
e-mail: lata.rani@chitkarauniversity.edu.in

A. L. Srivastav
Department of Applied Sciences, Chitkara University, Solan, Himachal Pradesh, India

© The Author(s), under exclusive license to Springer Nature Singapore Pte Ltd. 2026
A. L. Srivastav et al. (eds.), *Digital Twin Technology for Sustainable Agriculture*,
https://doi.org/10.1007/978-981-95-5915-2_10

Keywords

Climate change · Crop yield · Sustainable agriculture · Natural resource management · Artificial Intelligence

10.1 Introduction

Modern agriculture needs to adopt advance technologies to manage increasing global demand of food, as well as working efficiently to use more sustainable resources during production. Owing to intense population rise day by day, soil health deterioration, adverse climate change, etc., innovative and resilient farming techniques are required (Gonzalez 2011). Weather variations in certain areas are considered as the reasons for significant hazards to agriculture because they result in frequent food shortages and imbalance of ecosystems (Srivastav et al. 2022; Dubey et al. 2022). Noticeable enhancement of global temperatures is also a reason for significant loss of crop yields (Rathour et al. 2025). In the same way, unpredictable rainfall patterns are also responsible for droughts or floods which may influence irrigation procedures and cropping schedules (Agele 2021). The influence of these climatic variations is specifically seen in rain-fed agricultural areas, where strong climatic unpredictability; risk of crop failure, and food insecurity are significantly enhanced. To deal with the increasing challenges in farming, there's been a clear urge of a plethora of digital tools and automation. Various techniques like Artificial Intelligence (AI), Internet of Things (IoT), and Digital Twin (DT) are helpful to reshape procedures of agricultural works. These different tools support farmers to utilize resources more efficiently, minimize waste, and also to control environmental damage (Awais et al. 2025). Even though digital twins were initially introduced in industries like manufacturing, but now it is being applied in many other sectors, i.e., city planning, healthcare, and agriculture. Hence, DTs are more influential than the capability to develop virtual models of real systems, which can be evaluated in real time and utilized to run simulations or help for taking decisions. In farming, it facilitates better planning, prompt responses to problems, and a stronger push toward sustainability and climate flexibility (Peladarinos et al. 2023).

Owing to the boom of global population day by day, the need for food also increases significantly. To address these huge demands of food, the conventional farming procedures often result in soil health degradation, water loss, and increment in greenhouse gas emissions (Gomiero 2016). Climate alteration continues to cause a severe problem to agriculture with unpredictable weather, i.e., frequent incidence of droughts, floods, and growing outbreaks of pests. These obstructions are responsible for detrimental impacts on crop yields and farmer livelihoods. In this way, sustainability can be a game changer as it is about adopting innovative farming methods that are environment friendly, economical, practical as well as socially fair (Schaller 1993). To confirm sustainable agriculture, it is advised to use current resources very carefully, minimizing wastage concurrently try to inhibit dangerous emissions. Additionally, it also helps the farmers to address efficiently any minor or major alteration in weather conditions. The prominent aim is to make agriculture field more productive for long time without any negative impacts on the environment. Recent approaches like artificial intelligence (AI), big data analytics, Internet of Things (IoT) and digital twins, etc., facilitate stakeholders of agriculture to ensure this field is more innovative by installing smart techniques (Jain et al. 2025).

These all fields make it simple to analyse crop health, soil properties, weather conditions, and how resources, i.e., water, fertilizer, etc., are being consumed. With real-time tracking, farmers can deal with situation more precisely. Among all these recent and advanced methods, digital twin is superior because of its power to simulate agricultural environments with more accuracy (Cesco et al. 2023). By using combination of digital innovation and sustainability principles, it is possible to enhance flexible and better farming specifically with current environmental complications.

10.1.1 Digital Twin Technology

Digital Twin technology (DT) is a revolutionary approach that forms a virtual replica of a physical system, object, and process. This AI-based model is updated progressively on the basis of available date to facilitate the process of analysis and prediction of various environmental circumstances (Mihai et al. 2022). Initially, DT technology was developed to add smartness in industries for aerospace followed by involvement in healthcare system, agriculture, and smart cities (Grieves 2023). DT can also be connected to various models together to boost the performance. While each field may describe digital twins a little differently, the core idea stays the same—a digital twin is a virtual version of something real, whether it's a machine, a crop field, or even a full farming operation. DT is able to provide updated real-time data from the actual system. This integration of physical and digital world help people make better decisions by offering a clear picture of its working principle (Mazzetto 2024).

DTs play various functions in agriculture such as modelling farm environments, crop growth stages, irrigation systems, and even entire supply chains. For example, a digital twin of a crop field can combine weather forecasts, soil health data, and real-time sensor input to help farmers to take decisions about the best time to sow, the right amount of water to use, and effective strategies for managing pests (Getahun et al. 2024). Compared to traditional methods, DT is a sophisticated setup that can support both routine field analysis and enhancements to daily agricultural operations. Even under extreme environmental circumstances, digital twins can combine, store, and transport the original data that was collected. Effective resource management and climate planning are made possible by this creative integration. For a nation like India, where agriculture is regarded as a lifeline, DT therefore offers a valuable mcans of data monitoring, strategic planning, and quick decision-making (Ametefe et al. 2024).

10.2 Understanding Digital Twin Technology

10.2.1 Architecture of Digital Twin in Agriculture

Gathering real-time data from multiple sources, including soil sensors, weather stations, drones, and satellite imaging, is the responsibility of the data acquisition layer (Wu et al. 2023). To guarantee consistency and usability, the gathered data is processed, integrated, and organized utilizing cloud platforms or edge computing. Additionally, the data is processed to imitate real-world situations and forecast future events with the aid of AI algorithms, crop growth models, and simulation tools. This application can also be used to forecast future events, such as disease outbreaks or yield estimates (Javaid et al. 2023). Ultimately, an awareness can also be created with the help of interactive dashboards or decision-making support centre which facilitate farmers to evaluate the weather conditions as well as to assess the possible risks (Kumari et al. 2024).

10.2.2 Integration of DT with Sensors, IoT, and Cloud Systems

The use of different sensors, cloud computing systems, and IoT devices is able to support the digital twin brilliantly to enhance the production in agriculture (Nasirahmadi and Hensel 2022). Sensors can be used in the field for different purposes like weather monitors, soil moisture probes, and crop health cameras to uninterruptedly collect real-time environmental as well as operational data. Smooth transmission of data from the field to the digital environment is facilitated by various sensors allied with IoT networks that empower automation and communication between all systems (Al-Ali et al. 2020). Cloud storage is useful to store the data that will be used in processing, analysing as well as management of large datasets (Zhang et al. 2025; Peladarinos et al. 2023).

10.3 DT Applications in Sustainable and Climate-Resilient Farming

10.3.1 Crop Monitoring and Yield Prediction

Two important components of climate-resilient and sustainable agriculture are frequent monitoring and yield prediction of crop which are significantly improved by digital twin technology (Melesse 2025). The real-time information about different factors like nutrient levels, temperature, soil moisture, humidity, etc. and this information can be collected by using significant network of sensors. This information is really important and helpful to create virtual databank of the crop field can be owing to the involvement of digital twins, which is supportive to the farmers to keep a close and exhaustive check on their crops. Moreover, different diseases and other issues of fields like nutrient shortages, drought, flooding, soil pollutants, etc. can be early detected and addressed accordingly. Digital twins play its significant role in monitoring as well as to integrate sophisticated machine learning algorithms and other smart techniques to forecast crop yields in advance (Pylianidis et al. 2022). These procedures or technologies can provide information and management on the basis of past climate patterns and plant growth stages under current environmental conditions. Additionally, food waste can also be decreased by better harvest planning (Riaz et al. 2023). Hence, crop monitoring and yield prediction by using digital twin technology is an unmatchable advancement in this sector. It will make farming more productive, eco-friendly as well as resilient in the era of global warming.

10.3.2 Soil Health and Water Management

To achieve the goal of sustainable agriculture and to boost its climate resilience, two significant points to keep in mind are soil health and water management. These two pillars can also be managed in a magical way with the help of Digital Twin technology (Ametefe et al. 2024). The study of healthy soil is complex as it may be affected by many factors such as moisture, nutrients, pH, and even presence of microbes. With the help of smart sensors installed across the fields, digital twins gather real-time data on all these parameters of soil. This creates a virtual map that shows exactly what's going on under the surface. This helps in early detection of any issue and helps farmers to easily spot where nutrients are low, where the soil might be compacted, or which areas need special care (Tsakiridis et al. 2023). Using this method, farmers can give proper care and treatment to each part of the field whether the issue lies in finding the right amount of fertilizer or compost methods that protect the soil from erosion and degradation.

Water management is also a crucial factor as soil health, especially in present-day scenario of unpredictable weather. Too much or too little water can seriously harm the crops and to resolve this issue and make proper management of water, digital twins play a major role by integrating soil moisture data, weather forecasts, evapotranspiration rates, and how well irrigation systems are working or by matching crop water requirements accurately (Yousra 2023). Therefore, smarter irrigation systems can be developed that give crops the right amount of water without waste which will save water and energy and also protect crops from water stress (Kumar et al. 2023). Moreover, digital twins help farmers in adapting climate-induced challenges like extended drought periods or sudden heavy rainfall events by enabling simulation of various water management (Zhao et al. 2025). So that farmers can take decisions earlier like adjusting planting dates or selecting drought-tolerant crop varieties which would improve farm resilience.

This is very helpful to preserve ecosystem as well as it has its significant role in the detrimental effects of different illogical techniques of agriculture on the environment, specifically in the era of global warming (Ametefe et al. 2024).

10.3.3 Pest and Disease Forecasting

Crop health is affected by various factors, i.e., attack of pests, weeds and disease, and these issues can predicted early by using DT and managed accordingly. The real-time data from sources like field sensors, weather stations, satellite images, and past pest outbreak records together with digital twins can be an ideal and detailed virtual model of farms. These models analyse different environmental conditions such as temperature, humidity, rainfall, and the growth stage of crops and help to simulate how pests and diseases might spread under these conditions (Zhang et al. 2025). Integration of machine learning made these models smarter as predictions are possible by learning from new data over time. DT can send prompt intimation by sensors to farmers about the attack of weeds, pests, etc. Farmers can use different smart, targeted solutions, i.e., natural pest control methods, smart crop alteration, controlled pesticides use to preserve environmental health (Peladarinos et al. 2023).

10.3.4 Climate Impact Simulation and Adaptation Strategies

Digital twin technology plays an important role in tackling climate change challenges in agriculture through advanced climate impact simulation and adaptation strategies (Peladarinos et al. 2023). Digital twins are able to create a dynamic virtual environment by using previously collected records which influence both present and future climatic conditions (Riaz et al. 2023). These predictions help us to estimate how changes in temperature, rainfall, extreme weather, and rising carbon emission levels can influence crop growth, soil health, water supply, and the well-being of livestock badly. Digital twins can locate the issue and guide farmer for the development by adjusting various smart means, i.e., planting schedules, selecting climate-resilient crops, optimizing irrigation, and nutrient management (Uysal 2025). Farmers can first test these ideas virtually by seeing how they play out under different climate scenarios, then employ these suggestions for better outcomes in the real world. Therefore, this method has minimum risks, is smarter and creates more cost-effective decisions.

The data is collected from different sources including agriculture, climate science, AI, and is combined with digital twins to facilitate both farmers and policymakers to ensure proactive choices and smart solutions. As the climate becomes more unpredictable, taking this kind of forward-thinking approach helps farms become more resilient, use resources more efficiently, and adopt practices that are better for the environment (Batra et al. 2024). In the long run, it's not just about protecting harvests but also about ensuring food security and caring for the environment in a rapidly changing world.

10.4 Benefits and Opportunities

10.4.1 Enhanced Decision-Making and Resource Efficiency

Irrigation can be adjusted for different parts of a field by using up-to-date soil moisture data and weather forecasts. This helps to prevent overwatering and saves valuable water. In the same way, the use of fertilizer can be optimized by detecting nutrient deficiencies at an early stage and applying only what is necessary. This helps to minimize runoff and environmental harm (Agelli et al. 2024).

Digital twins help farmers to predict about weather and make various decisions as per problems like droughts, pest outbreaks, etc. that support farmers to ensure preventive steps to deal with challenges (Kalyani et al. 2024; Pires et al. 2023).

10.4.2 Increased Resilience to Climate Variability

Impulsive weather, global warming, etc. increase pressure day by day on agriculture. To deal with it, DT is a significant tool to boost the dependence of farming sector, supporting them acclimatize more successfully (Riaz et al. 2023). Weather is altered by different environmental conditions, i.e. droughts, floods, temperature rise, and drastic rainfall patterns. DT supports farmers to shift to adaptive strategies (Uysal 2025; Arulmozhi et al. 2024). This permits them to work more efficiently to unexpected weather alterations (Riaz et al. 2023). In this fashion, DT not only boosts productivity but also introduces the tools to the agricultural sector to adapt recent technology to handle the changing climate, which ultimately promotes stability, sustainability and food safety (Ametefe et al. 2024).

10.4.3 Support for Sustainability Certification and Compliance

Digital twin technology has a great advantage in facilitating sustainability certification and regulatory compliance within agriculture (Cesco et al. 2023). Nowadays, as environmental rules are getting stricter and global markets are pushing for more traceable, sustainable produce, digital twins help farmers to keep up with these changes by offering transparency, reliable data, and clear records (Omrany et al. 2024).

10.5 Challenges and Limitations

10.5.1 Data Privacy and Security Concerns

In the context of climate-responsive agriculture, digital twin technology demands extensive data collection from climate sensors, soil monitors, weather stations, drones, and IoT-enabled farm equipment (Khan et al. 2025). While this data is key to creating accurate simulations and helping farms adjust to changing climate conditions, it also raises real concerns about privacy and data security. The digital twin ecosystem often depends on cloud storage, AI analytics, and third-party platforms, which makes sensitive agricultural data vulnerable to breaches, unauthorized access, and misuse (Ali et al. 2020). In climate-smart farming, the data goes beyond just weather or environmental information as it also covers things like location, land use, crop yields, and quantity of water or fertilizer used. If this kind of information goes into wrong hands, it could be misused for unfair competition or even market manipulation. Smallholder farmers, especially in developing regions, are more susceptible to exploitation because they may lack awareness of data rights or the infrastructure to ensure secure data transmission.

Therefore, there is a need to build trust and also to ensure safe process for which digital twin solutions must embed strong cybersecurity protocols. DTs propose many safety measures such as encrypted communication, secure cloud access, user authentication, and compliance with data protection laws like General Data Protection Regulation (Wang et al. 2023). There's also an urgent need for clear rules and regulation for who owns the data, along with farmer-friendly agreements that focus on transparency, fair data sharing, and giving farmers control at the local level. These concerns are essential and need to be addressed properly not just for protecting privacy, but also for scaling digital twins as trustworthy tools in climate-resilient agricultural systems (Ahmad 2025).

10.5.2 Infrastructure and Scalability Issues

Digital infrastructure remains a significant barrier in many agricultural regions, particularly in developing countries which can be addressed by successful deployment of digital twin technology in climate-responsive agriculture (Omrany and Obaidi 2024). For effective functioning, the digital twins require seamless integration of IoT devices, high-speed internet connectivity, cloud computing, and data processing systems (Mashaly 2021). However, various obstacles while implementing digital twin at large scale are limited like internet availability, inconsistent power supply, lack of sensor networks, and inadequate data centres. Additionally, in many regions lack of technical workforce or institutional capacity is found as a hurdle to maintain and update digital twin systems regularly (George et al. 2023). If there aren't focused investments in rural internet access, sensors, and farmer training, digital twins might only benefit big agribusinesses or well-connected areas leaving others behind and missing the bigger goal of making climate-smart farming accessible to everyone. To overcome these challenges, collaborative efforts among governments, tech providers, and agricultural institutions are essential. Development of scalable, low-cost, and modular digital twin systems that are made to local conditions can help bridge the infrastructure gap and ensure broader, equitable access to this transformative technology (Mazzetto 2024).

10.5.3 Cost and Accessibility Barriers for Smallholders

Significant applications of digital twin technology to boost climate-resilient agriculture are restricted owing to its higher cost and difficult accessibility. It poses significant limitations to reach this technology at larger platform owing to its poor affordability for smallholder farmers, specifically in developing countries (Parra-López et al. 2024). For the Implementation of a digital twin system, substantial upfront investment is required for sensors, drones, IoT devices, data storage, analytics platforms, and connectivity infrastructure. The smallholders work on limited budgets and also at low profit margins, and these technologies are often costly and beyond their financial reach (Awais et al. 2025).

In addition to hardware and software costs, ongoing maintenance, data subscriptions, cloud service fees, and the need for technical support add to the financial burden. For the integration of complex multiple data sources such as soil sensors, weather forecasts, and crop growth models, specialized skills are required that smallholders may not possess or afford to hire. Hence, there is a need to develop affordable, farmer-friendly solutions, digital twin technology that could help small farmers as well (Ametefe et al. 2024). Digital literacy programs, technical training, or agricultural extension services can support the effective use of digital twin tools for many smallholders. Further adoption is hindered due to various issues such as language barriers, limited access to smart devices, and poor internet connectivity in rural areas (Gumbi et al. 2023).

Hence, to ensure digital twins are accessible and equitable, it's required to develop cost-effective and scalable models. To achieve this goal, support can be provided by government by ensuring subsidies, cooperative services, public-private partnerships, etc. Digital twins can become more accessible by using community or cooperative farming setups and depending on open-source tools. This may ensure the benefits for small farmers also, not only big ones (Buhler et al. 2023). Without such inclusive strategies, the transformative potential of digital twin technology may not be fully explored by everyone at every level in climate-responsive agriculture. There are chances that it will remain concentrated in technologically privileged segments only.

10.5.4 Interoperability and Standardization Gaps

There is lack of universal digital standards in agriculture, an urge to utilize smart devices by bringing together different brands or software made by various developers. This combination may be expensive to have an advanced custom setup because special tools may need to be installed to get them to work properly in combination (Avaria 2020). Fragmentation slows down smart implementation, makes systems less efficient, and may generate complications when trying to scale up digital twins across different regions for various types of farming. In climate-resilient agriculture, predicting extreme weather conditions, monitoring crop health, and optimizing resource consumption are critical to exchange data from multiple sources on time, and these time gaps can significantly limit the technology's impact (Sarma et al. 2024). Furthermore, unpredictable data semantics, lack of common ontologies, and variations in data granularity can lead to mistakes in model interpretation and decision-making. Wing to lack of common standards, it's difficult to check, share, or reuse digital twin models among different regions or organizations, which makes it difficult to work together on climate adaptation (Mazzetto 2024).

It is highly in demand to make collaboration between agricultural technology developers, policymakers, and standards organizations to develop common frameworks that support interoperability while guaranteeing data quality, privacy, and performance (Bahlo et al. 2019). Bridging these gaps will be a smart move to unlock the full potential of digital twins to ensure climate-smart agricultural transformation.

10.6 Future Perspectives

10.6.1 Integration with AI, Machine Learning, and Blockchain

In future, digital twins in combination with advanced technologies like artificial intelligence (AI), machine learning (ML), blockchain, etc. are to ensure sustainable and climate-aware farming (Hacker 2024; Sajadieh and Noh 2025). These technologies will enable digital twins to be more realistic to support farming as it is based on the study of past and real-time data which offer precise and advanced predictions (Eze et al. 2025). This level of transparency is specifically significant in the context of sustainable farming, where supply chain traceability from soil to shelf can affect consumer decisions as well as regulatory guidelines (Cordeiro et al. 2024; Mukherjee et al. 2022).

Digital twins of individual field, empowered by AI and linked through blockchain networks, could collaborate in real time to optimize regional water consumption, and utilize this information to decide climate adaptation strategies, and respond collectively to weather irregularities (Tyagi 2025). These developments can help enhance resilience to agriculture in climate change. These technologies enable farms to adapt more effectively to unpredictable weather patterns, shifting growing seasons, and evolving pest dynamics (Heeb et al. 2019).

Moreover, this technological evolution invites a broader reflection on ethical and inclusive design. It's important that these technologies also reach smallholder farmers and communities with fewer resources, so that digital agriculture becomes a tool for inclusion and fairness and not something that leaves people behind. Most likely, the combination of digital twins, artificial intelligence, machine learning, and blockchain could help to build a more intelligent, equitable, and sustainable future for agriculture (Jain et al. 2025).

10.6.2 Role in Policy Support and Global Food Security

Global food systems are facing increasing worries and pressure due to climate change, resource depletion, and population growth (Ametefe et al. 2024). In this situation, everyone expects digital twins to play a pivotal role in shaping agricultural policies and securing the future of food (Ametefe et al. 2024). Digital twins can model complex environmental and farming systems in real time so they give the policymakers a powerful way to test out different agricultural strategies and see the expected outcomes before taking any actual decisions (Purcell and Neubauer 2023). The decision made by digital twins balances the productivity environment stewardship because they can provide data-driven evidence by modelling many scenarios such as drought responses, fertilizer optimization, and land-use changes (Tsakiridis et al. 2023; Singh et al. 2023; Fankhauser 2024; Otsu and Maso 2024). In future, digital twins will likely become an important part of global food systems. Digital twins will help guide everything from tracking agriculture's carbon footprint to making sure water resources are shared fairly (Martindale and Lucas 2021). In the long run, digital twins are likely to play a vital role in securing our food supply and supporting sustainable policies. By making sense of real-world complexity through accurate digital models, they help turn big challenges into clearer, more manageable solutions both at home and around the world (Tzachor et al. 2022).

10.6.3 Opportunities for Collaborative Platforms and Open Innovation

The future of digital twins in agriculture is not limited to stand-alone tools or closed-off technologies. Instead, it is moving towards more open, collaborative spaces where ideas, data, and solutions can be shared and built together (Dal Mas et al. 2024). Today due to digital infrastructure, the farming has become more interconnected, therefore there is an urgent need for shared environments where researchers, technology providers, farmers, policymakers, and other stakeholders work on the challenge and co-create their solutions. Digital twins need a mix of data, ideas from different fields, and ongoing feedback to keep improving and stay useful (Mihai et al. 2022). This space can only be created by collaborative platforms which bring all these elements together to encourage inclusive, flexible innovation.

With the help of open-source digital twin frameworks, many institutions from different parts of the world share their unique farming practices and climate challenges by sharing local data, models, and algorithms (Peladarinos et al. 2023). Agricultural knowledge can be more accessible as the creation of solutions that truly fit local needs need to be speed up, especially in areas that are often overlooked or have less resources (Kuruppuarachchi and McGibney 2023). Moreover, cloud-based solutions make it easier for innovators to plug in new tools such as pest prediction modules, soil health indices, or sustainability scoring algorithms due to their support, compatibility, and data standards. One can use cloud-based technology instead of rebuilding entire systems (Johnraja et al. 2024).

10.7 Conclusion

Digital twins make it possible to monitor conditions in real time with the help of data-driven models of farming systems. With the emergence of early detection, one can predict the future challenges and can make plan for different scenarios. This would help stakeholders to use resources wisely also environmental harm would be reduced. Artificial intelligence, machine learning, and blockchain work together to make these systems much smarter and responsive. Due to the integration of AI, machine learning, and Blockchain, the model can learn and adjust on their own instead of just following fixed rules. Digital twin helps farmers to manage the pest, techniques of irrigation, different ways of planting, and give prior information about the weather conditions. This would increase the yield of the crops by minimized the risk. As the digital twins connect what's happening on the ground with big-picture planning, they're just as useful for small farmers working a few hectares as they are for policymakers designing national food strategies. The promise of digital twins in agriculture will only be realized through coordinated, cross-sector collaboration. For researchers, this is a call to move beyond theoretical models and engage with real-world agricultural contexts. This approach of researchers would ensure that digital twins reflect the ecological, cultural, and economic complexities of the systems.

References

Agele S (2021) Global warming and drought, agriculture, water resources, and food security: impacts and responses from the tropics. In: Handbook of climate change management: research, leadership, transformation. Springer International Publishing, Cham, pp 1–20

Agelli M, Corona N, Maggio F, Moi PV (2024) Unmanned ground vehicles for continuous crop monitoring in agriculture: assessing the readiness of current ICT technology. Machines 12(11): 750

Ahmad S (2025) AI-driven digital twin models for hybrid crop selection and climate-resilient yield prediction. Sci Res Timeline J 3(1):15–18

Al-Ali AR, Gupta R, Zaman Batool T, Landolsi T, Aloul F, Al Nabulsi A (2020) Digital twin conceptual model within the context of internet of things. Future Internet 12(10):163

Ametefe DS, Hussin N, John D, Dzorgbenya Ametefe G, Adozuka Aliu A, Abdi Ali Z (2024) Revolutionising agriculture for food security and environmental sustainability: a perspective on the role of digital twin technology. CABI Rev 19(1)

Arulmozhi E, Deb NC, Tamrakar N, Kang DY, Kang MY, Kok J, Basak JK, Kim HT (2024) From reality to virtuality: revolutionizing livestock farming through digital twins. Agriculture 14(12): 2231

Awais M, Wang X, Hussain S, Aziz F, Mahmood MQ (2025) Advancing precision agriculture through digital twins and smart farming technologies: a review. Agri Eng 7(5):137

Avaria RC (2020) Searching for sustainability in the digital agriculture debate: an alternative approach for a systemic transition. Teknokultura: Revista de Cultura Digital y Movimientos Sociales 17(2):225–238

Bahlo, C., Dahlhaus, P., Thompson, H. And Trotter, M., 2019. The role of interoperable data standards in precision livestock farming in extensive livestock systems: a review. Comput Electron Agric, 156, pp.459–466

Batra I, Sharma C, Malik A, Sharma S, Kaswan MS, Garza-Reyes JA (2024) Industrial revolution and smart farming: a critical analysis of research components in industry 4.0. TQM J 37:1497

Bühler MM, Calzada I, Cane I, Jelinek T, Kapoor A, Mannan M, Mehta S, Mookerje V, Nübel K, Pentland A, Scholz T (2023) Unlocking the power of digital commons: data cooperatives as a pathway for data sovereign, innovative and equitable digital communities. Digital 3(3):146–171
Cesco S, Sambo P, Borin M, Basso B, Orzes G, Mazzetto F (2023) Smart agriculture and digital twins: applications and challenges in a vision of sustainability. Eur J Agron 146:126809
Cordeiro M, Fernandes V, Curado J, Ferreira JC (2024) Enhancing agriculture products traceability towards sustainability. J Network Innov Comput 12:15–15
Dal Mas F, Valentina N, Massaro M, Bagnoli C (2024) Combining deep and digital technologies as a path towards twin transition: the "future farming" case study. IEEE Trans Eng Manag
Dubey AK, Kumar A, Narang SK, Khan MA, Srivastav AL (eds) (2022) Visualization techniques for climate change with machine learning and artificial intelligence. Elsevier
Eze VHU, Eze EC, Alaneme GU, Bubu PE, Nnadi EOE, Okon MB (2025) Integrating IoT sensors and machine learning for sustainable precision agroecology: enhancing crop resilience and resource efficiency through data-driven strategies, challenges, and future prospects. Discov Agric 3(1):1–34
Fankhauser K (2024) Fusion of in-situ and remote sensing data with machine learning toward global water and food security monitoring. Doctoral dissertation, University of Colorado at Boulder
George AS, George AH, Baskar T (2023) Edge computing and the future of cloud computing: a survey of industry perspectives and predictions. Partn Univers Int Res J 2(2):19–44
Getahun S, Kefale H, Gelaye Y (2024) Application of precision agriculture technologies for sustainable crop production and environmental sustainability: a systematic review. Sci World J 2024(1):2126734
Gomiero T (2016) Soil degradation, land scarcity and food security: reviewing a complex challenge. Sustainability 8(3):281
Gonzalez CG (2011) Climate change, food security, and agrobiodiversity: toward a just, resilient, and sustainable food system. Fordham Environ Law Rev:493–522
Grieves MW (2023) Digital twins: past, present, and future. In: The digital twin. Springer International Publishing, Cham, pp 97–121
Gumbi N, Gumbi L, Twinomurinzi H (2023) Towards sustainable digital agriculture for smallholder farmers: a systematic literature review. Sustainability 15(16):12530
Hacker P (2024) Sustainable AI regulation. Common Market Law Rev 61(2)
Heeb L, Jenner E, Cock MJ (2019) Climate-smart pest management: building resilience of farms and landscapes to changing pest threats. J Pest Sci 92(3):951–969
Jain G, Jain A, Jaiswal A, Ahmad S, Kapse VM, Bisht SS (2025) Transformative potential: IoT, cloud, big data, and AI-driven Blockchain digital twins in agriculture. In: Blockchain and digital twin applications in smart agriculture. Auerbach Publications, pp 92–111
Javaid M, Haleem A, Khan IH, Suman R (2023) Understanding the potential applications of artificial intelligence in agriculture sector. Adv Agrochem 2(1):15–30
Johnraja JI, Leelipushpam PGJ, Shirley CP, Princess PJB (2024) Impact of cloud computing on the future of smart farming. In: Intelligent robots and drones for precision agriculture. Springer, Cham, pp 391–420
Kalyani, Y., Vorster, L., Whetton, R. And Collier, R., 2024. Application scenarios of digital twins for smart crop farming through cloud–fog–edge infrastructure. Future Internet, 16(3), 100
Khan S, Sachan HK, Krishna D (2025) The role of smart farming technologies in mitigating climate change and enhancing agricultural sustainability. Int J Environ Climate Change 15(2):138–159
Kumar A, Srivastav AL, Dubey AK, Dutt V, Vyas N (eds) (2023) Innovations in machine learning and iot for water management. IGI Global, pp 51–70
Kumari S, Pandey GK, Tiwari S (2024) Data visualization techniques in smart agriculture implementation. In: AI applications for business, medical, and agricultural sustainability. IGI Global, pp 391–420
Kuruppuarachchi PM, Rea S, McGibney A (2023) Trusted and secure composite digital twin architecture for collaborative ecosystems. IET Coll Intell Manuf 5(1):12070

Martindale, W. And Lucas, K., 2021. Global resource flows in the food system. In Environment and climate-smart food production (pp. 219–257). Cham: Springer International Publishing
Mashaly M (2021) Connecting the twins: a review on digital twin technology & its networking requirements. Proc Comput Sci 184:299–305
Mazzetto S (2024) A review of urban digital twins integration, challenges, and future directions in Smart City development. Sustainability 16(19):8337
Melesse TY (2025) Digital twin-based applications in crop monitoring. Heliyon 11:e42137
Mihai S, Yaqoob M, Hung DV, Davis W, Towakel P, Raza M, Karamanoglu M, Barn B, Shetve D, Prasad RV, Venkataraman H (2022) Digital twins: a survey on enabling technologies, challenges, trends and future prospects. IEEE Commun Surv Tutor 24(4):2255–2291
Mukherjee AA, Singh RK, Mishra R, Bag S (2022) Application of blockchain technology for sustainability development in agricultural supply chain: justification framework. Oper Manag Res 15(1):46–61
Nasirahmadi A, Hensel O (2022) Toward the next generation of digitalization in agriculture based on digital twin paradigm. Sensors 22(2):498
Omrany H, Mehdipour A and Oteng D (2024) Digital twin technology and social sustainability: implications for the construction industry. Sustain 16(19):8663
Omrany H, Al-Obaidi KM (2024) Application of digital twin technology for urban Heat Island mitigation: review and conceptual framework. Smart Sustain Built Environ
Otsu K, Maso J (2024) Digital twins for research and innovation in support of the European green Deal data space: a systematic review. Remote Sens 16(19):3672
Parra-López C, Abdallah SB, Garcia-Garcia G, Hassoun A, Sánchez-Zamora P, Trollman H, Jagtap S, Carmona-Torres C (2024) Integrating digital technologies in agriculture for climate change adaptation and mitigation: State of the art and future perspectives. Comput Electron Agric 226:109412
Peladarinos N, Piromalis D, Cheimaras V, Tserepas E, Munteanu RA, Papageorgas P (2023) Enhancing smart agriculture by implementing digital twins: a comprehensive review. Sensors 23(16):7128
Pires F, Leitão P, Moreira AP, Ahmad B (2023) Reinforcement learning based trustworthy recommendation model for digital twin-driven decision-support in manufacturing systems. Comput Ind 148:103884
Purcell W, Neubauer T (2023) Digital twins in agriculture: a state-of-the-art review. Smart Agric Technol 3:100094
Pylianidis C, Snow V, Overweg H, Osinga S, Kean J, Athanasiadis IN (2022) Simulation-assisted machine learning for operational digital twins. Environ Model Softw 148:105274
Rathore N, Singh K, Kaur G, Sharma C, Sinha A (2025) Using recurrent neural networks to forecast climate change: a time series analysis of global temperature variability. In: 2025 First International Conference on Advances in Computer Science, Electrical, Electronics, and Communication Technologies (CE2CT). IEEE, pp 425–430
Riaz K, McAfee M, Gharbia SS (2023) Management of climate resilience: exploring the potential of digital twin technology, 3D city modelling, and early warning systems. Sensors 23(5):2659
Sajadieh SMM, Noh SD (2025) From simulation to autonomy: reviews of the integration of artificial intelligence and digital twins. Int J Precis Eng Manuf-Green Technol:1–32
Sarma HH, Paul A, Kakoti M, Talukdar N, Hazarika P (2024) Climate resilient agricultural strategies for enhanced sustainability and food security: a review. Plant Archives 24(1):787–792
Schaller N (1993) The concept of agricultural sustainability. Agric Ecosyst Environ 46(1–4):89–97
Singh G, Rajesh R, Daultani Y, Misra SC (2023) Resilience and sustainability enhancements in food supply chains using digital twin technology: a grey causal modelling (GCM) approach. Comput Ind Eng 179:109172
Srivastav AL, Dutt I, Sharma SR, Sharma A, Kanwar VS (2022, October) Global climate change: a culprit for the deterioration of nature and human life. In: AIP Conference Proceedings, vol 2451(1). AIP Publishing LLC, p 020001

Tsakiridis NL, Samarinas N, Kalopesa E, Zalidis GC (2023) Cognitive soil digital twin for monitoring the soil ecosystem: a conceptual framework. Soil Syst 7(4):88

Tyagi AK (2025) Blockchain and digital twin applications in smart agriculture. CRC Press, pp 29–47

Tzachor A, Sabri S, Richards CE, Rajabifard A, Acuto M (2022) Potential and limitations of digital twins to achieve the sustainable development goals. Nat Sustainability 5(10):822–829

Uysal İ (2025) Computational modeling for agriculture and climate change relation. In: Computational modeling applications for climate crisis. Morgan Kaufmann, pp 107–124

Wang Y, Su Z, Guo S, Dai M, Luan TH, Liu Y (2023) A survey on digital twins: architecture, enabling technologies, security and privacy, and future prospects. IEEE Internet Things J 10(17):14965–14987

Wu X, Lu G, Wu Z (2023) Remote sensing technology in the construction of digital twin basins: applications and prospects. Water 15(11):2040

Yousra HARABI (2023) Smart irrigation system for water distribution in arid areas using digital Twins. 1–10

Zhang R, Zhu H, Chang Q, Mao Q (2025) A comprehensive review of digital twins technology in agriculture. Agriculture (Basel) 15(9):903

Zhao T, Song C, Yu J, Xing L, Xu F, Li W, Wang Z (2025) Leveraging immersive digital twins and AI-driven decision support systems for sustainable water reserves management: a conceptual framework. Sustainability 17(8):3754

Evolution of Agriculture Technology: Assimilation of Biotechnology and Food Technology for Smart and Sustainable Agriculture

11

Manish Pant and Akanksha Kumari

Abstract

From ancient times, the use of hand tools like sickles, hoes, and wooden ploughs, to machineries involving tractors, seed drills, and threshers, as well as the shift from shifting cultivation to vertical and hydroponic farming, the agriculture sector has transformed into a modern, efficient, and sustainable industry. According to reports of the UN, the global population is currently increasing at a rate of approximately 0.85% per year, which translates to around 70 million people added annually. There is a constraint on agricultural production to keep pace with expanding global populations while avoiding the overexploitation of essential natural resources. The integration of biotechnology in the late twentieth century enabled genetically modified crops by enhancing traits. Concurrently, food technology assures processing innovation by modifying post-harvest preservation, enhancing nutritional quality, and reducing food loss. This book chapter discusses the integration of all relevant activities, such as sowing, harvesting, packaging, shipping, and new technologies available to farmers throughout the cropping cycle. This chapter also goes over the usage of other tools, such as unmanned aerial vehicles (UAVs), for agricultural monitoring and other helpful acts, such as crop yield optimization. Furthermore, advanced IoT programs are discussed. Furthermore, our research identified new prospects for IoT devices, which are crucial tools for sustainable agriculture.

M. Pant (✉)
Department of Chemical Engineering, Indian Institute of Science Education and Research, Bhopal, Madhya Pradesh, India

A. Kumari
Department of Food Technology and Nutrition (School of Agriculture), Lovely Professional University, Phagwara, Punjab, India

© The Author(s), under exclusive license to Springer Nature Singapore Pte Ltd. 2026

A. L. Srivastav et al. (eds.), *Digital Twin Technology for Sustainable Agriculture*,
https://doi.org/10.1007/978-981-95-5915-2_11

Keywords

Agriculture · Sustainable agriculture · Biotechnology · Food technology · Sustainability

11.1 Introduction

The increasing demands of the global population for greater amounts of food, water, and energy, the lack of arable land for growing food production, and the mounting strain on natural resources are some of the causes that have made the globe more complicated in recent years. Climate change, which is known to cause numerous environmental shifts, further exacerbates these factors. An agroecosystem consists of a set of interacting components that function together to support crop production and livestock rearing, with the aim of generating food, fiber, and energy from Earth's natural resources. Additionally, such systems might also have detrimental environmental consequences. Because field and farm experiments demand significant resources and may still not offer enough data in space and time to determine appropriate and effective management strategies, agricultural system models are playing an increasingly important role in the creation of sustainable land management across a variety of agro-ecological and socioeconomic contexts (Abiri et al. 2023). With projections showing that the world's population will reach approximately 9 billion by 2050, there will undoubtedly be an increase in the need for food. Agricultural land usage methods range from swidden farming to multi-cropping. This aligns perfectly with the concept of a land system, which considers the multifaceted aspects of how people utilize the land (Peter and Peng 2014). The total land area of the Earth, excluding Antarctica and inland waterways, is approximately 13 billion hectares (ha). A third or so of the world's total land area is made up of agricultural land, forest land, and other land (such as desert and barren areas, urban land, and infrastructure). 4.8 billion hectares of land were used for agriculture in 2021, compared to 4.1 billion hectares for forests and other land uses. The extent of permanent meadows and pastures was 3.2 billion hectares in 2021, whereas crops made up 1.6 billion hectares (12% of the world's total land area) (FAO 2023).

11.1.1 About Agricultural Evolution

The change from a nomadic hunter-gatherer existence to a relatively settled way of life in human behaviour greatly affected the agricultural landscape, hence influencing the qualities and features of flora and fauna favoured both consciously and subconsciously by people. Early farming sees Southwest Asia as both the most intensely researched site and one of the first regions in the world to practice agriculture. The field's environmental condition changed as agriculture progressed and paddocks grew more and more distinct from that among the natural surroundings in which animals and plants reside, evolved first. For plants, nutrient availability

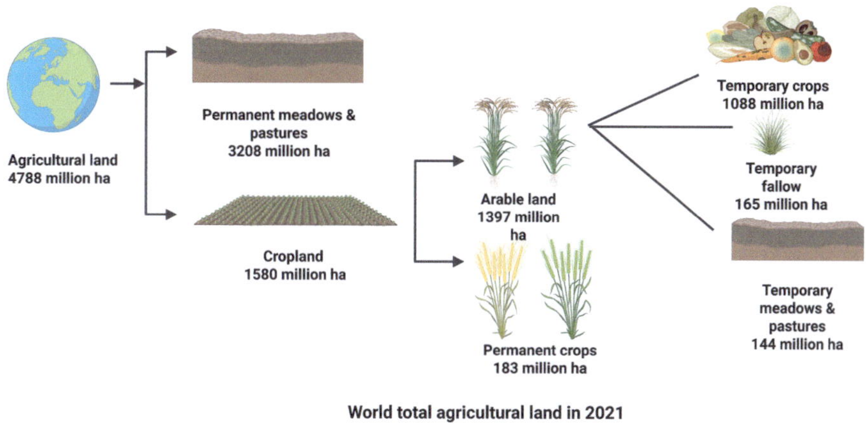

Fig. 11.1 Illustrates the world total land distribution in 2021

usually increased, genetic homogeneity and plant density rose, and the equilibrium in rivalry swung from intra-specific to inter-specific considerations (Flannery 1973; Thrall et al. 2010).

The farming revolution started in a much earlier period. The agricultural revolution is the collection of innovations that are more frequently associated with the years 1750 to 1850, while the other is the result of chemical and mechanical methods that have emerged during the current century. We shall use the term agricultural revolution to refer to this cluster of inventions rather than the promotion of temporary cultivation in place of permanent cultivation. Even though the effects of the agricultural revolution went far beyond considerations of space and wide prevent, we should hereinunder try more than just to answer the questions of what happened, when it occurred, and how, leaving the whys and wherefores for separate treatment. The development of computer and information technology has altered the way that people in the agriculture sector do business. The next step in the digitalization of the agriculture industry was the introduction of the Internet, which allowed digital data to be transmitted across personal and network-based computers at significantly quicker speeds than prior information dissemination systems (Bertoglio et al. 2021; Du et al. 2023). Global landscape of agricultural land in 2021 showed comparative distribution of cropland, permanent meadows, and pastures as shown in Fig. 11.1.

11.1.2 Relationship Between Biotechnology and Food Technology

Strong tools for enhancing plant alteration in ways that might benefit society are provided by genetic engineering. Biotechnology in agriculture refers to the use of live plant organisms or parts of them to produce food and feed products like insect-resistant maize, develop processes like tobacco's production of biologics, and provide services like the use of genetically modified poplar for the bioremediation of heavy metal contamination. The majority of the early biotechnology applications in the food sector focused on genetically modifying microbes to produce more enzymes for use in food production. At the moment, the food industry still uses genetically modified organisms (GM) in these areas the most. The creation of a diverse variety of plants designed for agronomic characteristics has been another significant field of use (Barling et al. 1999; Falk et al. 2002).

This chapter attempts to analyse the evolution of agricultural technology with a focus on the integration of food technology and biotechnology to attain intelligent and sustainable agriculture. This chapter aims to provide a comprehensive overview of the evolution of traditional agricultural practices into modern, technology-based strategies. It looks at the potential applications of biotechnology to boost food yields, create species that are adaptable to climate change, and enhance resistance to pests and diseases. To ensure food safety and nutritional value, it also highlights how important food technology is to enhancing post-harvest processing, storage, and value addition. By analysing their synergy, the chapter seeks to demonstrate how combining these two fields improves environmental sustainability, resource efficiency, and precision agriculture. It also looks at current challenges and real-world examples, and it makes recommendations for possible paths to building a strong and sustainable agri-food system in the face of rising global demand and climate uncertainty.

11.2 | History of Agriculture Evolution

India is regarded as an agricultural nation as 80% of its rural population makes their living from agriculture. In many Indian states, agriculture depends entirely on rainfall due to the lack of freshwater resources. A paradigm shift in water usage productivity is required since fresh water supplies are limited. In traditional agricultural irrigation methods, farmers either manually supervise the irrigation or use planned irrigation. Additionally, excessive water use can affect the ecosystem and cause water shortages, which is common in areas where irrigated agriculture is practiced. The current scenario calls for creative technologies to increase agricultural productivity while reducing irrigation water usage. This demonstrates how precision agriculture is replacing traditional agriculture (Gill and Chawla 2021).

11.2.1 Traditional Farming Methods

India's traditional farming methods, which have been refined over many centuries, provide a wealth of environmental expertise and sustainable land management. One essential element of these techniques is cropping rotation. The methods used to prevent erosion problems with pests and pathogens and maintain soil fertility. Crop rotation has historically been used by Indian farmers by switching between cereal crops and legumes, which fixes the soil's nitrogen needs and reduces the need for artificial fertilizers. Terracing is another age-old technique that is very popular in India's mountainous regions. These actions highlight the durability of old practices as well as their significance in light of current environmental issues and the need for a sustainable food production approach (Sekhar et al. 2024) (Table 11.1).

11.2.2 Green Revolution and Its Impacts

The primary promoters of the Green Revolution concept in India were C. Subramaniam, who was named agricultural minister in 1964, and M.S. Swaminathan, who was trained by Norman Borlaug and became the director of the Indian Agricultural Research Institute (IARI) in 1965. The increase in output from 35 million tonnes in 1960–61 to 54 million tonnes in 1980–81 and then to 112.9 million tonnes in 2017–18 indicates a substantial improvement in rice production. Furthermore, from 1960 to 2017–2018, the production per acre rose from 1013 kg to 2578 kg. Wheat production has once again increased significantly, rising from 11 million tonnes in 1950–1951 to 36 million tonnes in 1980–1981 and then to 99.7 million tonnes in 2017–2018. In addition, the production per acre jumped from 850 kg to 3371 kg within this time frame. Using a new agricultural technology to produce wheat and rice led to all of these improvements (Somvanshi et al. 2020).

11.2.3 Mechanization and Technological Advancement

Undoubtedly, advancements in technology have an effect on the expansion of the agriculture sector. Therefore, it is essential to look into and understand how different participants interpret technological advancement in order to have a more meaningful connection with new technologies. Modern civilization has been significantly affected by technological advancements, which many have hailed as the fourth industrial revolution. It is generally accepted that integrating new digital technology to automate procedures, optimize operations, and boost productivity is advantageous. Particularly notable is the debate of concepts like "digitalization" and "digitization." The process of transforming analog data into digital data is known as the former digitalization. Digitalization is the process of changing current business models and processes to make a company more competitive by utilizing and implementing digital technology (such drones and artificial intelligence) (Wakweya 2023). The standard for agricultural management is raised by increased agricultural mechanization. First, it elevates the subject matter.

Utilizing state-of-the-art agricultural machinery and technology, automating and intellectualizing the agricultural production process, and optimizing resource allocation efficiency in a more efficient manner are the methods used to reach the level of agricultural production automation. Second, it enhances the way the agricultural production process is managed. Precise monitoring of soil moisture, nutrient levels, and plant growth is made possible by the data collection capabilities and high operational efficiency of modern mechanized equipment. In order to help

Table 11.1 The effects of traditional farming methods on crop productivity and sustainability (Sekhar et al. 2024)

Traditional agricultural practices	Country	Cultivated crop	Description
Zero tillage	Indo-Gangetic plains of South Asia	Rice-wheat cropping	Compared to traditional methods under a rice–wheat cropping system, no-tillage techniques can enhance wheat yield up to 200–500 kg ha^{-1}
Conservation agriculture production systems (CAPSs)	Odisha (India)	Maize and cowpea	Reduced tillage, cover crops, and intercropping (maize and cowpea) for three years did not significantly boost maize output, but it did significantly increase cowpea production. The next horse gram and mustard crops grew by 37% and 25%, respectively
Conservation agriculture (CA)	Keonjhar district of Odisha (India)	Maize, cowpea and mustard	Compared to 227 US$ ha^{-1} using modern methods, a profit of 754 US$ ha^{-1} was achieved by reduced tillage and intercropping
Mixed cropping (crop diversification)	China	Rice	Four different types of rice were grown on a 3000-hectare plot of land, increasing yield by over 89% and reducing pest blast attacks by 44% without the use of pesticides
Small-scale integrated coconut cultivation with ruminants	Santa Cruz, Laguna, Philippines	Coconut	When goat and sheep grazing was combined with coconut farming (1991–1994), net earnings rose from $60 to $356 USD
Rubber plantations and food crops with integrated livestock	Butamarta, South Sumatra, Indonesia	Food crops and rubber plantation	Profit increased from 68 to 161 US dollars when food crops and rubber were grown alongside integrated livestock (one cow, three goats, and eleven poultry)
Agroforestry	Haryana (India)	Hordeum vulgare (barley)	In comparison to traditional cultivation, the plantation of plants such as *Prosopis cineraria, Azadirachta indica, Acacia albida*, and *Tecomella undulata* increased barley output by 16.8–86%
Agroforestry-based cultivation	Sahel, Sahara Desert, Africa	Maize and applering acacia	When compared to monocropping, intercropping with Faidherbia albida (*Applering acacia*) increased maize productivity from 1 to 3 tons per hectare
Agroforestry-based agriculture	Rajasthan (India)	Wheat, barley, gram	Mulching and leguminous plant agroforestry enhanced C, N, and P levels and better microbial density in comparison to non-plant cultivation
Optimized farming practices	Southern Italy	Durum wheat	Compared to current agriculture, crop rotation increased per kg wheat yield while minimizing the use of nitrogen fertilizer and lowering GHG emissions

agricultural production organizations provide more scientific crop planting and nurturing programs, increased crop resilience to weather-related hazards lowers the incidence of agricultural disasters. In an effort to lessen the negative effects that smallholders have on the environment and public health, developing nations have begun subsidizing alternate methods, like mechanical soil tillage, to minimize agricultural fires (Lu et al. 2024).

11.3 |Role of Biotechnology in Modern Agriculture

As the globe deals with more critical issues including resource scarcity, population growth, and climate change, there is a greater need than ever for innovative and sustainable agriculture solutions. In addition to being the mainstay of global food security, agriculture is also crucial for environmental preservation. Potential solutions to these problems can be found in the combination of biotechnology and advanced agricultural techniques. Precision farming, gene editing, and data-driven technologies offer chances to reduce agricultural impacts on the environment, optimize resource use, and increase crop yields (Pasupuleti 2024). Diverse applications of biotechnology in agriculture, including biofertilizers, biopesticides, selective breeding, CRISPR-mediated biofortification, GM crop development, and culture laboratory techniques, are shown in Fig. 11.2.

11.3.1 GM Crops and Genetic Engineering

The goal of genetic modification (GM), a branch of biotechnology, is to alter living things' genetic makeup so they can carry out certain tasks. Our ancestors employed "selective breeding" and "artificial selection," two Darwinian ideas that essentially refer to choosing parent organisms with desired qualities (such as tougher stems) and breeding them to spread those traits, more than 10,000 years ago. This was the initial idea of altering plants for domestication and human use. The use of genomics to agriculture represents a paradigm shift in our understanding of food production by offering a means of significantly boosting agricultural sustainability and productivity. Genomic analysis is the thorough examination of an organism's whole genetic makeup, including all of its genes. It offers a strong basis for finding and growing crop varieties with desired characteristics, like higher yield, disease resistance, and drought tolerance (Pasupuleti 2024). The most notable change in plant genetics using these methods was the artificial selection of corn, which transformed it from a weedy grass with few kernels and small ears (teosinte; earliest recorded growth: central Balsas river valley, southern Mexico 6300 years ago) to the modern varieties of maize and edible corn plants (Raman 2017).

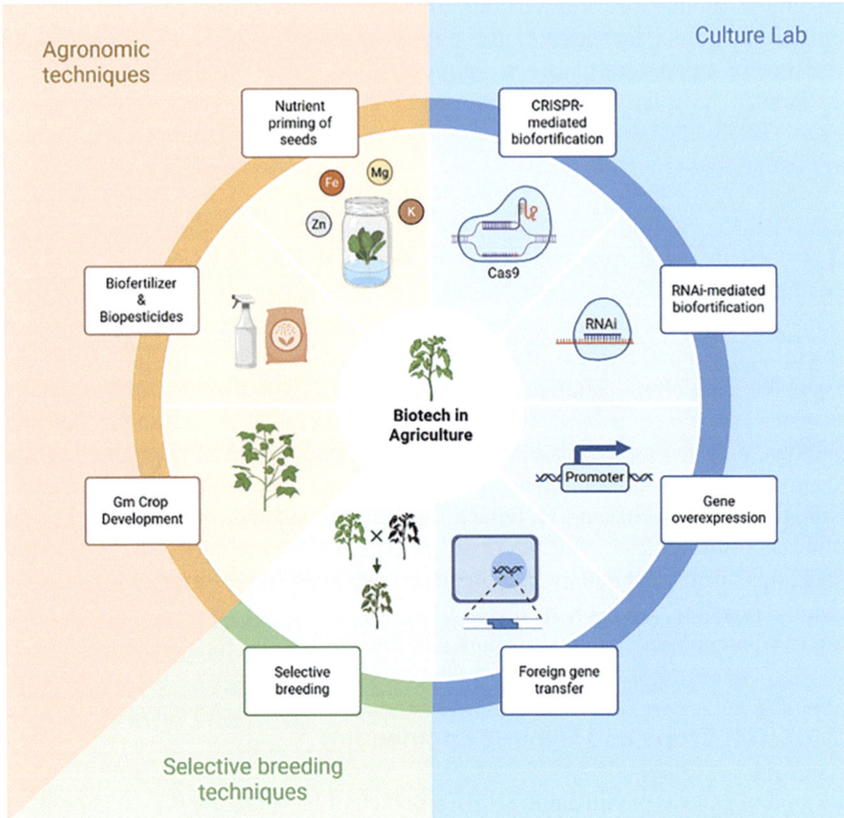

Fig. 11.2 Illustrates the major domains of biotech in agriculture

11.3.2 CRISPR Technology and Molecular Breeding

The development of genome editing using the clustered regularly interspaced short palindromic repeat (CRISPR) technology, combined with advances in computing and imaging technology, has ushered in a new era in which we can not only identify human diseases and even predict individual susceptibility based on personal genetics, but also respond to that knowledge. Our capacity to quickly identify and modify the genes governing plant characteristics has likewise transformed the pace of agricultural research and plant breeding. There are currently numerous and significant uses for this convergence of technologies. The CRISPR-Cas9 system has had a significant impact on biological research since it was first described as a genome editing tool ten years ago. It has affected not only people with genetic illnesses but also agricultural products and activities (Es et al. 2019). The 1980s witnessed the beginning of molecular marker development and use. A decade later, the creation of PCR-based DNA markers followed this significant advancement in plant genomics research. Since then, several uses of molecular markers in different facets of plant genomics and molecular breeding have been documented (Amiteye 2021).

11.3.3 Biopesticides and Biofertilizers

The biological health of the soil, environmental conditions, nutrient availability, and soil quality are other important aspects that contribute to boosting crop production per unit area in order to achieve the goal of food security. Chemical fertilizers (composed of N, P, or K) are overused in modern high-input farming techniques and technologies to meet plant nutritional requirements and increase agricultural output worldwide. However, due to ineffective fertilizer use, only 30–40% of these nutrients are absorbed by the plants; the rest is lost to the soil, contaminating the ecosystem. Chemical fertilizers also contain radionuclides and heavy metals, which are hard to manage and break down, making them long-term environmental contaminants. The eutrophication of water sources is another significant problem associated with the overuse of chemical fertilizers (Singh et al. 2025; Pant and Singh 2024). The creation of sustainable, environmentally friendly solutions that may lessen the use of synthetic fertilizers was spurred by these pollution issues, which posed a threat to public health. As a result, using beneficial microbiomes as biofertilizers in sustainable farming practices has become a cutting-edge and environmentally responsible way to increase soil fertility and plant development (Kumar et al. 2022) (Table 11.2).

Table 11.2 Benefits and limitations of biofertilizers and biopesticides (Ferreyra-Suarez et al. 2024)

Bioproduct	Benefit	Limitations
Biofertilizers	Promote plant growth and metabolites production Improve nutrient proximity and regulate and mobilize nutrient concentration in the soil. Improve soil fertility Promotes sustainable farming and circular economy Decrease in overall contamination	Ignorance of the product's quality on the part of the maker. No instantaneous visual effects conditions for special storage. There are crop-specific biofertilizers. A sufficient carrier resource constraint is not available. Since the nutritional levels are low, large amounts are needed for land application
Biopesticides	Promotes plant disease resistance Good and long shelf-life Ecofriendly Most biopesticides have a broad application range Increasing crop yields Considered as an effective biological control	Lack of quality assurance and hence limited production Mishandling techniques Some biopesticides Harmful consequences on biome interactions Lack of funding for research Field experiments are limited

11.3.4 Micropropagation and Plant Tissue Culture

Micropropagation is the term used to describe the aseptic cultivation of cells, tissues, organs, and their constituent parts in vitro under certain chemical and physical conditions. The basic concept of the plant body may be divided into smaller parts called "explants," each of which can develop into a whole plant. In order to produce multiple clone plantlets, micropropagation of plants is a comprehensive procedure that entails separating cells, tissues, or organs from selected plants and then incubating them in an aseptic, growth-promoting environment following sterilization. Cloning isolated cells that, under specific circumstances, can develop into a full plant demonstrated that somatic cells may indeed be cloned (Gupta et al. 2020). Under aseptic circumstances, plant tissue culture entails growing an entire plant from an explant or even a single plant cell. The phenomenon of cell totipotency—the ability of a single cell to differentiate into each of the unique cells that comprise an organ and to regenerate into a whole plant—is the foundation of this area of plant biotechnology. This essential property of plant cells is used in micropropagation to accomplish the following objectives: rapid mass replication of better genotypes on a wide scale in a comparatively short period of time. Because micropropagation produces healthy seedlings all year round and shortens the plant cycle, it now has a big influence on horticulture, business, and agriculture. Through genetic modification and somatic clonal variation, it is also a vital method for preserving plant genetic resources, improving harvests, and creating new cultivars (Soumare et al. 2021).

11.4 Assimilation of Food Technology in Agriculture

Addressing the widening gap between food availability and demand in a world population that is rising is crucial for human growth. Either reducing waste or increasing agricultural productivity might increase the demand for food. The problem of feeding the world's growing population is being addressed by a number of initiatives. Among these efforts, reducing postharvest losses is a particularly notable focus. This specific action has a direct and beneficial relationship to safeguarding our food supply and preserving important resources (Nath et al. 2024).

11.4.1 Post-harvesting Processing and Value Addition

Agriculture's main fields can be broadly divided into several categories, including horticulture, agricultural economics, agronomy, forestry, agricultural engineering, animal management, and fisheries. During the three critical stages of pre-harvest, harvest, and post-harvest, agriculture faces many difficulties. Unpredictable weather, inadequate irrigation, weed identification, and disease control are some of the problems that impact crop output at the pre-harvest stage. Crop size, maturity, and firmness must be determined at harvest time. To reduce losses and increase market value, post-harvest issues like grading, quality control, temperature control, and shelf-life forecasts must be addressed. These problems are difficult for traditional farming techniques to solve since they depend too much on manual labour and experiential judgment, which can lead to inefficiencies. The focus of this study is on applying AI and CV technologies to address these issues at every level of agriculture. Agricultural products must be handled, stored, and transported after harvest as part of the post-harvesting process. For fruits, vegetables, and other crops to maintain their nutritional content, shelf life, freshness, quality, grade, and temperature, several procedures are necessary. Post-harvest procedures that are effective can reduce rotting, increase marketability, and reduce losses. Some of the most crucial methods include using technology for grading, sorting, and quality control, as well as appropriate drying, freezing, and packing. ML and CV developments are also being applied to improve post-harvest procedures and guarantee that food is delivered to consumers in the best possible condition (Upadhyay and Bhargava 2025).

11.4.2 Food Preservation and Packaging Innovations

To increase food safety and quality while reducing environmental impacts, creative and sustainable food preservation techniques are crucial. A key element of food security and safety is food preservation, which attempts to increase the shelf life of goods while preserving their nutritional content, safety, and sensory appeal. To keep food available, traditional preservation methods including freezing, canning, refrigeration, and chemical preservatives have proved crucial. However, these processes can pose health and environmental risks, including the potential for toxic chemical residues, excessive energy consumption, and greenhouse gas emissions. The food industry faces the challenge of developing novel, environmentally friendly preservation methods that complement international initiatives to lessen the impact on the environment and safeguard public health in a time when sustainability is becoming more and more important. Sustainable food preservation refers to the use of techniques and equipment that not only successfully preserve food but also reduce environmental impact and enhance food safety and quality. This study seeks to advance the field by highlighting innovative, effective, and environmentally sound preservation techniques. Two of the cutting-edge technologies that are garnering interest are cold plasma and pulsed light, which provide viable substitutes for conventional methods. Additionally, there is potential to reduce the use of artificial chemicals and improve sustainability through the use of natural preservatives and innovative packaging methods like modified atmosphere packing (MAP) and biodegradable materials (Lisboa et al. 2024).

11.4.3 Functional Food and Nutraceuticals

Probiotics are live bacteria that offer the host health benefits. However, a number of concerns about their use have surfaced in recent years. Particularly, because probiotic products have the potential to drastically reduce cell viability, their industrial production and storage remain technological challenges. However, the safety of living microorganisms should be taken into account when delivering them to vulnerable groups, such as elderly or immunocompromised people. Due to these drawbacks, the emphasis has switched to innovative solutions that employ probiotics that are not viable, including postbiotics and paraprobiotics. One such example is "inactivated microbial cells," or para probiotics. Functional Foods and a New Perspective contain anti-inflammatory, antiproliferative, and antioxidant properties; they can regulate the innate and adaptive immune systems; and they have an antagonistic impact on non-viable pathogens, which benefits the consumer's health. Furthermore, paraprobiotics provide enhanced safety, provide technological and practical benefits, and may be used in products intended for the elderly and those with weakened immune systems. These qualities offer a great opportunity to launch novel, safer, and more stable functional foods or nutraceuticals onto the market. In order to guide regulatory agency decisions, this study looks at the main issues surrounding paraprobiotics and highlights the urgent need for more research into the drugs' safety, effectiveness, and mechanisms of action. Finally, a concept that distinguishes between paraprobiotics and postbiotics is proposed (Siciliano et al. 2021).

11.4.4 Waste Reduction and By-product Utilization

Around US$ 310 billion is lost in developing countries due to food waste and its consequences, whereas US$ 680 billion is lost in wealthier countries. The annual percentage of fruit and vegetable residue for legumes, fruits, and tubers was 40–50%, but the percentages for cereals and oleaginous fruits were 30% and 20%, respectively. An estimated 1.3 billion tons of food are lost or wasted year worldwide, according to FAO estimates. Fruits and vegetables generate a lot of waste that is either thrown away or used as animal feed because they are widely used in the production of beverages, minimally processed foods, and other products. For instance, during the cutting and pressing stage of processing, the pulp, seeds, peel, and stems—which are frequently found at the outside and extreme regions of the fruit and vegetable—are discarded. These components are rich in biopolymers and bioactive chemicals that can be used to create additional products. Numerous businesses may leverage these cheap and abundant renewable resources to develop new functional ingredients, natural additives, nutraceuticals, and medications to make value-added products and reduce negative environmental effects. Waste from the agri-food sector is abundant, particularly in the form of horticulture leftovers. This waste offers the potential for a wide range of industrial uses, including the production of end goods with higher nutritional value, given the diverse spectrum of biomolecules that may be recovered from these leftovers. A growing number of studies have focused on the benefits that bioactive chemicals, such as antioxidants, provide for the body (Brito et al. 2022).

11.5 Introduction of Digital Innovations in Smart Agriculture

The agriculture sector is now dealing with several challenges on a local and worldwide scale. Food and other ecosystem services provided by the agriculture sector are in greater demand. The underlying agricultural output capacity is also severely strained as a result of problems like soil degradation and climate change. Whole societies, including the agri-food sector, are rapidly and simultaneously being revolutionized by digitalization, and agriculture is presently undergoing its "fourth revolution." This development might assist in addressing the main problems confronting the agriculture sector, including increasing productivity, reducing environmental impact, and safeguarding natural resources—all of which would contribute to the achievement of the Sustainable Development Goals. For example, new digital technologies such as precision farming, the internet of things, remote sensing, unmanned aerial vehicles, data-driven applications, artificial intelligence, digital twins, robots, and more are increasingly being used in the agricultural sector. Digital technologies also alter communication routes, administrative processes, and exchanges in the food value chain. This innovation might simultaneously promote improvements in productivity, sustainability, resilience, and animal welfare while also improving and revolutionizing management and production decisions and reducing trade-offs. It is anticipated that digital advancements would be "key to developing sustainable agriculture" (Finger 2023).

11.5.1 Smart Agriculture

Every industry is changing and evolving as a result of Industry 4.0. This strategic plan makes it possible for the industry to become more digital by integrating disruptive and new digital technologies like the Internet of Things (IoT), big data and analytics (BDA), system integration (SI), cloud computing (CC), artificial intelligence (AI), cyber-physical systems (CPS), simulation, autonomous robotic systems (ARS), augmented reality (AR), wireless sensor networks (WSN), digital twin (DT), and additive manufacturing (AM). The next stage of industrial agriculture is called "agriculture 4.0," sometimes known as "smart agriculture," "smart farming," or "digital farming." It is fuelled by the integration of contemporary technologies into the agricultural industry. A number of issues pertaining to agricultural food production, including crop losses, sustainability, food security, environmental effects, and farm productivity, can be resolved by farmers through the application of smart agriculture (Singh et al. 2022).

11.5.2 Use of IoT and AI

The Internet of Things (IoT) and artificial intelligence (AI) are poised to bring about a dramatic change in the agricultural sector. Previously linked to consumer and industrial usage, these technologies are now found in farmhouses and farms, offering the possibility of increased agricultural sustainability, production, and efficiency. This change is a necessary adaptation to the increasingly complex environment we live in, as well as a reaction to the technological advancements of our time. The effect of combining artificial intelligence and machine learning in agriculture is about a paradigm change in the way people think about agriculture. Forecasting crop yields, detecting diseases, and generating useful data for livestock and agricultural management today need the use of AI algorithms that can analyse enormous data sets from sensors and satellite images. Precision agriculture, a strategy made possible by these advancements, enables farmers to make decisions based on precise, data-driven information rather than on gut feeling or conventional approaches. Better resource allocation results, which not only lessens environmental effects but also boosts productivity. The development of intelligent agricultural practices has benefited greatly from the Internet of Things. Real-time monitoring of a number of characteristics, such as soil moisture, temperature, and weather conditions, is made possible by the Internet of Things' sensor network spread throughout agricultural fields. This continuous flow of information enables farmers to make snap decisions about whether to harvest or irrigate, increasing the effectiveness of agricultural activities. Large agricultural businesses especially benefit from the opportunities for remote monitoring and management that the Internet of Things in agriculture provides (Hussein et al. 2024). Figure 11.3 illustrates IoT-enabled smart agriculture, highlighting soil pH sensors, environmental monitoring parameters, direction of winds, and equipment monitoring.

11.5.3 Drone Technologies and Remote Sensing

Examples of biotic risks include weeds, insects, fungi, viruses, and others. Bacteria can have a major influence on crop quality and productivity. The biggest problem among them is weeds, which cause a large loss in productivity worldwide. The most well-known consequence of weeds is competition for resources, such as nutrients, light, water, and space. Additionally, among the many environmental pests like bacteria, fungi, and insects that might be detrimental to crops, weeds interact with other living things and serve as a haven for plants.

In order to maximize agricultural yields while reducing environmental impacts, precision agriculture uses technology that combines sensors, information systems, and trained management. Nowadays, precision farming is employed. It is utilized in many different agricultural contexts, including sowing, harvesting, fertilizing, irrigation, and insect control.

In IWM, precision agriculture can also be incorporated. Over the past decade, precision farming has advanced rapidly due to technological developments in sensors, computer hardware, nanotechnology, unmanned vehicle systems, and robotics. These innovations enable the accurate detection and management of weeds in the field. Unmanned aerial vehicles (UAVs) are among the most efficient technologies used in precision agriculture. For conducting and observing agricultural activities, mobile vehicle systems—whether aerial (UAV) or terrestrial (UTV)

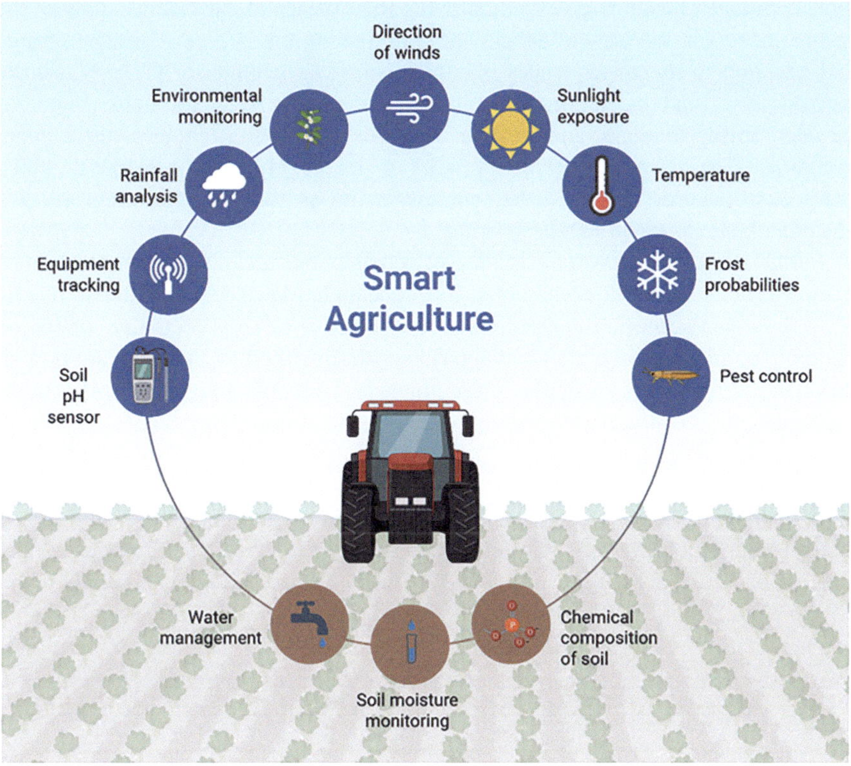

Fig. 11.3 Depicts the smart agriculture approaches for real-time monitoring of water, soil, and climate

platforms—offer a number of advantages. UAVs might be very helpful since they enable Site Specific Weed Management (SSWM) (Esposito et al. 2021). Drones are among the most challenging systems to assist. Farmers and governments should increase trade growth. Drones are used, among other things, for agricultural assistance and soil analysis. Drone technology provided priceless precision management because of its cost-effectiveness and operational flexibility (Meivel and Maheshwari 2021).

11.5.4 Climate Smart Agricultural Practices

Although the biophysical and socioeconomic impacts of human-caused climate change exhibit considerable spatial variation, regional and national policies' reactions to shifts in the global food security discourse centre on consistent technological solutions. There is a great deal of scientific and political discussion around farmers' capacity to adjust to shifting climate circumstances and opportunities to reduce climate change without jeopardizing their means of subsistence. One of the main concepts in the present debate about mitigating the effects of climate change and adapting to its consequences is climate-smart agriculture, or CSA. It may be defined as a kind of farming that lowers greenhouse gas emissions, increases farmers' ability to adapt, and sustainably raises agricultural output to raise living standards. The primary critique of CSA is its largely technical focus, which overlooks concerns pertaining to land ownership as well as other elements like customer behaviour and labour reforms.

Furthermore, climate-smart farming Misconceptions and civil society resistance hamper it in the scientific community. The concept has been demonstrated to live up to its three-win scenario internationally, with positive benefits on the environment, agriculture, and socioeconomic advancement, despite the fact that they are now in the restructuring phase. Popular CSA practices including integrated soil fertility management, no tillage, alternating watering and drying, and others have been shown in recent studies to increase agricultural productivity.

Drying and efficient nitrogen use may have global advantages. The necessity for CSA to prioritize including spatial and temporal dimensions in order to alleviate the strain of agricultural work for women and satisfy the demands of producers and customers is further highlighted by the fact that CSA technology can greatly lower (Autio et al. 2021).

11.6 Aspects of Sustainability in Agri-Biotech and Food Tech

Environmental factors are growing more dynamic and unpredictable in the present context, making it difficult for businesses to maintain their competitive advantages during uncertain times. Businesses must thus modify their resources to react to shifts in the market if they are to maintain an advantage over time.

They must thus be adaptable, and this adaptability calls for the ability to manage a variety of data that has to be processed, understood, and produced, like the recently developed Blockchain, which is an illustration of information technology (IT) management. Additionally, the agriculture sector is seeing a rise in the use of technology like the Internet of Things (IoT). Beginning with the input suppliers and concluding with the consumers, the effects of Big Data (BD), Blockchain, Artificial Intelligence (AI), and robots are transforming the whole value chain. By rapidly detecting food that is harmful, digital technologies are improving food safety, reducing the use of pesticides and water, and generating new businesses and employment, enhancing environmental, social, and economic sustainability. First of all, these technologies have relevant economic implications, such as boosting the competitiveness of already-existing businesses and creating new jobs or businesses (economic sustainability). Second, the use of these technologies—like precision farming systems that reduce the use of pesticides and water—is improving environmental sustainability with regard to climate or habitat problems. Third, in terms of social considerations (social sustainability), they are helping to improve certain rural regions with disadvantaged populations. Consequently, by pushing the sector toward a more intelligent and sustainable future, these technologies are aiding in the fight against climate change. At least five of the Sustainable Development Goals of the UN are related to the concept of smart agriculture: Objective 8: Good Work and Economic Development Objective 9: Infrastructure, Industry, and Innovation Goal 11: Communities and Cities That Are Sustainable Goal 13: Climate Action, and Goal 12: Conscientious Production and Consumption (Manvel Ciruela-Lorenzo et al. 2020).

11.7 Case Studies

The shift to more sustainable business practices is crucial given the environmental problem, which is mostly caused by pollution and the depletion of natural resources. As an alternative to the conventional linear model, which uses material recovery, recycling, and reuse to reduce waste and increase industrial productivity, the circular economy (CE) has received a lot of attention. In this context, biotechnology is essential because it provides novel approaches to converting waste into products that can be used, such as sustainable wastewater treatment that lessens reliance on fossil fuels (Carmona Marques et al. 2025).

11.7.1 Successful Biotech Crop Implementation

Agriculture was very different before the scientific and technical revolution thanks to biotechnology. By increasing the diversity of food production in terms of both quantity and quality, recent scientific and technological developments have assisted in meeting the demands of the growing world population for food security. These alterations are explained by molecular immunization, which strengthens a plant's resistance to disease. By successfully managing emerging pests and providing established crops with defence against abiotic stressors and disease-causing pathogens, these innovative technologies have shown promise in the establishment of a plant protection system. Growing plants and animals has historically been referred to be both an art and a science in agriculture. Its goals were to increase profit margins, enhance quality, and increase yields. The farmers' expertise, which came from their experience, determined the accuracy of procedures including tillage, irrigation, fertilizer application, plant health management, and crop harvesting (Sinha et al. 2024) (Table 11.3).

11.7.2 Food Tech Innovations

In addition to being great to the taste, food must be visually pleasing. Some of the purported "nutraceutical" substances taste harsh or repulsive when added to cooked foods. Therefore, despite their nutritional benefits, many consumers find it difficult to incorporate such components into their daily diets. Developing foods that provide

Table 11.3 Advances in agriculture using biotechnology to improve and protect crops (Shikha et al. 2024)

Advancement	Description	Impact	Example
Genetic modification	Transgenic techniques are used to introduce novel features into crops in order to improve nutritional quality, productivity, and pest resistance	Increased crop productivity and reduced reliance on chemical inputs	Bt cotton, GM soybean, Golden Rice
CRISPR-Cas9 gene editing	Precise genome editing of crops to enhance characteristics without adding extraneous DNA	Development of crops with improved nutritional profiles and stress tolerance, as well as accelerated breeding programs	CRISPR-edited tomatoes with extended shelf life and drought-tolerant rice
Biopesticides	Pest and disease control is achieved by using natural organisms or compounds	Environmentally friendly pest control that decreases chemical pesticide use while also promoting biodiversity	Bacillus thuringiensis (Bt) sprays, fungal-based biopesticides
Biofertilizers	Microorganisms are used to promote soil health by increasing nutrient availability	Increased soil fertility, lower chemical fertilizer consumption, and higher agricultural yields	*Rhizobium* inoculants for legumes and mycorrhizal fungi for different crops
Nanotechnology	Nanomaterials are being developed for tailored nutrient and herbicide delivery	Increased agrochemical efficiency, decreased waste, and minimized environmental effect	Nano-encapsulated fertilizers, nano-pesticides

additional health advantages without compromising flavor, texture, appearance, or functionality remains a major challenge. Encapsulation technologies offer a promising solution for delivering active ingredients. These techniques can minimize undesirable nutrient interactions, mask unpleasant flavors associated with many nutraceutical compounds, and regulate the timing, rate, and targeted site of release. Moreover, encapsulation helps maintain the stability and bioactivity of these components under harsh processing or storage conditions. However, in order to create better final products, some natural food constituents that are thought to be undesirable or harmful to one's health might be decreased or deleted. For example, several processing techniques have been developed to reduce antinutritive compounds and to remove caffeine from caffeinated beverages within the natural food matrix. Synthetic lipids have also been engineered to mimic the functional properties of natural oils and fats while providing fewer calories. Enzymatic and fermentation-based processes are widely applied to degrade toxic, allergenic, and antinutritional substances in natural food ingredients, while simultaneously enhancing flavor and improving the bioavailability of essential nutrients. Despite these advancements, a major challenge in developing healthier food products is maintaining desirable taste, which remains a significant barrier in food technology. (Hsieh and Ofori 2007).

11.8 Challenges and Future Prospects

Recent research highlights the growing role of nanotechnology in agriculture. Nanomaterials are being explored for their potential to enhance crop productivity, improve soil health, and support sustainable farming practices. Key developments include nanobiosensors, nanopesticides, nanofertilizers, and nano-enabled approaches for soil pollution remediation. Nanofertilizers play an especially important role by improving nutrient-use efficiency, increasing crop yield and quality, and lowering production costs. Their controlled and slow-release properties reduce nutrient losses to the environment, contributing to environmentally sustainable plant nutrition. Nanotechnology has found several uses in agriculture, such as environmental remediation agents, nanobiosensors, nanopesticides, and nanofertilizers. In agricultural and environmental sciences, a solid grasp of the fate of nanomaterials and their effects on the environment is still a significant obstacle. For the development of effective, multifunctional, stable, economical, and environmentally friendly nanomaterials, cooperation across institutions investigating various applications of nanomaterials would be essential (Pant et al. 2024, 2025; Usman et al. 2020).

Researchers, farmers, and other stakeholders may make educated decisions on soil fertility, moisture content, and nutrient content by using machine learning techniques to anticipate soil properties. In order to optimize crop output and yield, proactive adjustments to agricultural practices may be made using the insights that machine learning models can provide from the analysis of data from a variety of sources. Furthermore, ML provides real-time information about crop health, growth phases, and potential environmental stressors, which simplifies the monitoring of crops and soil conditions by using data from computer vision and remote sensing.

Machine learning algorithms have proven to be highly effective in improving water-related operations through the merging of cutting-edge sensor techniques with IoT technology. One well-known application where ML models use real-time data processing to provide accurate schedules is precision irrigation. Additionally, these models excel at closely monitoring water quality and guaranteeing that plants are given water with the ideal balance of nutrients. Additionally, ML-driven predictions of crop evapotranspiration rates provide valuable insights into water needs, enabling more environmentally friendly irrigation techniques. There are still obstacles to overcome despite the potential of ML integration in agriculture. Important issues to solve include increasing the degree of digital literacy among regular farmers, promoting human-machine collaboration, and improving interpretability. Creating ML apps with intuitive user interfaces that don't require a lot of technical know-how is essential. Simple dashboards and clear visualizations might be helpful for farmers who struggle with reading. Additionally, outreach and training initiatives might be created to cater to the unique requirements of farming communities. Farmers may be able to get the knowledge and abilities necessary to successfully integrate ML-based technology into their daily tasks through seminars, demonstrations, and instructional programs. Furthermore, creating lightweight models and investigating distributed computing techniques are essential first steps in successfully integrating machine learning into farming in resource-constrained settings. Resolving these issues in the agriculture industry would result in a more thorough and efficient application of ML technology (Araujo et al. 2023; Pant and Singh 2025; Pant et al. 2025b).

11.9 Conclusion

To feed the growing world population, more food must be generated from agricultural land systems. The issues of global food security require a number of solutions that address both the supply and demand sides of food availability, accessibility, and consumption. Although the problem of food security is primarily caused by agricultural land usage practices, solving it requires the recently developed field of Land System Science, which views the architecture of land systems in human-environment interactions more holistically and views food production as one aspect of consequences among other ecological and economic outcomes. Future farming will therefore focus on producing more food with fewer resources while minimizing adverse ecological and environmental repercussions. In order to contribute to sustainable solutions, land systems must systematically analyse land availability and assess the trade-offs related to activities such as growing, using farmland, assigning crops, and improving management. Therefore, a variety of selective pressures originate from agriculture, and modern agriculture, including its species and practices, is the result of a continuous process of change that has profoundly changed all facets of agricultural ecosystems, both genetically and physically. Future advancements in the genetic potential of plants and animals will mostly come from conventional breeding techniques, which are becoming more and more successful because of advancements in marker technology. Another important advancement is the incorporation of genes from other species using molecular manipulation (GM) technologies, which will gain traction but will be used to make modifications that are not possible with conventional techniques.

Overall, the development and administration of agriculture will depend more and more on the identification and use of ecological and evolutionary concepts, particularly in light of climate change (new plants and animals) and the need for greater sustainability. A comprehensive collection of agri-food technology is needed to produce safe, healthy, sustainable meals, and technologies branded 'disruptive' are considered as part of the solution to food security in the period of climate change and population expansion. The majority of disruptive ideas that radically alter an industry come from sectors other than food.Disruptive technologies are obviously needed to address the problems caused by population expansion, climate change, and food chain disruptions. Early on in the development process, it is necessary to assess society's approval in order to successfully introduce such innovations.

References

Abiri R, Rizan N, Balasundram SK, Shahbazi AB, Abdul-Hamid H (2023) Application of digital technologies for ensuring agricultural productivity. Heliyon 9(12). https://doi.org/10.1016/j.heliyon.2023.e22601

Amiteye S (2021) Basic concepts and methodologies of DNA marker systems in plant molecular breeding. Heliyon 7(10)

Araujo SO, Peres RS, Ramalho JC, Lidon F, Barata J (2023) Machine learning applications in agriculture: current trends, challenges, and future perspectives. Agronomy 13(12):2976

Autio A, Johansson T, Motaroki L, Minoia P, Pellikka P (2021) Constraints for adopting climate-smart agricultural practices among smallholder farmers in Southeast Kenya. Agric Syst 194: 103284. https://doi.org/10.1016/j.agsy.2021.103284

Barling D, De Vriend H, Cornelese JA, Ekstrand B, Hecker EF, Howlett J, Jensen JH, Lang T, Mayer S, Staer KB, Top R (1999) The social aspects of food biotechnology: a European view. Environ Toxicol Pharmacol 7(2):85–93. https://doi.org/10.1016/S1382-6689(99)00009-5

Bertoglio R, Corbo C, Renga FM, Matteucci M (2021) The digital agricultural revolution: a bibliometric analysis literature review. Ieee Access 9:134762–134782

Brito TBN, Ferreira ML, Fai AE (2022) Utilization of agricultural by-products: bioactive properties and technological applications. Food Rev Intl 38(6):1305–1329

Carmona Marques P, Fernandes PC, Sampaio P, Silva J (2025) Advances in biotechnology in the circular economy: a path to the sustainable use of resources. Sustainability 17(14):6391

Du X, Wang X, Hatzenbuehler P (2023) Digital technology in agriculture: a review of issues, applications and methodologies. China Agric Econ Rev 15(1):95–108. https://doi.org/10.1108/CAER-01-2022-0009

Eş I, Gavahian M, Marti-Quijal FJ, Lorenzo JM, Khaneghah AM, Tsatsanis C, Kampranis SC, Barba FJ (2019) The application of the CRISPR-Cas9 genome editing machinery in food and agricultural science: current status, future perspectives, and associated challenges. Biotechnol Adv 37(3):410–421. https://doi.org/10.1016/j.biotechadv.2019.02.006

Esposito M, Crimaldi M, Cirillo V, Sarghini F, Maggio A (2021) Drone and sensor technology for sustainable weed management: a review. Chem Biol Technol Agric 8(1):18

Falk MC, Chassy BM, Harlander SK, Hoban TJ IV, McGloughlin MN, Akhlaghi AR (2002) Food biotechnology: benefits and concerns. J Nutr 132(6):1384–1390. https://doi.org/10.1093/jn/132.6.1384

FAO (2023) Land use statistics and indicators 2000–2021. Global, regional and country trends. FAOSTAT Analytical Briefs Series No. 71. Rome. https://doi.org/10.4060/cc6907en

Ferreyra-Suarez D, García-Depraect O, Castro-Muñoz R (2024) A review on fungal-based biopesticides and biofertilizers production. Ecotoxicol Environ Saf 283:116945. https://doi.org/10.1016/j.ecoenv.2024.116945

Finger R (2023) Digital innovations for sustainable and resilient agricultural systems. Eur Rev Agric Econ 50(4):1277–1309

Flannery KV (1973) The origins of agriculture. Annu Rev Anthropol 2:271–310

Gill R, Chawla P (2021) A review on various techniques to transform traditional farming to precision agriculture. Turk J Comput Math Educ 12(2):131–135

Gupta N, Jain V, Joseph MR, Devi S (2020) A review on micropropagation culture method. Asian J Pharm Res Develop 8(1):86–93. https://doi.org/10.22270/ajprd.v8i1.653

Hsieh PYH, Ofori JA (2007) Innovations in food technology for health. Asia Pac J Clin Nutr 16 (S1):65–73

Hussein AHA, Jabbar KA, Mohammed A, Jasim L (2024) Harvesting the future: AI and IoT in agriculture. E3S Web Conf 477:00090. EDP Sciences

Kumar S, Sindhu SS, Kumar R (2022) Biofertilizers: an ecofriendly technology for nutrient recycling and environmental sustainability. Curr Res Microb Sci 3:100094. https://doi.org/10.1016/j.crmicr.2021.100094

Lisboa HM, Pasquali MB, dos Anjos AI, Sarinho AM, de Melo ED, Andrade R, Batista L, Lima J, Diniz Y, Barros A (2024) Innovative and sustainable food preservation techniques: enhancing food quality, safety, and environmental sustainability. Sustainability 16(18):8223

Lu F, Meng J, Cheng B (2024) How does improving agricultural mechanization affect the green development of agriculture? Evidence from China. J Clean Prod 472:143298. https://doi.org/10.1016/j.jclepro.2024.143298

Manuel Ciruela-Lorenzo A, Rosa Del Aguila-Obra A, Padilla-Melendez A, Jose Plaza-Angulo J (2020) Digitalization of agri-cooperatives in the smart agriculture context. Proposal of a digital diagnosis tool. Sustainability 12(4)

Meivel S, Maheswari S (2021) Remote sensing analysis of agricultural drone. J Ind Soc Remote Sens 49(3):689–701

Nath B, Chen G, O'Sullivan CM, Zare D (2024) Research and technologies to reduce grain postharvest losses: a review. Foods 13(12):1875. https://doi.org/10.3390/foods13121875

Pant M and Singh J (2024) Distribution of heavy metals contamination in the environment and their toxicological impacts. In Heavy metal contamination in the environment, pp 95–113

Pant M, Singh J (2025) Integrating AI, machine learning, and nanotechnology: shaping the future of water quality management. In: Computational automation for water security. Elsevier, pp 111–134. https://doi.org/10.1016/B978-0-443-33321-7.00006-8

Pant M, Singh J, Bisen D, Manori S, Shukla RK (2024) Nanocomposite-based photocatalysis: tackling pharmaceutical and personal care products (PPCPs) pollutants in environmental remediation. In: Sustainable development goals towards environmental toxicity and green chemistry: environment and sustainability. Springer, Cham, pp 83–105

Pant M, Singh P, Landge AS, Bisen D, Singh J, Manori S, Shukla RK (2025a) Graphene oxide-p-phenylenediamine nanocomposite for efficient dye degradation and bacterial eradication in wastewater treatment. J Water Process Eng 75:108009. https://doi.org/10.1016/j.jwpe.2025.108009

Pant M, Bisen D and Singh J (2025b) Biomass residue utilization: harnessing unconventional resources. In Bioenergy: sustainable solutions for a greener planet. American Chemical Society, pp 67–86

Pasupuleti MK, Volume-14| Issue-3| March| Year-2024 IJAIRI

Peter VH, Peng YANG (2014) How could agricultural land systems contribute to raise food production under global change? J Integr Agric 13(7):1432–1442. https://doi.org/10.1016/S2095-3119(14)60819-4

Raman R (2017) The impact of genetically modified (GM) crops in modern agriculture: a review. GM Crops Food 8(4):195–208

Sekhar M, Rastogi M, Rajesh CM, Saikanth DRK, Rout S, Kumar S, Patel AK (2024) Exploring traditional agricultural techniques integrated with modern farming for a sustainable future: a review. J Scientific Res Report 30(3):185–198

Shikha D, Sindhura KAV, Rastogi M, Saritha B, Satapathy SN, Srivastava S, Kurdekar AK (2024) A review on propelling agricultural practices with biotechnology into a new era. J Adv Biol Biotechnol 27(3):99–111

Siciliano RA, Reale A, Mazzeo MF, Morandi S, Silvetti T, Brasca M (2021) Paraprobiotics: a new perspective for functional foods and nutraceuticals. Nutrients 13(4):1225

Singh G, Kalra N, Yadav N, Sharma A, Saini M (2022) Smart agriculture: a review. Siberian J Life Sci Agric 14(6):423–454

Singh J, Pant M, Ghoshal T, Srivastava AL and Singh S (2025) Spatio-seasonal observations of water quality index, heavy-metal pollution index and microbial contagion in springs of lower Indian Himalayan zone. Environ Claims J 1–36

Sinha T, Sharma A, Das A, Kumar A, Rahman T, Panotra N, Anbarasan S (2024) The impact of biotechnology in shaping the modern agriculture. J Adv Biol Biotechnol 27(11):844–851

Somvanshi PS, Pandiaraj T, Singh RP (2020) An unexplored story of successful green revolution of India and steps towards ever green revolution. J Pharmacogn Phytochem 9(1):1270–1273

Soumare A, Diédhiou AG, Arora NK, Tawfeeq Al-Ani LK, Ngom M, Fall S, Hafidi M, Ouhdouch Y, Kouisni L, Sy MO (2021) Potential role and utilization of plant growth promoting microbes in plant tissue culture. Front Microbiol 12:649878

Thrall PH, Bever JD, Burdon JJ (2010) Evolutionary change in agriculture: the past, present and future. Evol Appl 3(5–6):405. https://doi.org/10.1111/j.1752-4571.2010.00155.x

Upadhyay N, Bhargava A (2025) Artificial intelligence in agriculture: applications, approaches, and adversities across pre-harvesting, harvesting, and post-harvesting phases. Iran J Comput Sci 8: 1–24

Usman M, Farooq M, Wakeel A, Nawaz A, Cheema SA, ur Rehman H, Ashraf I, Sanaullah M (2020) Nanotechnology in agriculture: current status, challenges and future opportunities. Sci Total Environ 721:137778. https://doi.org/10.1016/j.scitotenv.2020.137778

Wakweya RB (2023) Challenges and prospects of adopting climate-smart agricultural practices and technologies: implications for food security. J Agric Food Res 14:100698

Practical Implementation of Digital Twins in Agriculture: Case Studies and Future Directions

12

Saurabh Kumar, Naveen Patel, Akansha Patel, Vinod Kumar Chaudhary, Anurag Singh, Arun Lal Srivastav, and Neeraj Kumar Verma

Abstract

The expanding human population, degradation of soil quality, climate change, and food security are the issues that have become critical concern for sustainable agriculture. In this regard, the Digital Twin (DT) is in a unique position to overcome these obstacles and can help to achieve the goals of sustainability. Though there have been a number of technical developments in agriculture recently, DT technology, because of its ability to enhance sustainability, productivity, and decision-making processes, has become an emerging technology in several industries, including agriculture. This book chapter discusses the basic architecture of digital twin and various applications in agriculture sector such as crop growth monitoring and management, pest and disease management, smart irrigation system, aquaponics, impact of climate on crops, as well as livestock management. Moreover, some case studies related to the application of this

S. Kumar · A. Patel · V. K. Chaudhary
Department of Environmental Sciences, Dr. Rammanohar Lohia Avadh University, Ayodhya, Uttar Pradesh, India

N. Patel (✉)
Department of Civil Engineering, Dr. Rammanohar Lohia Avadh University, Ayodhya, Uttar Pradesh, India

A. Singh
Department of Mechanical Engineering, Dr. Rammanohar Lohia Avadh University, Ayodhya, Uttar Pradesh, India

A. L. Srivastav
Department of Applied Sciences, Chitkara University, Solan, Himachal Pradesh, India

N. K. Verma
Department of Data Science, Manipal University, Jaipur, Rajasthan, India

© The Author(s), under exclusive license to Springer Nature Singapore Pte Ltd. 2026

A. L. Srivastav et al. (eds.), *Digital Twin Technology for Sustainable Agriculture*,
https://doi.org/10.1007/978-981-95-5915-2_12

emerging technology have also been discussed. Limitations associated with this technology like cost issue and lack of proper knowledge of technique have also been covered in this chapter.

Keywords

Digital Twin · Agriculture · Emerging techniques · Smart water irrigation

12.1 Introduction

According to the Food and Agriculture Organization (FAO), the population of world is expected to exceed 10 billion by 2050, necessitating an increase in food production to meet the rising demand (Falcon et al. 2022). Ensuring food security in the future and demonstrating sustainable development is a major challenge for humanity. Due to this, the economy is exerting more pressure on the agricultural sector primarily due to labor shortages, environmental issues, and challenges related to climate change (Prause 2021). Agriculture is a highly dynamic and complex process where various environmental factors such as climate, soil conditions, agricultural operations, and the presence of weeds, pests, and diseases significantly affect crop production (Trienekens et al. 2014). Moreover, without affecting other environmental concerns such as food security, ecological balance, and sustainability, meeting the demand of consumers from farmer's fields has become an important goal. There is a need of increasing agricultural productivity without creating enormous pressure on natural resources. Climate change, resource scarcity, population growth, and changing consumer preferences are causing severe challenges in agriculture sector that threaten productivity and affect crop production and the sustainability of the food system. Hence, it is becoming important to address these issues in a sustainable and effective manner.

Data analytics, machine learning (ML), digital twin (DT), artificial intelligence (AI), and Internet of Things (IoT) devices are recent technologies that can play a significant role in solving the issues related to agriculture. Digital twin (DT) is one of the emerging technologies that allows for managing agricultural operations remotely using real-time observation data, with low manual labor cost. DT is virtual representation of real-world physical things, where the virtual description of physical object or system reflects its behavior and current state. This technique develops a strong foundation of physical objects to study their dynamic behavior. Nowadays, DT successfully demonstrates the possibilities of edits in various industrial sectors, including the automotive and energy sectors. Additionally, DT is transforming agriculture by providing innovative solutions in various areas, including crop monitoring, irrigation management, evaluation of soil ecosystems, livestock production systems, and supply chain optimization (Escribà-Gelonch et al. 2024).

Moreover, application of digital twins in industrial sector has expanded the boundaries of industrial activities by predicting and optimizing the environment for achieving physical products, assets, and processes. In this way, this technique is particularly important in the mechanical transformation of manufacturing industries, enhancing sustainability, increase in production, and management of product lifecycle. Digital twins are crucial for intelligent manufacturing and play a significant role in the digital transformation of enterprises (Liu et al. 2021). With the help of this emerging technology, various types of crop management are being done. For example, rice cultivation, which is crucial as a staple food globally, faces serious challenges every year due to diseases that present biological barriers that can lead to a reduction in production of 20–100% (Shivappa et al. 2021). Among different issues related to agriculture, rice blast is one of the most prominent diseases caused by fungus *Magnaporthe oryzae* in rice crop and is considered as one of the most impactful diseases with an annual production loss of about 10–30%. Under normal conditions, this disease has the potential to rapidly destroy entire rice crops within 15–20 days, leading to significant losses up to 100% in some areas. One of the most effective techniques for early detection of leaf fungal and nutritional deficiencies in rice plants is digital twin by using regular visual inspections. With its assistance, it can be further improved to provide real-time monitoring and visual representation of the field to aid in timely disease detection and management (Asibi et al. 2019). In addition to crop management, livestock management can also be done using DT that provides farmers with tools to monitor animal health, breeding cycles, and productivity, thereby improving animal welfare and farm efficiency (Neethirajan and Kemp 2021). In recent days, the application of DT with machine learning (ML) and IoT devices can further enhance predictive accuracy and decision-making capabilities. In agricultural irrigation, digital twins can assess the behavior of irrigation system before it is implemented in the field, and its management can be made easier by evaluating different irrigation strategies in parallel to the current farm practices (Goldenits et al. 2024). The integrated application of digital twin technology, sensor data, and machine learning algorithms has gained enormous attention in agricultural revolution. These components provide an innovative approach to monitoring and managing nutrient deficiencies, drought conditions, soil health, and disease control in agricultural fields. In this book chapter, application of digital twin technology in agriculture sector like crop health monitoring, soil health assessment, climate impact studies, irrigation improvement, aquaponics, vertical farming, cropping systems, and livestock farming has been discussed along with the agriculture disaster monitoring and assessment.

12.2 Digital Twin Technology

Digital twin is a virtual description of an object or system that is developed to accurately reflect the physical object that has to be analyzed. It utilizes real-time data, machine learning based modeling, and simulation for taking logical decisions. It combines data analytics, artificial intelligence (AI), and sensors to facilitate performance optimization, simulation, and monitoring while offering predictive information and well-informed decision-making (Awais et al. 2025). In order to mirror real-world difficulties, digital twins are updated frequently with real-time data. Consequently, this technology is widely employed in order to enhance the performance of the system, reduce the possible hazards, and improve efficiency in various industries, including manufacturing, healthcare, and agriculture. The idea of digital twin technology was introduced first by David Gelernter in his book "Mirror Worlds" in 1991. However, in 2002, Dr. Michael Grieves is known for the first application of the digital twin concept in manufacturing, where digital twin software concept was introduced formally. In 2010, principal technologist of NASA, John Vickers further described the digital twin. Since then, utilization of DT has increased gradually but no official definition of DT was accepted till then each expert has their own definition to describe it. The fundamental principles of digital twin emphasize the role of integrating modeling, simulation, and real-time data to enhance decision-making in complex aerospace systems (Wang 2024). Later, this concept was adopted in various fields such as manufacturing sector, healthcare, urban planning, and agriculture sector for prediction, maintenance, optimization, and data-driven decision-making. Moreover, the application of digital twins in recent years has increased gradually in different sectors such as transportation, manufacturing and services, food and beverage industry, oil and gas production, construction, agriculture field, aviation, and energy sectors. Furthermore, the recent availability of various business software tools such as Predix and Simcenter 3D has proven the growing interest in the application of DT (Negri et al. 2017).

Agricultural digital twin contains four stages including sensing, updating, predicting, and management. The first stage is sensing that includes collecting live data through various equipments and techniques such as soil moisture sensors, temperature and humidity sensors, satellite image as well as UAV (Unmanned Aerial Vehicle) image. Subsequently, in the updating stage, the collected data is integrated, cleaned, and stored in cloud platforms' databases to ensure digital twin is updated properly with accurate information. The third stage is the forecasting stage that uses machine learning algorithms, simulation models, and trend analysis techniques to predict future agricultural crops, soil, and climate conditions. Finally, in the management stage, decisions are made on the forecast to utilize decision support system for optimizing agricultural practices such as irrigation scheduling, fertilization systems, and pest control strategies (Manivasagam 2025). Nowadays, cutting-edge technologies such as AI, big data, and IoT are reported to have great potential of application in various fields. The advantages of such integration have also been seen in DT applications in different areas. However, the application of DT in agriculture sectors is still limited, and extensive discussion regarding its cost value is still needed. Due to this, questions arise about the benefits of DT in agriculture, what characteristics differentiate them from current technological practices, and how their design and implementation can be done and managed at the grassroots level (Pylianidis et al. 2021).

The architecture of DT is basically made from three-unit parts that include physical space, virtual space, and feedback mechanism. When it comes to its operational function, the physical and virtual spaces are connected through feedback mechanism, transfer data mutually, and serve as the core of digital twin architecture. The physical space includes the actual entity or system that collects the sensor data

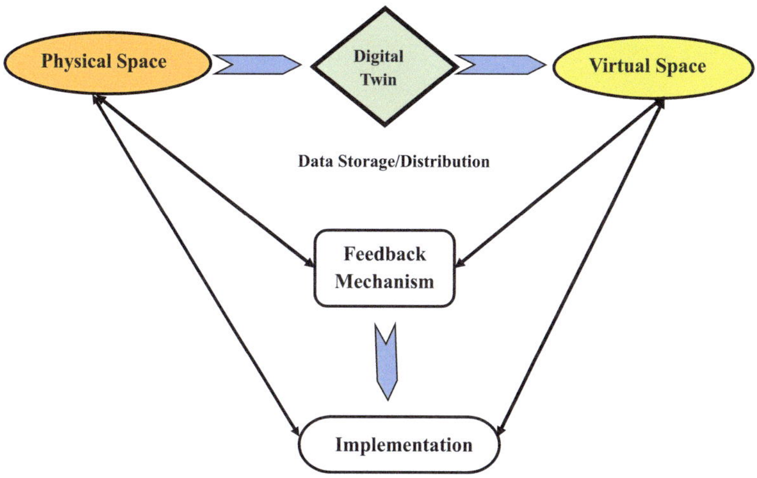

Fig. 12.1 Simple architecture diagram of Digital Twin Technology

from equipment, machines, or agricultural fields. On the other hand, the virtual space receives feedback from sensor data, stored in the real world, based on which digital models, simulations, and analytical tools, hence helping in identification of problem. Figure 12.1 represents the architecture of digital twin technology. The main challenge of digital twin architecture is to ensure accurate and real-time coordination between physical and virtual counterparts (Zhang et al. 2025).

12.3 Application of Digital Twin Technology in Agriculture

The application of DT in the field of agriculture is vast and includes various sectors of agriculture like soil management, water irrigation management, and crop monitoring as well as fertilizer utilization. It also plays an important role in livestock farming, pest and disease control on crops. Figure 12.2 represents how the agriculture sector digital twin can play an important role.

12.3.1 Application of DT in Crop Monitoring

Digital twins utilize real-time data for assessment of statistics, simulation models, and machine learning models that helps in predicting the crop's growth pattern. However, the accuracy of the predictions depends on the adequacy of the model that describes the actual process during the development of the plant necessary for growth using digital twin (Slob et al. 2022).

Physical models simulate the behaviors of crops under different environmental conditions like photosynthesis, evaporation, and nutrient absorption. Whereas, mathematical models commonly describe crop characteristics using data such as

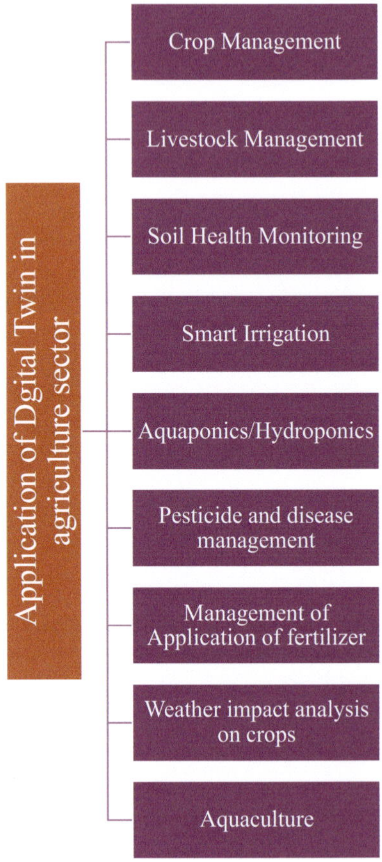

Fig. 12.2 Application of digital twin agriculture sector

plants' dry weight, soil moisture, temperature, and composition. Models like CERES (Crop Environment Resource Synthesis) and APSIM (Agricultural Production Systems Simulator) are broadly applied to simulate crop growth model and predict responses to climatic stresses (Ibrahimov et al. 2023).

Furthermore, hybrid models integrate data-driven and physical-based approaches to develop precise and adaptable models as well as methodologies for crop production prediction, anomaly detection, and environmental resource improvement. While spatial models are developed using Geographic Information System (GIS) to describe the basic process and properties for a specific set of spatial characteristics of agriculture field, and integration of DT and GIS can help to simulate crop growth under various factors. In land-use planning, these models are especially essential as they help in determining the best locations for crop rotation and hence maximizing the yield by reducing the environmental damage (Melesse 2025).

12.3.2 Application of DT in Pest and Disease Control

Insect prediction models are being developed in recent days to improve the operational efficiency and accuracy of Integrated Pest Management (IPM) in the agriculture field. DT techniques were used to develop a relationship between environmental factors including; humidity, CO_2 concentration, temperature, wind speed, sunlight intensity, rainfall, and insects. Dai et al. have developed a pest management system that compiles relevant data in the perception layer and convolution layer through RGB (Red, Green, Blue) cameras, hyperspectral cameras, various sensors, environmental acquisition tools, and established an AI-based DT model for pest prediction by setting up IoT architecture in a controlled agricultural environment for successful management of pests and diseases.

12.3.3 Application of DT in Management of Livestock Farming

Animal-based emissions are a major source of greenhouse gas, and farming of livestock has been identified as a significant contribution to global warming. It is an essential component in farming. Although, application of DT in livestock farming is limited as compared to other fields, but finds its importance in areas of supplying chain transparency, welfare enhancement, labor reduction, and decision support. Optimizing energy use and environmental conditions are essential for achieving the objectives of higher sustainability and quality standards in the animal husbandry sector. Application of DTs along with Internet-of-things, cyber-physical systems, and machine learning in the livestock sector has been reported to have great potential in livestock management (Neethirajan and Kemp 2021). DT technology can offer numerous advantages in livestock management including real-time monitoring, prediction analysis, enhancement in decision making, and management of personalized animal care. DT can provide a comprehensive report about the animal health, feeding patterns, and environmental interactions by using virtual image of the livestock and its nearby environment.

12.3.4 Application of DT in Management of Impact of Weather on Crops

In agriculture, DT is mainly focused on managing the effects of climate change by managing the appropriate use of water and soil resources in order to ensure to mitigate the impacts of extreme weather events, or handling multiple contemporary stresses. DTs are now being used for climate adaptation and other areas of environmental science. For weather and climate forecasting on a global scale and overall time period, DTs can be used as an interactive model. Application of DT for disaster management and climate change was a key component that was adopted by the European Commission under Destination Earth (DestinE) initiative (Hoffmann et al. 2023).

In addition, weather monitoring stations primarily collect regional meteorological data such as temperature, rainfall, humidity, wind direction, and wind speed with the help of weather sensors. The integration of collected data from these sensors into a digital twin (DT) enables farmers and agricultural scientists to play a pivotal role in providing information about crop growth and health, leading to efficient and sustainable management of agricultural crops by minimizing climatic disasters. However, such DT technique demands high cost for sensors, advanced analysis system, and processing power units. For example, simulating the impact of rising temperatures or drought conditions on crop health can help farmers take proactive steps like adjusting planting times or choosing more drought-resistant crops (Prakash et al. 2023). Regular monitoring of real-time weather data with the help of DT helps in creating model for the prediction of the impact of weather conditions on crops and provides information that can be helpful in managing crop production even in extreme climatic conditions.

12.3.5 Application of DT in Soil Health Management

Soil is an important natural renewable resource as crop health depends on the health of the soil. By assessing the properties of soil, in order to promote sustainable crop production, various factors like nutrient balance, water balance, sustainable fertilization, and irrigation method to prevent soil degradation are selected. For example, integration of digital twin (DT) technology with soil sensors to monitor the soil moisture, soil temperature, organic matter, and also contaminants present in the soil can play an important role in mapping the soil profile of the field. The digital technologies used in soil-related digital twin include wireless system networks, IoT, edge computing, local weather-based controllers, and soil sensors (Parewai et al. 2025). Alves et al. (2023) designed DT technique for smart water management system that gathers information from sensors installed for collection of data regarding temperature and humidity, soil moisture, sun radiation, as well as geospatial information. Models obtained are included in the cloud server to simulate the behavior of soil and crops. Moreover, DT helps in managing and guiding the water consumption patterns in order to minimize the water loss.

12.3.6 Application of DT in Smart Water Irrigation System

In agriculture, water management is a major challenge, particularly in areas of limited water supplies. A real-time irrigation tracking and management system along with Digital Twin (DT) technology helps in managing the water irrigation system. DT model receives data from sensor network installed in the field that helps in tracking temperature, rainfall, and soil moisture levels (Alves et al. 2023). In order to help farmers to minimize over-irrigation and maintain the sufficient level of water availability for crops, the real-time data integration with DT helps to replicate irrigation requirements. Digital twin is a computerized model of the actual irrigation system that is presently being used. The crop growth models were produced by gathering information from sensors and additional sources including soil moisture sensors, and weather forecasts. The models developed using DT can help in achieving the maximum crop productivity while minimizing water usage by optimizing irrigation timings and providing better decisions. Moreover, DT can further assist farmers in identifying problems associated with irrigation system problems, such as leaks or malfunctions, and help us in taking necessary steps in order to maintain the crop yield (Kalyani et al. 2024).

Digital twin technology also helps in simulating the implementation of different irrigation methods such as drip, sprinkler, or flood irrigation, which further helps in determination of most effective irrigation system based on the characteristics of agricultural fields and local climate condition.

12.3.7 Application of DT in Fertilizer Management on Crops

Application of fertilizer in agriculture fields plays crucial role in increasing crop yields and reduce the impact of lack of growth regulator nutrients like nitrogen, phosphorus, potassium, zinc, etc. present in soil. DTs technique on integration with real-time data from physical process of crops and soil like photosynthesis, respiration, water holding capacity, and other environmental factors helps in suggesting efficient fertilizer application practices in agriculture field. This technique will help in minimizing the risks of overuse and runoff of water from fields and thereby improving the quality of soil and crops. Furthermore, with advancement in digital twin technology, increase in expectation for automating tasks such as fertilizer spraying on crops has been expected to grow, i.e., independent drones equipped with digital twins can spray fertilizers based on real-time environmental data (Peladarinos et al. 2023).

12.4 Case Studies Related to Application of Twin Technologies

There are some case studies that provide proof of the utilization of emerging technologies in the area of agriculture. In 2023, Zhang along with his research team investigated the application of digital twins in agriculture for the purpose of perceiving and simulating dairy cow eating behavior. The classification of cow feeding and non-feeding behaviors, digital twin perception and modeling techniques, and the creation of a digital cow shadow are the parameters that were used in the study. Five healthy non-lactating Holstein cows were used to collect neck movement and real-time location data using a bespoke collar equipped with inertial measurement units (IMUs) and ultra-wide band (UWB) chips. The positioning anchor received the gathered data via UWB signals, which were then sent to a local server. The experimental results indicated that all three classification techniques were successful, but LSTM (Long-Short Term Memory) performed better than the others. The accuracy of detecting cow grazing patterns using only

IMU data along with LSTM neural network was obtained to be 91.05% with 92.23% precision and recall of 91.35%. The accuracy gets improved to 94.97% while the precision rate was obtained to be 99.99% with recall rate of 93.86% by combining the data obtained from UBW and IMU devices with the application of LSTM. The findings of this study showed accuracy rate of 95.07% while determining the feeding and non-feeding behaviors of cows with combination of the local server, anchor points, and custom-made collar. 3.92% greater accuracy rate was achieved with IMU data alone. Furthermore, this work offers a way for investigating digital twins for cows, particularly by creation of their digital shadows.

Alves et al. (2023) have developed a digital twin for smart water irrigation system in agriculture fields for sustainable application of water on crops. In this research, digital twin technology was used for irrigation system in which the physical components, such as sensors and actuators, are connected to each other to represent data virtually. Siemens Plant Simulation software was integrated with FIWARE-based IoT platform and simulation model. The IoT platform was applied to store and process data of soil quality, weather, and related to crop in order to calculate the daily irrigation requirement. The application of simulation model software was used to simulate the behavior of irrigation technique based on application scenario in agriculture area. Thereby, communication between IoT platform and simulation model was easily established for real-time analysis through OPC UA server. The advantage of the proposed systems is twofold. The first advantage is to evaluate the behavior of IoT platform and irrigation system before its implementation in the agriculture field. The second advantage is associated with the technical evaluation of different irrigation strategies in parallel with recent farm practices. Within the proposed system, agriculture professionals can evaluate the characteristics of the system before implementing in the agriculture area so that drawbacks associated with this technique can be identified.

Dai et al. (2024) have applied the digital twin technology for pest control on agriculture crops for minimizing the impact of attack of pest and to minimize the application of pesticide on crops too. Digital twin framework to manage insect pests in the life cycle of crop growth was first presented. Then, the integrated application of random forest algorithm and genetic algorithm along with digital twin model is used to predict the number of pests based on twin data searching technique for pest control interventions. The study was conducted to verify the feasibility of proposed system for growth state and pattern of pepper plant (*Capsicum annuum*) and pepper pests. In this experimental study, virtual and real interactive response was obtained for pepper aphid insect management system. The obtained result prediction reported an accuracy of insect attack on pepper plant is 88.01% with the training set while 85.73% with the experimental set in the field. The application of the prediction model in decision-making objective in pest management can reduce 20% of economic loss. Present study was found to be better than manual regulatory method in pest management of agriculture crops.

12.5 Limitations

The complete application of DTs in agriculture still remains subjected to some limitations. The high expenses for developing, implementing, and maintaining agricultural DT systems are the key challenge. The establishment of fully functional DT requires significant investments in software (such as AI, ML, and big data analytics), hardware (such as IoT sensors, tools of remote sensing, and high-performance computing tools), and expertise including data scientists, agronomists, and system engineers. Particularly for small-scale farmers and agricultural enterprises with little funding, these expenses may be unaffordable (Tao and Qi 2019). In addition to limitations of budget, the relationship between the potential advantages of DT implementation and its complexity must be carefully considered for determining whether adoption of this technology in agriculture activities is economical and practically feasible or both. The cost of implementing DTs frequently and inadequate digital infrastructure and data governance frameworks make other affordable smart farming options like basic IoT-based monitoring systems more viable (Rajeswari et al. 2024). Furthermore, in rural communities, where internet bandwidth is often limited, crop monitoring and synchronization with a digital twin are difficult tasks. In addition to these, the digital divide limits the adoption of technology by exacerbating knowledge gaps. These problems can be resolved by better internet infrastructure, and cutting-edge computing for local data processing, which will lead to application of digital twins more widely in remote locations (Manivasagam 2025). The absence of advanced IoT device sensors required for detailed, real-time data collection further proves a major challenge. The technological gap can be solved by creating and implementing inexpensive sensor technology and applying open-source data platforms (Khan et al. 2022).

12.6 Conclusion

Although the application of DT technique in agriculture has gained momentum in recent years but its widespread implementation is still missing mainly due to the lack of availability of a huge volume of data associated with the agriculture field. The success and sustainability associated with this technique in the agriculture sector does not only depend upon technology, skill, and data infrastructure but also on the economic aspect and ability to demonstrate its relevance. Farmers and other stakeholders associated with the agriculture sector will adopt digital twin only when positive return on agricultural economic investment can be obtained. Moreover, collaborative DT systems allow many farmers to share real-time data that can reduce agricultural disaster risk, maximize supply chain efficiency, and enhancement of collective aptitude. Governments and research organizations need to actively support the adoption of DTs in the agriculture sector from a strategic viewpoint. The comprehensive adoption of smart farming technology will be aided by open-access system, research funding, and subsidies. Therefore, the application of DT in integration with machinery, AI, and IoT in the agriculture sector is essential for managing natural resources sensibly and production of food in a sustainable manner to feed the huge population of the world.

References

Alves RG, Maia RF, Lima F (2023) Development of a digital twin for smart farming: irrigation management system for water saving. J Clean Prod 388:135920

Asibi AE, Chai Q, Coulter JA (2019) Rice blast: a disease with implications for global food security. Agronomy 9(8):451

Awais M, Wang X, Hussain S, Aziz F, Mahmood MQ (2025) Advancing precision agriculture through digital twins and smart farming technologies: a review. Agri Eng 7(5):137

Dai M, Shen Y, Li X, Liu J, Zhang S, Miao H (2024) Digital twin system of Pest management driven by data and model fusion. Agriculture 14(7):1099

Escribà-Gelonch M, Liang S, van Schalkwyk P, Fisk I, Long NVD, Hessel V (2024) Digital twins in agriculture: orchestration and applications. J Agric Food Chem 72(19):10737–10752

Falcon WP, Naylor RL, Shankar ND (2022) Rethinking global food demand for 2050. Popul Dev Rev 48(4):921–957

Goldenits G, Mallinger K, Raubitzek S, Neubauer T (2024) Current applications and potential future directions of reinforcement learning-based digital twins in agriculture. Smart Agric Technol 8:100512

Hoffmann J, Bauer P, Sandu I, Wedi N, Geenen T, Thiemert D (2023) Destination earth—a digital twin in support of climate services. Clim Serv 30

Ibrahimov F, Rzayeva U, Balayev R (2023) Opportunities and perspectives of the digital twins' conception: the case in agriculture. East-Eur J Enterp Technol 1(13):121

Kalyani Y, Vorster L, Whetton R, Collier R (2024) Application scenarios of digital twins for smart crop farming through cloud–fog–edge infrastructure. Future Internet 16(3):100

Khan LU, Han Z, Saad W, Hossain E, Guizani M, Hong CS (2022) Digital twin of wireless systems: overview, taxonomy, challenges, and opportunities. IEEE Commun Surv Tutor 24(4): 2230–2254

Liu M, Fang S, Dong H, Xu C (2021) Review of digital twin about concepts, technologies, and industrial applications. J Manuf Syst 58:346–361

Manivasagam VS (2025) From bytes to farm: transferability of industrial digital twins in agricultural systems. J Biosyst Eng 50(1):130–144

Melesse TY (2025) Digital twin-based applications in crop monitoring. Heliyon 11(2):e42137

Neethirajan S, Kemp B (2021) Digital twins in livestock farming. Animals 11(4):1008

Negri E, Fumagalli L, Macchi M (2017) A review of the roles of digital twin in CPS-based production systems. Procedia Manuf 11:939–948

Parewai I, Köppen M (2025) A digital twin approach for soil moisture measurement with physically based rendering simulations and machine learning. Electronics 14(2):395

Peladarinos N, Piromalis D, Cheimaras V, Tserepas E, Munteanu RA, Papageorgas P (2023) Enhancing smart agriculture by implementing digital twins: a comprehensive review. Sensors 23(16):7128

Prakash C, Singh LP, Gupta A, Lohan SK (2023) Advancements in smart farming: a comprehensive review of IoT, wireless communication, sensors, and hardware for agricultural automation. Sensors Actuators A Phys 362:114605

Prause L (2021) Digital agriculture and labor: a few challenges for social sustainability. Sustainability 13(11):5980

Pylianidis C, Osinga S, Athanasiadis IN (2021) Introducing digital twins to agriculture. Comput Electron Agric 184:105942

Rajeswari D, Parthiban AV, Ponnusamy S (2024) Digital twin-based crop yield prediction in agriculture. In: Harnessing AI and digital twin technologies in businesses. IGI global, Hershey, pp 99–110

Shivappa R, Navadagi DB, Baite MS, Yadav MK, Rathinam PS, Umapathy K et al (2021) Emerging minor diseases of rice in India: losses and management strategies. In: Integrative advances in rice research. IntechOpen

Slob N, Hurst W (2022) Digital twins and industry 4.0 technologies for agricultural greenhouses. Smart Cities 5(3):1179–1192

Tao F, Qi Q (2019) Make more digital twins. Nature 573(7775):490–491

Trienekens JH, Van der Vorst JGAJ, Verdouw CN (2014) Global food supply chains. In: Encyclopedia of agriculture and food systems, 2nd edn. Academic Press, pp 499–517

Wang L (2024) Digital twins in agriculture: a review of recent progress and open issues. Electronics 13(11):2209

Zhang Y, Zhang Y, Gao M, Dai B, Kou S, Wang X et al (2023) Digital twin perception and modeling method for feeding behavior of dairy cows. Comput Electron Agric 214:108181

Zhang R, Zhu H, Chang Q, Mao Q (2025) A comprehensive review of digital twins technology in agriculture. Agriculture (Basel) 15(9):903

13 Digital Twin Applications in Agriculture: Emerging Prospects and Opportunities

Arti, Garima, Yogesh Mohan, and Sandeep Kumar

Abstract

Agriculture practices are undergoing a significant transformation through the integration of digital technologies to address global challenges such as climate change, population growth, and resource scarcity. Digital Twin (DT), among these innovations, has gained importance as a transformative concept for agriculture, enabling real-time virtual representations of possessions, processes, and equipment. DTs support predictive modeling, scenario analysis, and data-driven decision-making across domains such as crop cultivation, livestock management, controlled environment farming, and water resource optimization through the integration of Internet of Things (IoT) sensors, cloud-fog-edge computing, artificial intelligence, and big data analytics. These applications boost productivity, promote utilization of sustainable resource, and strengthen the resilience of agricultural systems. However, the widespread implementation of DTs faces multiple barriers including data heterogeneity, insufficient infrastructure, high deployment costs, and socio-economic inequalities. This research paper also discusses federated DT networks, Industry 5.0 interfaces, blockchain-enabled data ecosystems, and quantum-assisted simulations. As a foundation stone of Agriculture 5.0, Digital Twins (DTs) have the potential to transform global food systems by driving smart, sustainable, and inclusive farming practices.

Arti (✉) · Garima
Sanatan Dharma College, Ambala Cantt, Kurukshetra University, Kurukshetra, India
e-mail: artisachdeva@sdcollegeambala.ac.in

Y. Mohan · S. Kumar
Department of Computer Science, Himachal Pradesh University, Shimla, India

© The Author(s), under exclusive license to Springer Nature Singapore Pte Ltd. 2026

A. L. Srivastav et al. (eds.), *Digital Twin Technology for Sustainable Agriculture*,
https://doi.org/10.1007/978-981-95-5915-2_13

Keywords

Digital Twin · Federated Digital Twin · Internet of Things · Smart agriculture · Precision farming

13.1 Introduction to Emerging Technologies in Agriculture

13.1.1 Evolution and Importance of Digital Technologies in Agriculture

The agriculture sector has seen a revolutionary transformation in the last few decades due to the excessive use of digital technology. The traditional methods of farming depend on human labor and primitive equipment but with the use of the latest technologies make the farming techniques more sustainable and productive. This revolution is driven by the need to cure complex problems like a changing climate and growing population that is putting the world's food security at risk due to scarce resources. The aim is to develop smart agricultural systems using the digital technologies such as the Internet of Things (IoT), Artificial Intelligence (AI), Big Data analytics, and Digital Twins to ensure maximum efficiency while reducing their impact on the environment.

The fourth industrial revolution or Industry 4.0 technologies will allow the integration of the real and virtual worlds and make follow-ups possible in real time, scale predictive analytics, and automation. Since IoT and robotics to precision agriculture and smart farming techniques, precision agriculture has been a core taproot of industry 4.0 enabling technologies that have contributed to improve quality of produce with implications on both quality and utilization. Digital solutions have also become more popular in the agri-food sector due to international issues like the climate crisis or the COVID-19 pandemic (Hassoun et al. 2023).

Also, the development of new technologies makes it possible to capitalize on agricultural practices in a sustainable manner which in turn makes it possible to meet the growing food demand thus safeguarding the integrity of ecosystems (Klerkx et al. 2019; Saikanth et al. 2024). Digital transformation will encompass some technological gains and also socioeconomic changes that are related to management of knowledge in farming communities, decision-making process, and market access. In this way, the technologies are fully productive as well as promoting rural development and promoting inclusive growth (Klerkx et al. 2019; Singh et al. 2021; Saikanth et al. 2024).

13.1.2 Concept and Definition of Digital Twin Technology

Industry 4.0 is built around the Digital Twin (DT) technology, which consists of real-time synchronized virtual models of physical assets and processes. DTs, which started in manufacturing, can be used to simulate, monitor, and proactively maintain, connecting the digital and physical worlds, to support descriptive, prescriptive, and predictive modeling to achieve increased operational efficiency.

Digital twins do not emulate digital threads, which are systems that handle the flow of data across life cycles, or cyber-physical systems, which carry out computational components, and are virtual models of the real-time physical objects. They are critical in complex fields such as agriculture because both biotic and abiotic parameters have to be simulated and supported by these types of tools (Singh et al. 2021).

Digital twins encompass a series of attributes such as bidirectional data transfer, real-time synchronization with the physical assets, autonomous control, and scenario-based simulations. These attributes enable digital twins not only to reflect the current state but also to forecast future states or estimating their possible response to hypothetical changes thereby supporting evidence-based decision-making. By integrating sensor-based inputs, historical records, and computational models ranging from physics-based approaches to artificial intelligence algorithms, digital twins collaboratively manage, govern, and optimize complex systems.

13.1.3 Scope and Objectives of Digital Twin in Agriculture

The adoption of digital twin technology within agriculture including crop, livestock, and forest farming has gained significant momentum, although its development is still in the early stages compared to its use in manufacturing and industrial application. In agriculture, digital twins enable the convergence of physical and virtual spaces through the convergence of data from diverse sources, hence improving operational efficiencies that enhance productivity, optimize the use of resources, and secure sustainability. Their uses extend across a wide range of activities, such as crop planting and monitoring, precise irrigation, pest control, livestock health monitoring, and managing environment agriculture systems.

Two key uses of digital twins in the agro-industry are: first, optimizing the production systems through reduced waste of resources and second is to enhance real-time detection of abiotic and biotic stressors. It can make predictions and prevent risks like disease or climatic stresses and support good decision-making by running simulations on various agricultural set-ups. Also, digital twins support sustainable use of natural resources and boost the circular economy by optimizing inputs such as water, fertilizer, and energy (Tagarakis et al. 2024).

In the general domains of precision agriculture and smart agriculture, digital twins are sophisticated systems supplementing sensor networks and data analysis methods. Smart agriculture is generally the integration of information and communication technologies to improve farming practices, while precision agriculture is related to the management of specific areas by taking into account spatial variability. In addition, digital twins provide an integrated and comprehensive model that simulates dynamic interactions and feedback loops in agricultural systems and thus offer a more comprehensive and interactive digital platform for management (Cesco et al. 2023; Melesse 2025).

13.1.4 Stages of Evolution

The evolution of Digital Twin (DT) technology has been a continuous progression driven by computing, connectivity, and artificial intelligence. Initially developed with aerospace sector, the innovation has evolved gradually into a revolutionary medium used across multiple sectors, including agriculture. This chronology of evolution offers insight into the transition of Digital Twins from theoretical representations to tangible, real-time decision-support tools in agriculture and food systems.

Stage 1: Conceptualization (1960s–2000s)
Originally, Digital Twin technology can be linked to NASA's Apollo missions during the 1960s and 1970s, when engineers developed mirrored spacecraft simulations to guarantee mission safety, identify risks, and test solutions from Earth. This was the initial concept of forming a digital duplicate of a physical system. However, due to constraints in computing power and the lack of sophisticated networking, these simulations were limited to static models and manual data entry (Allen 2021). In 2002, Dr. Michael Grieves officially presented the notion of the Digital Twin in relation to manufacturing, highlighting the potential for developing virtual replicas of physical assets to manage product lifecycles. At this point, the idea mostly stayed theoretical and academic, representing more of a vision for future industrial advancement rather than a technology commonly utilized. It provided the conceptual groundwork but was deficient in real-time integration, high-speed computing capabilities, and practical scalability (Grieves 2016).

Stage 2: Early Adoption (2000–2010)
The Digital Twin technology initially found application in precision-based industries such as in aerospace and automotive, where it facilitated design validation and evaluation of performance, as well as maintenance planning. The major benefit of this was its ability to test the designs virtually in order to reduce costs and risk but the first designs were sedentary and would only be updated periodically. Also, they relied heavily on pre-existing data and not live data feeds. Without IoT connectivity, these systems were unable to be flexible and responsive. Despite these difficulties, this phase was nonetheless important as it demonstrated the applied value of Digital Twins for engineering and product optimization, which in turn led to the real-time integration in advanced stages (Fuller et al. 2020).

Stage 3: Integration with IoT (2010–2015)
Digital Twins have been developed to work with IoT, cloud computing, and big data to evolve common models into active and real-time systems. They were deployed in sectors like energy, manufacturing, and logistics for tracking the assets and performing predictive analysis. Also, access to cloud computing enabled them to do computations on large chunks of data, bringing the possibility of integrating AI and ML to make high-level decisions (Sharma et al. 2020).

Stage 4: Expansion Across Sectors (2015–2020)
Digital Twins also ventured into other areas such as energy, healthcare, agriculture, urban planning, and supply chains to become increasingly autonomous with AI and ML-driven predictive and prescriptive analytics. Their applications included uses in personalized healthcare and precision farming, traffic and disaster simulations, and sustainability and resource efficiency became key objectives toward economic and environmental gains (Jeong et al. 2022).

Stage 5: Ecosystem Stage (2020–Present)
The transition to single asset models to integrated system-wide digital ecosystems is the next step in Digitizing. In this respect, it is necessary to promote cross-industrial collaboration of numerous twins which communicate and integrate in order to reproduce simulated complex spaces (Jeong et al. 2022). Smart cities are now modeled as Digital Twin ecosystems incorporating the energy sector, transport sector, healthcare, and infrastructure with a focus on governance, interoperability, and data security. They can be improved with AI, real-time analytics, and edge computing, thereby facilitating predictive resilience, sustainability, and climate change adaptation solutions (Fuller et al. 2020).

Stage 6: Future Vision (2030 Onwards)
The future of Digital Twin technology will become more interconnected with emerging or future technologies such as 6G networks, blockchain, quantum computing, and the metaverse. These advances will allow high level of safety, speed, and complex Digital Twin systems to operate with almost zero human input. Digital Twins will eventually be able to simulate whole societies, entire industries, or ecosystems, providing decision makers with new levels of predictive and strategic insight. For example, climate-smart agriculture could be revolutionized to model weather, soil conditions, and crop growth. Similarly, blockchain integration would improve data security and trust among multiple parties, while quantum computing would unlock simulations of unparalleled complexity and speed. Ultimately, the goal is to create self-sustaining, intelligent, and autonomous digital ecosystems with transformative potential across industries, governance, and sustainability on a global scale (Sharma et al. 2020; Jeong et al. 2022) (Figs. 13.1 and 13.2; Table 13.1).

13.2 Architectural Foundations and Enabling Technologies of Digital Twins in Agriculture

13.2.1 IoT and Sensor Technologies as Data Acquisition Backbone

The implementation of digital twin technology in agriculture consists of Internet of Things (IoT) sensors and devices that continuously collect real-time data from the physical world. The devices track parameters of crop like soil moisture, temperature, humidity, pH levels, nutrient levels, and sun exposure.

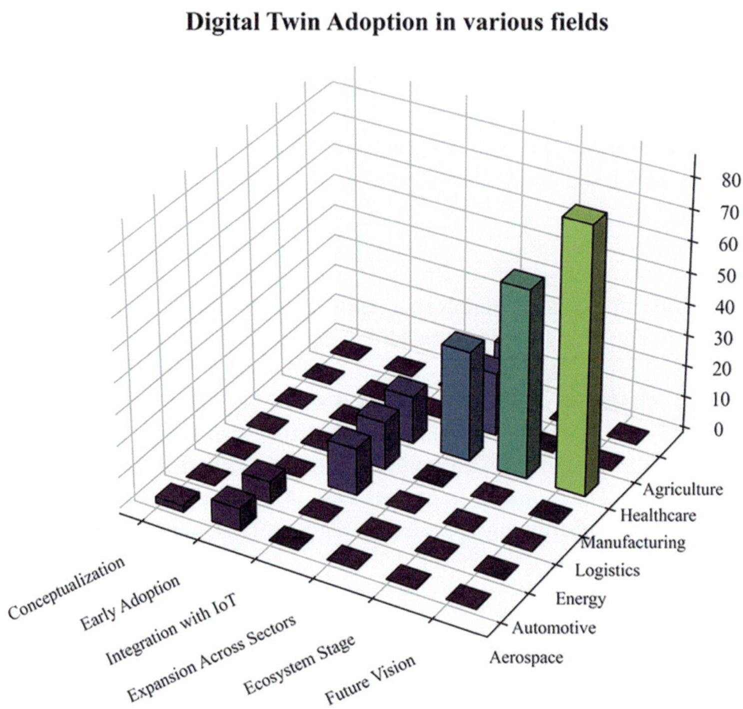

Fig. 13.1 Digital twin adoption in various fields

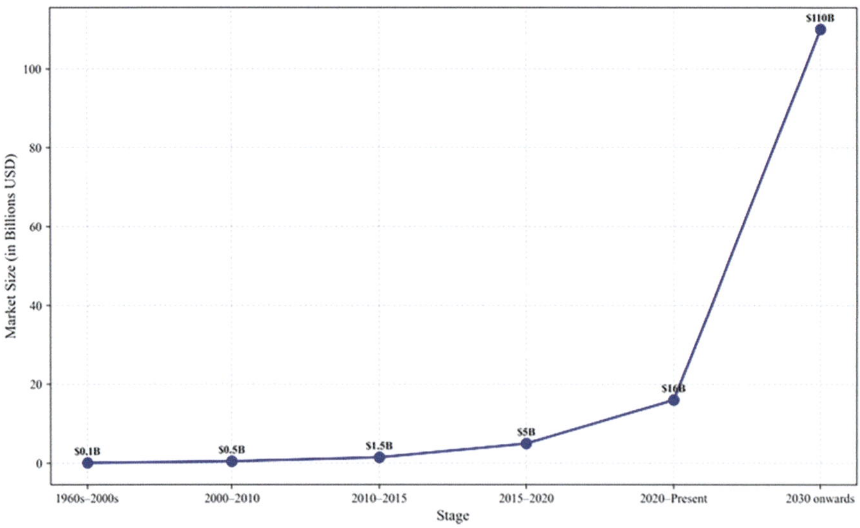

Fig. 13.2 Digital twin market growth over time

Table 13.1 Evolution of Digital Twins in agriculture across stages, key technologies, and relevance (Klerkx et al. 2019; Tao et al. 2019; Jones et al. 2020; Saikanth et al. 2024)

Stage	Key technologies	Agriculture relevance
Conceptualization	Simulation models, CAD, early PLM	Early crop growth and soil simulation models; static DSS (Decision Support Systems) for irrigation and weather prediction
Early adoption	CAD/CAM, PLM, static simulations	Use of computer-based tools in greenhouse climate control and basic crop yield forecasting; isolated digital models without IoT integration
Integration with IoT	IoT, cloud, big data, visualization	Precision farming begins with IoT sensors in soil and crops; drone-based monitoring; real-time farm data collection for decision-making
Expansion across sectors	AI, ML, IoT, big data, edge analytics	Digital twins simulate farm ecosystems (soil–crop–weather); predictive models for pest/disease outbreaks; livestock monitoring; focus on sustainability
Ecosystem stage	AI, IoT, Blockchain, edge computing	Integration of farm-level DTs with regional weather, supply chains, and smart farming platforms; blockchain for traceability; resource efficiency optimization.
Future vision	6G, Blockchain, quantum computing, autonomous AI	Global-scale agricultural DTs for climate-smart farming; quantum-powered crop-climate simulations; metaverse-enabled virtual farm testing; food system resilience planning

IoT sensors in agriculture are afflicted with various issues like different sensors, heterogeneous hardware and software platforms and unreliable data communication. However, to capture the richness and the diversity of agricultural landscapes, it is essential to properly integrate heterogeneous data sources to support the development of holistic models (Purcell and Neubauer 2023).

Real-world applications involve sensor networks in fields and controlled spaces that collect data and feed it as input into digital twin systems. For example, sensors are used for measuring soil moisture and nutrients that deliver crucial information to inform irrigation and fertilization planning. With the combination of different sensors with digital twin models, crop models can be continuously updated and farmers can actually see in real-time how crops are growing and developing. This enhances prediction accuracy and recommendations, enabling farmers to make data-driven, more informed decisions that enhance productivity and promote sustainability.

The IoT device integration with digital twins also maximizes resource utilization efficiency through accurate input management and enabling early detection of anomalies. However, sensor network scalability and reliability remain an area of research especially with the volume of data generated presenting storage, processing, and management issues (Ramachandran et al. 2022).

13.2.2 Cloud, Fog, and Edge Computing for Data Processing and Management

To address the data deluge from IoT devices and maintain real-time responsiveness, the Cloud-Fog-Edge computing paradigm is fundamental for digital twin deployment in agriculture. Cloud computing provides extensive storage and computational resources to manage large datasets, run complex simulations, and store historical records. However, due to latencies and connectivity constraints typical in agricultural environments, purely cloud-based solutions may cause delays detrimental to real-time decision-making.

Fog and edge computing supplement cloud capabilities by the ability to offload processing capacity close to points of data origin. Edge computing does initial processing of data close to or at the sensor node point, reducing latency and bandwidth usage, whereas fog computing is an intermediate layer providing localized data processing, aggregation, and analysis before forwarding to the cloud. This hierarchical architecture supports scalable models, with primary decisions made at the edge while providing rich analytical capabilities to centralized cloud infrastructures.

Recent proposals for digital twins in agricultural applications use multi-agent systems based on microservices architectures implemented on cloud, fog, and edge computing layers. Such frameworks can be useful for building scalable, modular, and context-aware systems where different agents/microservices are used to perform specific functions (data collecting, resources scheduling, cognitive decision-making, etc.) (Kalyani et al. 2023).

Because of the strengths of all computing paradigms, the cloud, fog, and edge layers coordinate with one another and lend overall robustness and efficiency to the system. The architecture addresses the problem of unreliable network conditions for rural farming areas, while at the same time supporting the flow requirements critical for real-time operation inherent to precision agriculture applications (Fuller et al. 2020; Habibi et al. 2020). Further, federated computing allows for coherent digital twin systems to be populated and interlinked with multiple farms (or agricultural units) for collective learning and widespread knowledge exchange (Vergara et al. 2023).

13.2.3 Artificial Intelligence and Machine Learning Integration

AI and ML capacities of agricultural digital twins provide simulation, forecasting, optimization, and automation capabilities for crop growth forecasting, disease, pest, and resource modelling, and deep learning is capable of accurately modeling the environment at dynamic conditions and identifying anomalies (Taha et al. 2025).

Utilizing real-time data from various sensors, AI-powered digital twin ecosystems are being used to monitor the health of animals, forecast the occurrence of disease, and optimize feeding and breeding cycles in livestock farming ventures. Integration of DT systems with AI refers to autonomous decision making and intelligence of systems through automated anomaly detection, intelligent guidance support, adaptive control techniques, etc. (Mesías-Ruiz et al. 2023).

Further, the cognitive ideologies in AI-based digital twin architectures are adopted for extending knowledge acquisition and reasoning process in order to extend these systems for higher accuracy. Such intelligent digital twins are the next generation of smart agricultural systems able to tackle complex biological and environmental interrelations (Rathore et al. 2021).

13.3 Applications of Digital Twins in Crop Production and Management

13.3.1 Crop Monitoring, Growth Simulation, and Health Assessment

Digital twins provide a strong platform that can integrate various streams of data to track the physiology of plants and simulate the growth dynamics under changing environmental conditions. DT models dynamically model the interactions between crops and their environment by utilizing continuously streaming sensor data to model the crop.

Such models can simulate physiological changes, for example, in photosynthetic rate, nutrient uptake, water stress and phenological stages and thus capture events of early adverse conditions such as drought stress, nutrient deficiency, or pathogen attack. Predictive simulations allow farmers to estimate the course of crop growth over time and plan the allocation of resources in advance. For example, for improving grape quality, vineyards can use DTs to track soil moisture and nutrient levels to enable precision viticulture (Faluomi 2023).

Digital twins can integrate complex plant-environmental correlations and allow simulation of scenarios such as the impact of changing weather patterns or treatment options that can be applied to increase or decrease crop yields. Data-driven simulation models come with decision support systems and reduce uncertainty to offer farm and field-level actionable information (Awais et al. 2025).

In addition, digital twins are contributing to the sustainability of the agricultural ecosystem through surveillance which allows us to reduce the excessive use of chemicals and preventative treatment and real-time feedback loops to health monitoring status for dynamic adjustments, timely relief against stressors, minimizing crop losses, and guaranteeing higher productivity (Zhang et al. 2025).

13.3.2 Precision Irrigation and Resource Optimization

Digital twin technology can enhance precision irrigation with real-time soil, weather, and crop data being coupled with simulation models to predict water availability, evapotranspiration, and crop stress. When used with GIS and remote sensing, DTs allow site-specific water management to maximize yield and quality (Bwambale et al. 2022).

Based on the case studies, it can be seen that the DT-guided irrigation had significant efficiencies in water conservation, wastage reduction, and demand-supply equilibrium. Such innovations are justified not only on the grounds of sustainability, but also economically to those farms where water is not freely available (Ramachandran et al. 2022).

Integration of IoT-based sensors along with DT-enabled irrigation systems enables for automated, data-based irrigation management to reduce human errors and to adapt to environmental changes. Water conservation is further improved by using predictive analytics that determine the need for water based on weather forecasts and soil moisture trends (Purcell and Neubauer 2023).

13.3.3 Pest, Disease Control, and Crop Protection Using DT

Precision crop protection holds revolutionary prospects in bringing together digital twins and AI/ML technologies. Digital twin refers to a combination of sensors and predictive models to perform simulations of pest population dynamics and disease accumulation that could assist in the early warning and control decisions on pests and diseases.

This could involve applying AI-enabled models to DTs to conduct anomaly detection regarding pest/pathogen infestation based on sensor data collected by devices such as spectral cameras, microclimate sensors, and crop stress sensors. These systems give early warning and suggestions for targeted interventions thus reducing chemical pesticide application and environmental effects (Mesías-Ruiz et al. 2023).

Additional solutions to these issues include sensorization of the DT with automated surveillance, spraying, and mechanical treatment. Such integration ensures the highest possible accuracy and sustainability in crop protection, while simultaneously reducing chemical residues and maintaining the presence of organisms which have a positive impact on an ecosystem (Kalyani et al. 2023).

Further, digital twins can help to evaluate and simulate crop protection strategies and experiment with alternative strategies, including integrated pest management and biological control. By continuously monitoring efficacy and environmental conditions, DT-enabled systems support sustainable agriculture and protect productivity (Zhang et al. 2025).

13.4 Digital Twins in Livestock and Animal Farming

13.4.1 Virtual Modeling of Animal Housing and Environmental Control

In livestock production, Digital Twins are being used to simulate animal housing and optimize energy consumption and animal welfare by using sensor data from HVAC and climate systems and monitoring devices. Studies about Pig house indicated that when combined with optimal control, DT frameworks could be employed to analyze energy needs to support heating and ventilation and lead to an optimal system. These models also ensure to get valuable air quality and thermal feedback and will benefit healthier working environments and drive productivity (Jeong et al. 2023).

Grassland management service-based systems can be used to provide minimal viable implementations of DT that can be extended as needed and adapted to accommodate other kinds of farms, while increasing the utilization in animal agriculture (Taubert et al. 2024). Smart agri-systems and digital twins can improve precise livestock housing control resulting in improved energy efficiency, livestock welfare, and sustainability (Collins and Smith 2022).

13.4.2 Health Monitoring, Disease Prevention, and Management

Animal health management can leverage digital twin technology for real-time monitoring, simulation, and disease dynamic prediction. Filtration systems, electrostatic precipitators, and ultraviolet inactivation are just a few of the control technologies installed in DTs that reduce pathogen spread through ventilation systems. Emerging technologies such as ionization and reactive oxygen species generators are an area of active research and DT-based testing frameworks provide a way of safety and efficacy testing before rollout into the field (Ouyang et al. 2023).

Digital twins can be used to predict disease outbreaks early, by correlating environmental, physiological, and behavioral data for intervention. The unique capability to simulate different mitigation strategies in the digital environment increases responsiveness and allocation of resources, while improving animal health outcomes (Gupta et al. 2020).

Furthermore, these systems contribute to veterinary biosecurity by modeling pathogen transmission pathways and providing real-time alerts, ultimately reducing economic losses and improving livestock resilience (Ouyang et al. 2023).

13.4.3 Precision Feeding and Reproduction Management

Digital twins enable the optimization of nutrition and reproductive management by modeling metabolic needs and reproductive cycles based on continuous data from sensors. DTs have the potential to synchronize feeding regimes with physiological states in animals, leading to increased feed efficiency and decreased waste.

Predictive models are integrated in DTs for forecasting optimal timing of breeding and early warning of reproductive problems. These traits increase the sustainability and profitability of livestock operations through improved reproductive rates while reducing production cycles.

Although there is little in the literature in terms of specific detailed studies of precision feeding with DTs, reviews of digital twin applications in livestock suggest the need for increased research in this area in order to achieve the DT-enabled decision support's potential (Tagarakis et al. 2024).

13.5 Controlled Environment Agriculture (CEA) and Digital Twin Integration

13.5.1 DT Architectures for Climate Control and Crop Management in CEA

Digital twin architectures used in Controlled Environment Agriculture achieve climate and crop management simulation and optimization by modelling microclimate conditions such as temperature, humidity, light, and CO2 in a greenhouse or plant factory.

By enabling the simulation of climate control strategies and treatment schedules, digital twins help optimize the balance between crop yields and resource consumption, chiefly energy and water. Various case studies validate that these architectures hold significant potential to enhance production efficiency while maintaining crop quality (Chaux et al. 2021).

Infrared photovoltaic panels and saltwater cooling are implemented in new greenhouse designs in extreme areas, such as the Middle East and North Africa, and digital twins combine renewable energy and climate control to make agriculture more sustainable (Lefers et al. 2020).

Furthermore, plant factory control systems based on DT technology provide intuitive interfaces to show power usage and environmental parameters that facilitate operational adjustments to improve sustainability and productivity (Liu et al. 2023).

13.5.2 Emerging Technologies for Extreme Environment Agriculture

Digital twins also have applications in agriculture in hard environments—where agricultural systems are built and run to fit the local conditions. These new processes represent one part of the answer to an increase in sustainable biomass production on marginal lands and progress toward circular carbon economies. As DT-driven models are best able to compute flows of resources and environmental connections for extreme climates, these might lead to the maximization of input and energy consumption (Lefers et al. 2020).

The integration of new technologies such as renewables with controlled environment agriculture systems supported by digital twins creates new frontiers for food production security and sustainable biomass production in geographical locations that were previously unusable for agriculture (Lu et al. 2023).

13.5.3 Digital Twins and Algae-Based Biotechnologies

Algae-based biotechnology is a new frontier in sustainable agriculture where digital twins help optimize algae cultivation for biofertilizer and biomass production. DT systems integrate data on algae growth conditions, nutrient dynamics, and wastewater quality that can simulate and control algae-based processes.

Digital twins in this area address productivity limitations, contamination risks, and resource consumption in algal biofertilizer production. By allowing precise management of cultivation parameters, DTs overcome technical barriers and improve system efficiency. This integration will enable environmentally friendly, resource recycling agricultural practices that form the basis of an algae-based bioeconomy supporting sustainable agriculture (Fabris et al. 2020; Lu et al. 2023).

13.6 Digital Twin-Driven Smart Irrigation and Water Management

13.6.1 Advances in IoT-Enabled Smart Irrigation Systems

IoT devices and digital twins smart irrigation systems adjust irrigation schedules in real time based on data inputs. Sensor networks measure soil moisture, climate and crop water requirements, feed digital twin models that simulate irrigation needs and deliver water.

DT-supported dynamic irrigation scheduling reduces water waste and improves crop water productivity by matching supply to demand. Automation minimizes human error and adapts to changing conditions (Ramachandran et al. 2022).

Case studies show that smart irrigation systems integrated with DTs save water and maintain or increase yields, emphasizing their key role to water sustainability (Zhang et al. 2025).

13.6.2 GIS and Remote Sensing Integration with Digital Twins

Geographic Information Systems (GIS) and remote sensing provide spatially explicit data that enhance the accuracy and scalability of digital twin models in water management. Integrating these technologies with DTs enables precise mapping of irrigation zones, soil properties, and crop water requirements.

Remote sensing contributes satellite and aerial imagery data layers that feed into DTs, improving real-time sensing and modeling capabilities. This integration facilitates site-specific water management but also helps to assess the performance of irrigation at larger scales (Bwambale et al. 2022).

These three (GIS, remote sensing, and digital twins) enable resource optimization in irrigation management to enable policymakers, researchers, and farmers to make data-driven decisions to enhance sustainability (Purcell and Neubauer 2023).

13.6.3 Machine Learning and Predictive Analytics in Irrigation Management

Even more advanced machine learning models incorporated into a digital twin framework can help better predict the need for irrigation for both historical and real-time data. These models can be used for predicting irrigation scheduling, future climate, soil moisture modelling, and crop water stress simulation.

Integrating information from other data sources, such as climate models and weather forecasts, is also changing the way we manage water; not only does this make better use of information, but it also reduces the risk of over- or under-supply (Mesías-Ruiz et al. 2023).

Predictive Analytics are also being used in the development of decision-support systems to provide guidance on optimal irrigation schedules and amounts required in complex and variable agricultural environments, to support sustainable water resource management (Pawar et al. 2022).

13.7 Challenges and Limitations in Implementing Digital Twins in Agriculture

13.7.1 Data Acquisition, Integration, and Standardization Issues

However, the application of digital twins within agriculture still encounters major challenges in data acquisition, integration, and standardization. Technical issues are raised about multi-source datasets that can be combined into coherent DT models, if different data formats, sampling rates, and spatial-temporal resolutions are encountered. Further complicating the integration efforts are the lack of standardized protocols for 3D crop modeling and agricultural data representation that prevent interoperability among systems and domains (Bwambale et al. 2022).

Scaling of DT solutions is also impeded by these data-related barriers, because without harmonization, the fidelity and reliability of digital twins are compromised, limiting their utility for decision support and optimization (Tagarakis et al. 2024).

13.7.2 Technological and Infrastructure Barriers

Adaptive infrastructure is usually lacking in farmland, particularly in rural or remote regions, influencing the adoption level of advanced digital twin technologies. Insufficient network coverage, unstable power output, or the complexity and high cost of sensor deployment and maintenance may be the causes of those limitations.

Edge, fog, and cloud computing infrastructures for low latency on a large scale data processing also require the investment in lots of capital and technical skills that are not available to resource poor or smallholders farmers Generally speaking, plethora of capital and technical skills are required to deploy with edge, fog and cloud computing infrastructures for low latency massive data processing, which normally are not at the reach of resource discriminated or smallholder farmers (Stakens et al. 2023).

Furthermore, the energy intensive use cases of sensor use and use of the communication modules have to be limited by the availability of energy, which affects the continuity in the acquisition of data, and the communicating abilities of the systems.

Addressing these infrastructure constraints is considered to be important in ensuring equitable access and to maximize usage of digital twin technologies in agriculture (Kalyani et al. 2023).

13.7.3 Socioeconomic and Organizational Constraints

In addition to technology challenges, it becomes increasingly important to take into account the organizational structures and socio-economic factors that influence the adoption of digital twins in agriculture. It also suffers problems on high cost of investment, inadequate technical expertise at the farm household level, and problems of data ownership, privacy, and security.

The ethical issues usually concern privacy of the data and the possible exclusion of the food security scenarios of small-scale farmers who might not be in a position to be able to use high-tech solutions. Technical solutions, models of participatory governance, and enabling policy regimes are needed to overcome these challenges (Gupta et al. 2020).

Secondly, tailored approaches to technology diffusion that are contextually respectful and promote a balanced inclusion are required due to the fragmented agricultural value chains and multi-stakeholder involvement (Klerkx et al. 2019; Saikanth et al. 2024) (Fig. 13.3).

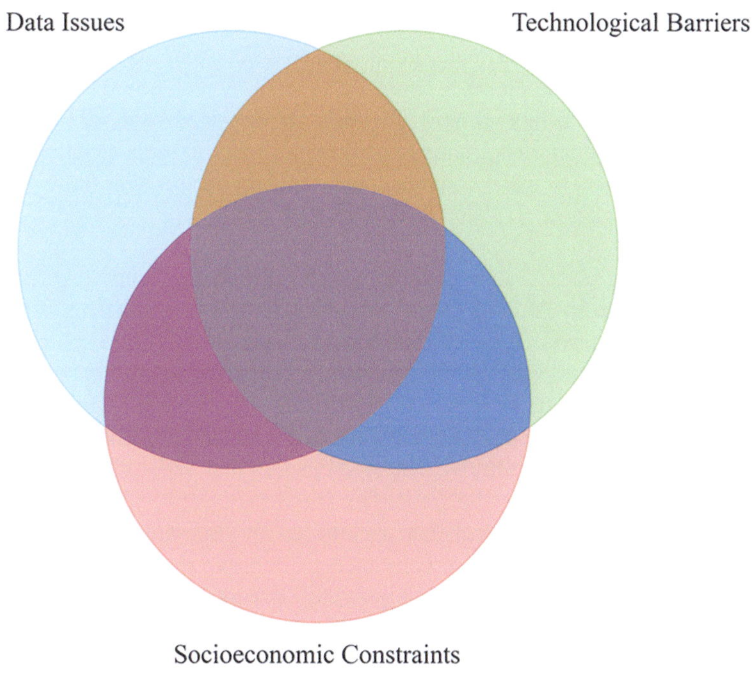

Fig. 13.3 Challenges in implementing Digital Twins in agriculture

13.8 Future Directions and Research Opportunities

13.8.1 Federated and Collaborative Digital Twin Networks

As we visualize this union as being fulfilled in the future, we can imagine a federated digital twin which means the process of connecting a number of digital twins. They are federated systems capable of learning, knowing, making decisions, and acting synergetically around certain overlapping subsets of agri-ecosystems.

FDT infrastructures enable every digital twin to exercise autonomous operation, but with the collective intelligence and information resources of a storied use case. This would enable large-scale agricultural landscape scaling, resilience, and adaptability, enabling the collaboration of multi-stakeholders and improve optimization of the system overall (Vergara et al. 2023).

13.8.2 Industry 5.0 and Extended Reality Technologies

Emerging frameworks of Industry 5.0 focus on human-centric technologies enabling operators to use improved decision-support tools. Digital twins are paired with Extended Reality (XR) such as virtual, augmented, and mixed reality interfaces, to create immersive and interactive environments to engage with digital models.

This connection provides improved user interaction, support for training, and personalized farming solutions as operators can access and manipulate digital twin data. Hypermedia multi-agent systems can enhance cognitive flexibility, semantic scalability, and the ability to make good adaptive decisions for the digital twin ecosystems (Kolekar et al. 2023).

This convergence offers a new direction in the landscape of agriculture, and digital twins hold a transformative potential as powerful tools enabling farmers and other stakeholders to make better decisions and connect to their farms virtually and in real time (Fuller et al. 2020; Akundi et al. 2022).

13.8.3 Advances in AI, Big Data, and Quantum Computing for DT Enhancement

The evolution of digital twins is also closely linked with the evolution of Artificial Intelligence, big data analytics, and new computing paradigms such as quantum computing. The intelligent, adaptive, and automated digital twins powered by AI can learn independently and understand complex scenarios.

Large-scale, rapid agricultural data can be easily ingested into the big data platforms, and multiple inputs are combined to improve their model results and decision support. Quantum computing provides a computational benefit and can perform simulations in a much faster and more accurate way than is computationally feasible with classical systems (Hubbard et al. 2020).

These technological developments enable digital twins to transform into autonomous agents of smart agro-ecosystems whose optimization and prediction capabilities can be scaled and utilized in real time and with previously unprecedented speed (Rathore et al. 2021; Taha et al. 2025).

13.9 Comparative Analysis of Digital Twin with Other Emerging Agricultural Technologies

13.9.1 Digital Twin in Relation to Smart Agriculture and Precision Agriculture

While smart agriculture is a broader term for the use of digital and communication technologies in the whole farming value chain, precision agriculture is about site-specific management. Using mass data, digital twins add a dynamic integrative layer of simulative images to improve real-time response, monitoring, and testing scenarios.

Digital twins are supplementary to sensor networks and data analytics, providing the ability to predict and prescriptively analyze with a detailed view of the complex interactions within a holistic, virtual world. Unlike other methods that emphasize more on descriptive analytics or measurement, DTs alone facilitate virtual experimentation and optimization, which offer actionable insights beyond traditional frameworks (Cesco et al. 2023).

13.9.2 Integration of Internet of Things and Blockchain with Digital Twins

The Internet of Things (IoT) constitutes the foundational layer for digital twins by capturing physical data essential to virtual modeling. IoT devices continuously feed environmental, crop, and machine data that digital twins transform into meaningful insights.

Blockchain technology, while not inherently part of DT architecture, offers potential in securing DT data, ensuring transparency and immutability, particularly beneficial for supply chain management and data provenance within agricultural ecosystems. However, combining these technologies poses challenges related to interoperability, scalability, and data privacy (Purcell and Neubauer 2023; Saikanth et al. 2024).

13.9.3 Robotics and Automation Coupled with Digital Twin Systems

Digital twins increasingly integrate with intelligent robotics systems, facilitating sensing, actuation, and manipulation in precision agriculture. Collaborative robots working within DT frameworks enable automated field tasks such as planting, weeding, and harvesting, enhancing operational efficiency and precision.

AI-enhanced machinery coordinated through digital twins allows fine-grained control and real-time adjustment, reducing labor needs and increasing productivity. This integration is vital for realizing the full potential of smart, automated farming systems under the emerging Agriculture 5.0 paradigm (Mesías-Ruiz et al. 2023; Licardo et al. 2024).

13.10 Conclusion

Digital twins have emerged as transformational technologies in agriculture, enabling optimized production systems, real-time monitoring, sustainable resource management, and enhanced decision support. Evidence from diverse sectors including crop production, livestock farming, and controlled environments demonstrates substantial benefits in productivity and sustainability. The evolving maturity of digital twin architectures, incorporating cloud-fog-edge computing, AI and multi-agent systems, supports flexible, scalable deployments that address the complexity of agricultural ecosystems. These advances position DTs as foundational components in the digital agriculture landscape, contributing to smarter, more resilient farms. Despite promising progress, several barriers persist. Data acquisition and integration complexities, infrastructural and technological limitations, and socio-economic challenges including privacy, ethics, and equitable adoption impede widespread and effective DT deployment.

Looking ahead, the convergence of digital twin technology with Industry 5.0 principles, AI advancements, IoT expansion, and environmental sustainability promises to revolutionize agricultural systems. DTs are positioned to become cornerstones of resilient, adaptive, and precision agriculture, facilitating data-driven, environmentally responsible farming. Emphasizing equitable access, ethical technology adoption, and multi-stakeholder engagement is fundamental to maximizing societal benefits. Through continued innovation and collaboration, digital twins hold the potential to support global food security and sustainable rural development within a smart, interconnected agricultural future.

References

Akundi A, Euresti D, Luna S, Ankobiah W, Lopes A, Edinbarough I (2022) State of industry 5.0—analysis and identification of current research trends. Appl Syst Innov 5(1)

Allen BD (2021) Digital Twins and living models at NASA, Washington, DC

Awais M, Wang X, Hussain S, Aziz F, Mahmood MQ (2025) Advancing precision agriculture through digital twins and smart farming technologies: a review. Agri Eng 7(5)

Bwambale E, Naangmenyele Z, Iradukunda P, Agboka KM, Houessou-Dossou EAY, Akansake DA, Bisa ME, Hamadou AAH, Hakizayezu J, Onofua OE, Chikabvumbwa SR (2022) Towards precision irrigation management: a review of GIS, remote sensing and emerging technologies. Cogent Eng 9(1)

Cesco S, Sambo P, Borin M, Basso B, Orzes G, Mazzetto F (2023) Smart agriculture and digital twins: applications and challenges in a vision of sustainability. Eur J Agron 146:126809
Chaux JD, Sanchez-Londono D, Barbieri G (2021) A digital twin architecture to optimize productivity within controlled environment agriculture. Appl Sci (Switz) 11(19)
Collins LM, Smith LM (2022) Review: smart agri-systems for the pig industry. Animal 16:100518
Fabris M, Abbriano RM, Pernice M, Sutherland DL, Commault AS, Hall CC, Labeeuw L, McCauley JI, Kuzhiuparambil U, Ray P, Kahlke T, Ralph PJ (2020) Emerging technologies in algal biotechnology: toward the establishment of a sustainable, algae-based bioeconomy. Front Plant Sci 11
Faluomi V (2023) A digital twin application for vineyards sustainable management, BIO Web of Conferences, vol 68. EDP Sciences
Fuller A, Fan Z, Day C, Barlow C (2020) Digital twin: enabling technologies, challenges and open research. IEEE Access 8:108952–108971
Grieves M (2016) Origins of the Digital Twin concept
Gupta M, Abdelsalam M, Khorsandroo S, Mittal S (2020) Security and privacy in smart farming: challenges and opportunities. IEEE Access 8:34564–34584
Habibi P, Farhoudi M, Kazemian S, Khorsandi S, Leon-Garcia A (2020) Fog computing: a comprehensive architectural survey. IEEE Access 8:69105–69133
Hassoun A, Marvin HJP, Bouzembrak Y, Barba FJ, Castagnini JM, Pallarés N, Rabail R, Aadil RM, Bangar SP, Bhat R, Cropotova J, Maqsood S, Regenstein JM (2023) Digital transformation in the agri-food industry: recent applications and the role of the COVID-19 pandemic. Front Sustain Food Syst 7
Hubbard SS, Varadharajan C, Wu Y, Wainwright H, Dwivedi D (2020) Emerging technologies and radical collaboration to advance predictive understanding of watershed hydrobiogeochemistry. Hydrol Process 34(15):3175–3182
Jeong DY, Baek MS, Lim TB, Kim YW, Kim SH, Lee YT, Jung WS, Lee IB (2022) Digital twin: technology evolution stages and implementation layers with technology elements. IEEE Access 10:52609–52620
Jeong DY, Jo SK, Lee IB, Shin H, Kim JG (2023) Digital twin application: making a virtual pig house toward digital livestock farming. IEEE Access 11:121592–121602
Jones D, Snider C, Nassehi A, Yon J, Hicks B (2020) Characterising the digital twin: a systematic literature review. CIRP J Manuf Sci Technol 29:36–52
Kalyani Y, Bermeo NV, Collier R (2023) Digital twin deployment for smart agriculture in cloud-fog-edge infrastructure. Int J Parall Emergent Distrib Syst 38(6):461–476
Klerkx L, Jakku E, Labarthe P (2019) A review of social science on digital agriculture, smart farming and agriculture 4.0: new contributions and a future research agenda. NJAS—Wageningen J Life Sci 90:90–91
Kolekar A, Shalgar S, Malawade I (2023) Beyond reality: a study of integrating Digital Twins. J Phys Conf Ser 2601. Institute of Physics
Lefers RM, Tester M, Lauersen KJ (2020) Emerging technologies to enable sustainable controlled environment agriculture in the extreme environments of Middle East-North Africa coastal regions. Front Plant Sci 11
Licardo JT, Domjan M, Orehovački T (2024) Intelligent robotics—a systematic review of emerging technologies and trends. Electronics 13(3)
Liu J, Wang L, Wang Y, Xu S, Liu Y (2023) Research on the interface of sustainable plant factory based on digital twin. Sustainability 15(6)
Lu Q, Xiao Y, Wu P (2023) Emerging technologies of employing algae and microorganisms to promote the return-to-field of crop straws: a mini-review. Front Bioeng Biotechnol 11
Melesse TY (2025) Digital twin-based applications in crop monitoring. Heliyon 11(2):e42137
Mesías-Ruiz GA, Pérez-Ortiz M, Dorado J, de Castro AI, Peña JM (2023) Boosting precision crop protection towards agriculture 5.0 via machine learning and emerging technologies: a contextual review. Front Plant Sci 14

Ouyang H, Wang L, Sapkota D, Yang M, Morán J, Li L, Olson BA, Schwartz M, Hogan CJ, Torremorell M (2023) Control technologies to prevent aerosol-based disease transmission in animal agriculture production settings: a review of established and emerging approaches. Front Vet Sci 10

Pawar RS, Nema S, Jawale DR, Joshi K, Debnath S, Singh SP (2022) The role of innovative data mining approaches for analyzing and estimating the crop yield in agriculture among emerging nations, 2022 2nd International Conference on Advance Computing and Innovative Technologies in Engineering, ICACITE 2022, 2339–2342. Institute of Electrical and Electronics Engineers Inc.

Purcell W, Neubauer T (2023) Digital Twins in agriculture: a state-of-the-art review. Smart Agric Technol 3:100094

Ramachandran V, Ramalakshmi R, Kavin BP, Hussain I, Almaliki AH, Almaliki AA, Elnaggar AY, Hussein EE (2022) Exploiting IoT and its enabled technologies for irrigation needs in agriculture. Water 14(5)

Rathore MM, Shah SA, Shukla D, Bentafat E, Bakiras S (2021) The role of AI, machine learning, and big data in digital twinning: a systematic literature review, challenges, and opportunities. IEEE Access 9:32030–32052

Saikanth DRK, Ragini M, Tripathi G, Rajesh K, Giri A, Pandey SK, Verma LK (2024) The impact of emerging technologies on sustainable agriculture and rural development. Int J Environ Climate Change 14(1):253–263

Sharma A, Kosasih E, Zhang, J., Brintrup, A. Calinescu, A., 2020, 'Digital twins: state of the art theory and practice, challenges, and open research questions'

Singh M, Fuenmayor E, Hinchy EP, Qiao Y, Murray N, Devine D (2021) Digital twin: origin to future. Appl Syst Innov 4(2)

Stakens J, Mutule A, Lazdins R (2023) Agriculture electrification, emerging technologies, trends and barriers: a comprehensive literature review. Latv J Phys Tech Sci 60(3):18–32

Tagarakis AC, Benos L, Kyriakarakos G, Pearson S, Sørensen CG, Bochtis D (2024) Digital Twins in agriculture and forestry: a review. Sensors 24(10)

Taha MF, Mao H, Zhang Z, Elmasry G, Awad MA, Abdalla A, Mousa S, Elwakeel AE, Elsherbiny O (2025) Emerging technologies for precision crop management towards agriculture 5.0: a comprehensive overview. Agriculture (Basel) 15(6)

Tao F, Qi Q, Wang L, Nee AYC (2019) Digital Twins and cyber–physical systems toward smart manufacturing and industry 4.0: correlation and comparison. Engineering 5(4):653–661

Taubert F, Rossi T, Wohner C, Venier S, Martinovič T, Khan T, Gordillo J, Banitz T (2024) Prototype biodiversity Digital Twin: grassland biodiversity dynamics. Research Ideas and Outcomes 10

Vergara C, Bahsoon R, Theodoropoulos G, Yanez W, Tziritas N (2023) Federated Digital Twin, Proceedings—2023 IEEE/ACM 27th International Symposium on Distributed Simulation and Real Time Applications, DS-RT 2023. Institute of Electrical and Electronics Engineers Inc., pp 115–116

Zhang R, Zhu H, Chang Q, Mao Q (2025) A comprehensive review of digital twins technology in agriculture. Agriculture (Basel) 15(9)

Digital Twin Applications for Air Pollution-Resilient Rice Agriculture

14

Ritesh Rana, Naveen Kumar, Rakesh Kumar, and Girdhar Gopal

Abstract

Air pollution, particularly fine particulate matter ($PM_{2.5}$), is a vital yet inadequately researched hazard to Northern Indian crop productivity. Rice residue burning seasons in Punjab and Haryana deliver record $PM_{2.5}$ levels, with potential impacts not only on human health but also on the physiology and yield of crops. This chapter investigates the processes by which $PM_{2.5}$ affects rice cultivation and proposes a digital twin (DT) approach to detect, model, and mitigate these effects. A conceptual DT structure is introduced that integrates real-time air quality monitoring, satellite/UAV imagery, and agro-environmental simulation models. Empirical studies from remote sensing and ground-based sensor networks are summarized, showing high levels of pollution exposure in rice farming areas. The chapter addresses the prospects of DTs for augmenting precision agriculture, enlightening policy-making, and encouraging resilient agricultural systems. Lastly, it outlines research and technical innovations required to upscale DTs in the Indo-Gangetic Plains so as to enhance sustainable agriculture under increasing air pollution.

R. Rana (✉) · N. Kumar
Department of Computer Science, Himachal Pradesh University, Shimla, Himachal Pradesh, India
e-mail: ritesh.rana53@gmail.com; nkjaglan@hpuniv.ac.in

R. Kumar
Department of Computer Science & Applications, Kurukshetra University, Kurukshetra, Haryana, India
e-mail: rakeshkumar@kuk.ac.in

G. Gopal
Sanatan Dharma College, Ambala Cantt, Kurukshetra University, Kurukshetra, Haryana, India
e-mail: girdhar.gopal@gmail.com

© The Author(s), under exclusive license to Springer Nature Singapore Pte Ltd. 2026

A. L. Srivastav et al. (eds.), *Digital Twin Technology for Sustainable Agriculture*,
https://doi.org/10.1007/978-981-95-5915-2_14

Keywords

Digital Twins · Smart Agriculture · Air Pollution · Crops

14.1 Introduction

Agriculture in the Indo-Gangetic Plains, especially in Northern India, stands at the intersection of opportunity and environmental vulnerability. Although states such as Punjab and Haryana have led India's agricultural revolution, increased farming activity has brought unforeseen consequences, with major contributions to local air pollution. Of specific interest is $PM_{2.5}$, which is a fine particulate air pollutant that is emitted in huge quantities through post-harvest stubble burning. This chapter delves into the implications of $PM_{2.5}$ for rice cultivation and looks at how DT technologies already proven within other sectors can be applied to monitor and measure these impacts for agriculture.

14.1.1 Defining Digital Twins in the Context of Agriculture

A Digital Twin (DT) is a dynamic virtual replica of an existing asset or process in the physical world. Practically, it combines real-time data and models such that the virtual equivalent reflects the state of the physical system. The Digital Twin Consortium describes a DT as *"a virtual representation of real-world entities and processes, synchronised at a specified frequency and fidelity"* (Digital Twin Consortium 2025). Mourtzis (2023) also points out that DTs broaden traditional simulations by facilitating ongoing, two-way communication between the physical and virtual environments.

In architectural terms, a DT system will usually consist of several layers, such as a perception layer of IoT devices and sensors gathering data from the physical asset, a network/cloud layer for data storage and transit, and an application/analytics layer where computational models (AI, ML, or physics-based simulators) consume the data to produce insights and control actions. These components work together to provide real-time synchronization. Sensor data from the field continually updates the virtual model, and analytics or control signals from the twin can be fed back to the physical system (Fig 14.1).

Historically, the digital twin concept emerged in manufacturing and cyber-physical systems (CPS). Sharma et al. (2022) observe that Michael Grieves first introduced the DT idea in the early 2000s within product lifecycle management. It was later popularized in the Industry 4.0 era of smart factories, where pervasive sensors and CPS integrate physical processes with virtual models for enhanced monitoring and maintenance. DTs were useful in such industrial environments for real-time monitoring (machine performance) and predictive maintenance.

DTs have since been utilized in numerous industries. Surveys identify DT uses in automotive, aerospace, energy, urban planning, and healthcare. DTs are increasingly

Fig. 14.1 Key benefits of Digital Twin technology in agriculture

being applied in agriculture to simulate plant growth, observe environmental data, and aid precision farming. Tagarakis et al. (2024) mention open-field DTs that allow farmers to observe and even remotely adjust parameters like irrigation through integrated sensor feedback. These applications underscore that any data-intensive, complex system, such as farms, would be aided by a continuously updated virtual twin.

DTs offer real-time monitoring, simulation, and forecasting capabilities in agro-environmental systems. With the integration of data from IoT sensors, weather stations, and satellite sensors, a DT can ascertain whether the soil is wet, pests are infesting, and exposed to pollution. This feature is highly relevant for regions like Northern India, where timely decision-making and localization are essential to sustainability.

14.1.2 Agricultural Landscape of Northern India

Punjab and Haryana are the leading rice-producing states of India, commonly known as the nation's "food granary" (Panwar et al. 2018). Punjab supplied close to 10% of

India's rice (about 12–13 million tonnes) (Vivek Gupta 2025; Economic Survey of Haryana 2025), collectively contributing a significant percentage of the national production. Despite this high contribution, Punjab cultivates only 6–8% of India's total rice area. Both states employ intensive double-cropping systems, primarily rice during the Kharif season and wheat in Rabi, supported by widespread irrigation and mechanization (Bhatia and Singh 2024).

The cropping intensity in Punjab is around 200%, and the rice–wheat rotation occupies roughly 95% of its food crop area (Bhatt et al. 2021). Over the past five decades, rice cultivation in Punjab expanded significantly from just 7% of the state's cropped area in 1970–1971 to approximately 40% in 2021–22 (Prabhjyot-Kaur et al. 2021). Haryana followed a similar pattern. Most farmlands in both states are irrigated using tubewells, and Punjab alone hosts over 1.4 million of them (Pandey 2016). These extract an estimated 4.8 × 10^11 liters of groundwater every day, leading to overexploitation in almost 80% of administrative blocks of Punjab (Amandeep Kaur et al. 2025).

Rice production in the state is extremely water-intensive, where Punjab needs approximately 5337 liters of water per kilogram of paddy, more than double the consumption in West Bengal. This is compounded by free power and lenient regulations on borewell digging (Groundwater Governance in the Indo-Gangetic and Yellow River Basins: Realities and Challenges 2009). Fertilizer usage is also high in Punjab; it averages 380 kg/ha, almost triple the national average (Nayak et al. 2023). Pesticide use similarly exceeds national levels.

The widespread use of combined harvesters, while effective, dumps 25–30 cm of paddy straw, giving rise to a significant residue management problem (Dutta et al. 2022). Only about 20% of this straw is repurposed, with the rest (~15 million tonnes) typically burnt in the fields. This practice continues because of the small window for sowing rice and wheat, even though legal prohibitions were enacted in 2013 (Punjab) and again strengthened nationwide in 2015.

Recent satellite images indicated a decrease in fire incidents. Punjab noted the lowest count of agricultural fires between September and November 2024 (Jethva 2022). Nonetheless, levels of atmospheric pollution are still high, indicating underreporting or non-enforcement. As a countermeasure, central and state governments are encouraging no-burn technologies (happy seeders, rotavators) and inducing crop diversification. Punjab plans to deal with all (~20 million tonnes) of its paddy straw through sustainable means by 2024, although utilization is affected by the rental cost of machinery and logistical challenges (Anand and Kaur 2024) (Table 14.1).

14.1.3 Air Pollution Crisis and Its Overlap with Rice Cultivation

Air quality in Northern India has reached critical levels, particularly during winter and post-monsoon periods. One of the key reasons for this peak during seasons is open burning of agricultural residues, most notably rice straw in Haryana and Punjab. The emissions released are high in fine particulate matter ($PM_{2.5}$), which

Table 14.1 Changes in cropping patterns in Punjab and Haryana between 1970–71 and 2021–22, showing increased dominance of rice and wheat cultivation

Crop	Punjab (1970–71)	Punjab (2021–22)	Haryana (1970–71)	Haryana (2021–22)
Rice (Paddy)	7%	40%	5%	23%
Wheat	41%	45%	23%	36%
Other cereals	52%	15%	72%	40%

enters deep into the respiratory tract as well as remains suspended in the lower atmosphere, impinging on visibility, solar radiation, and even plant health (Kaushal 2020; Rathore et al. 2025).

Evidence from research and satellites indicates that agricultural fires in Punjab and Haryana are major contributors to regional air quality deterioration (Saxena et al. 2021; Lan et al. 2022; Neelam et al. 2025). Estimates suggest that stubble burning contributes up to 30–40% of Delhi's $PM_{2.5}$ levels during peak fire periods (Chetna et al. 2023; Khan et al. 2023). The transported smoke plume affects not only urban areas but also adjacent agricultural zones, exposing crops to sustained pollution episodes during critical growth or post-harvest periods.

The overlap of intensive rice production and seasonal contamination is especially significant. Peak burning in October–November coincides with wheat sowing, which necessitates quick field clearance. Burning is the quickest and least expensive method used by farmers, even after the enactment of controls, such as fines and subsidies for equipment. Ground enforcement is still weak. Persistent fire hotspots, especially in Sangrur, Ludhiana, Karnal, and Kurukshetra districts, are evident from the remote sensing analysis (Vadrevu and Lasko 2018; Kumar et al. 2019).

Moreover, $PM_{2.5}$ levels during these months frequently exceed the national safe limit of 60 $\mu g/m^3$, with spikes above 300 $\mu g/m^3$ observed in some areas (Singh et al. 2024; Joshi et al. 2025). This has health consequences for urban dwellers and also for rural workers who are directly exposed in the fields. Studies have shown a rise in respiratory diseases and impaired worker productivity when it is burn season (Chakrabarti et al. 2019).

Notably, although the consequences on human health have been widely documented, $PM_{2.5}$ effects directly on plants remain still to be fully investigated. Potential mechanisms include suppression of photosynthetically active radiation (PAR), blockage of stomata, and changed transpiration through particulate deposition on surfaces (Rai 2016). There is new evidence that long-term pollution can reduce yield and quality in rice and wheat, although studies are limited.

To address these gaps, there is growing interest in applying sensor networks and modeling techniques that are capable of monitoring pollutant exposure at the farm level. These involve stationary air quality sensors and low-cost particulate matter with pollution payloads. Upon integration with digital twins, such data streams have the potential to provide high-resolution information on spatial variability and cumulative pollution stress on crops.

14.1.4 Motivation for Applying Digital Twins in Agri-Environmental Monitoring

Industrialization of Indian agriculture within the breadbasket states at a rapid rate has been followed by new problems along with it. Production has risen, but environmental damage, in this case, air pollution, has become a major concern. Fine particulate matter ($PM_{2.5}$), mainly from stubble burning and road transport, not only constitutes a health risk for humans but also impacts the physiology of rice crops through minimized photosynthesis, surface deposition, and changed microclimates (Nair et al. 2020).

Despite this growing risk, most surveillance systems are still disconnected from farm-level decision-making. Conventional farm models rarely incorporate real-time air quality information, and environmental policies aimed at reducing air pollution rarely consider agronomic susceptibility (Jones et al. 2017). This disconnection constrains both timely action and systems planning.

Digital twins (Sharma et al. 2022; Cowell et al. 2025) provide an exciting prospect for closing this gap. Their power is in the dynamic, real-time integration of physical and virtual environments. By providing environmental sensing intrinsically within crop systems, DTs can sense, mimic, and anticipate pollution-caused stress at field-level resolution. This responsiveness is particularly important in the case of Punjab and Haryana, where winter months' peak air quality coincides spectacularly with rice growth and harvest patterns.

Employing DTs in agriculture thus holds tremendous potential not only for the protection of crops against air pollution stress but also for balancing agronomic management with environmental health. The technology has the potential to enhance decision-making and offer early signals and more targeted interventions and thus serve as a strategic intervention in India's march toward sustainable and pollution-tolerant agriculture.

14.2 Air Pollution and Crop Systems: Scientific Basis

Rice farming in Northern India is increasingly exposed to air pollution, specifically fine particulate matter ($PM_{2.5}$), and this has become an agronomic as well as an environmental issue. Exposure to $PM_{2.5}$ can disrupt all phases of plant development, including germination and grain filling (Yadav et al. 2022; Teng et al. 2023). The difficulty is that different facets of plant biological and chemical interactions with airborne particulates need to be understood. While epidemiological evidence has conclusively established the health risks of $PM_{2.5}$, its direct and indirect effects on crop production systems, particularly in intensively cultivated rice belts, are poorly understood. In this section, the physiological mechanisms, empirical evidence, and existing knowledge gaps related to $PM_{2.5}$ and rice cultivation are outlined.

14.2.1 Impact of $PM_{2.5}$ on Crop Physiology

Fine particulate matter ($PM_{2.5}$) may have adverse impacts on the health and productivity of crops through a number of direct and indirect physiological processes (EPA 2025). Although research on the specific impact of $PM_{2.5}$ on rice is ongoing, inferences from similar crops and controlled studies constitute the basis for conceptualizing the mechanisms underlying the impacts.

14.2.1.1 Reduction in Photosynthetically Active Radiation (PAR)

High concentrations of atmospheric $PM_{2.5}$ decrease incoming solar radiation, especially the radiation available for photosynthesis by plants. Reduction in PAR can impair photosynthetic efficiency during peak stages such as tillering and panicle initiation in rice. Long-term exposure to haze or smog can lead to stunted growth and reduced biomass accumulation (Sankhla et al. 2024).

14.2.1.2 Deposition on the Leaf Surface

The $PM_{2.5}$ particles tend to settle on the leaf and establish a physical layer on a leaf surface that interferes with light absorption and gas exchange. The coating will clog stomatal openings, preventing transpiration and slowing the absorption of CO_2, weakening the plant's carbon assimilation and nutrient transport mechanisms (Ayala-Cortés et al. 2023).

14.2.1.3 Oxidative Stress and Cellular Damage

$PM_{2.5}$ often contains reactive chemical components such as heavy metals, PAHs, and sulphates. During deposition, these chemicals may form reactive oxygen species (ROS) in the plant tissues. The oxidative stress produced by these ROS can cause disruption of the cellular membrane, chloroplast damage, and lipid peroxidation. In rice, such damage can occur as chlorosis, grain filling deficiency, and decreased panicle weight (Mansoor et al. 2023).

14.2.1.4 Microclimate Alterations

Airborne particulates affect the microclimatic conditions by modulating radiative heat transfer, temperature, and humidity. Dense $PM_{2.5}$ concentrations can suppress daytime temperatures and lower diurnal variability, which may interfere with the thermal regime of the crop. This is especially important at anthesis and grain-filling stages when temperature sensitivity is increased (Wu et al. 2024).

14.2.1.5 Soil–Plant Interface Impacts

Particulate fallout may also impact the rhizosphere through the modification of pH and loading of toxicants in the soil environment. Microbial community modification and alteration in enzymatic activities, which play a crucial role in nutrient cycling, are indirect effects of particulate fallout (Thepbandit and Athinuwat 2024).

Collectively, these physiological stressors induce sublethal damage that is not apparent at the time but leads to quantifiable reductions in yield, quality, and resistance to secondary stressors. These processes make it worthwhile to have integrated real-time monitoring tools such as digital twins that can identify and interpret the early indicators of pollution stress.

14.2.2 Empirical Studies on Crop Yield Loss Due to $PM_{2.5}$

More recent studies (Von Schneidemesser et al. 2020; Dong and Wang 2023) have started measuring the negative impacts of $PM_{2.5}$ on farm productivity, though most of the evidence is based on extrapolation from general air pollution and crop exposure research because direct measurements are sparse.

14.2.2.1 Observational and Remote Sensing Evidence

In India, combined satellite-based estimates and crop production have indicated a negative trend between $PM_{2.5}$ levels and yields of major crops. Burney and Ramanathan (2014) estimated the reduction in rice and wheat yields in the Indo-Gangetic Plains due to air pollution between 1980 and 2010 by 10–20%, attributing virtually half of the loss to high $PM_{2.5}$ levels. Remote sensing studies using MODIS and AVHRR also report suppressed vegetation indices such as NDVI and EVI in high-pollution zones (Jarchow et al. 2018; Huang et al. 2019). More recent studies are by Sinha et al. (2015) and Grover and Chaudhry (2019).

14.2.2.2 Experimental and Controlled Environment Studies

Experiments in controlled chambers on rice plants have shown yield decreases of 10–30% under high $PM_{2.5}$ exposure, mainly attributed to lowered photosynthetic rate and shorter panicle length (Das et al. 2024). One study (Cheabu et al. 2018) exposed rice seedlings to simulated urban air pollution and observed a 23% reduction in grain weight and spikelet fertility. While such studies may overstate real-world exposure, these confirm the plausibility of physiologically disruptive effects.

14.2.2.3 Spatial Patterns in India's Rice Belt

District-level yield data from Punjab and Haryana overlaid with satellite fire counts (Jethva et al. 2018; Liu et al. 2020) and aerosol optical depth (AOD) show consistent yield penalties in high-burn areas like Sangrur, Ludhiana, Karnal, and Fatehabad. Although confounding factors exist, such as irrigation variability or pest pressure, statistical models adjusting for these still show a significant pollution signal.

These empirical findings reinforce the concern that $PM_{2.5}$ is not merely a health hazard but also an agronomic one. However, exact attribution will still be difficult without datasets with high spatio-temporal resolution that connect pollution measurements to crop yields. This makes it important to have systems such as digital twins that can track exposure to pollution in real time and match that against the response of plants and yield measurements.

14.2.3 Knowledge Gaps in Rice Crop Growth Under Pollution Stress

Despite growing evidence of $PM_{2.5}$ adverse effects on rice, there are notable gaps in knowledge. Relatively few studies clearly delineate the levels of exposure that cause physiological damage to rice, giving little practical guidance to farmers. Field measurements of $PM_{2.5}$ in cropping environments where rice is grown in rural areas are scarce, which give low-resolution data often lacking critical exposure duration during sensitive crop stages. In addition, the complex interactions between air pollution and other stressors, like heat, water restriction, and nutrient shortages, are not often incorporated into existing models, but these interactive effects also play significant roles in regulating crop performances. Socioeconomic determinants and farmer behavioral adaptations to pollution, like changed sowing seasons or residue management, are also seldom taken into account, which constitute blind spots for separating real impacts.

Moreover, most research is limited to short-term studies, with few longitudinal datasets available to assess the cumulative and multi-year effects of pollution on rice phenology and yield. Addressing these gaps through integrated approaches like digital twins, which combine environmental, agronomic, and social data in real time, holds promise for more effective and adaptive management of pollution stress in rice agriculture.

14.3 Digital Twin Frameworks in Agriculture

Digital Twins (DTs) (Sharma et al. 2022) are virtual, dynamic copies of real-world systems that constantly incorporate real-time data to mimic, predict, and optimize processes. DTs originally developed for aerospace and manufacturing have seen growing uptake in agriculture, where they can contribute to greater precision and sustainability. By their simulation of crop growth, environmental factors, and management practices, agricultural DTs (Abiri et al. 2023) enable stakeholders to monitor crop health, predict stress impact, and test intervention strategies without physical testing. The integration of digital and physical settings is especially important for complicated systems such as rice production systems where environmental variables, such as air pollution, intersect with the biology of the crop as well as managerial choices.

In agricultural contexts, DTs have data from a range of sources, including remote sensing, ground sensors, weather stations, and farmers inputs, to construct an integrated dynamic model of the plot or farm. The model supports decision-making at scales ranging from single plots to regional landscapes. In particular, DTs can simulate environmental stressors like $PM_{2.5}$ pollution, both the temporal behavior and spatial heterogeneity. Such capability is crucial for regions like Punjab and

Haryana, where air pollution episodes are episodic but of high intensity and simultaneous with vulnerable crop phases. The following subsection outlines the necessary building blocks that constitute an effective agricultural digital twin.

14.3.1 Core Components of a Digital Twin (Sensing, Modeling, and Feedback)

A robust digital twin integrates three essential elements:

14.3.1.1 Sensing and Data Acquisition

This is the sustained acquisition of high-precision data through sensors continuously. For rice cultivation under pollution stress, these are air quality sensors monitoring $PM_{2.5}$ levels, meteorological stations measuring temperature, humidity, and solar radiation, soil moisture sensors, and remote sensing platforms like satellites and drones collecting canopy health and phenology. Heterogeneity and data granularity constitute the basis for DT's truthfulness.

14.3.1.2 Modeling and Simulation

The virtual twin applies computational models to simulate crop development, physiological response, and environmental interactions. Crop models such as DSSAT or APSIM can be adapted to include air pollution effects, using data-driven machine learning models or process-based simulations. Pollution impact models incorporate mechanisms such as photosynthetic inhibition, stomatal closure, and oxidative stress induced by $PM_{2.5}$. Dynamic feedback loops enable updating predictions as new data arrives, allowing near real-time monitoring and scenario analysis.

14.3.1.3 Decision Support and Feedback

The DT provides actionable insights for farmers, agronomists, and policymakers through user-friendly interfaces, alerts, and visualization tools. For instance, early warning of pollution peaks coinciding with delicate growth phases can prompt adaptive measures such as realigned irrigation or fertilizer use. DTs also enable regulatory compliance by recording pollution exposure and agricultural practices. Ongoing feedback loops promote repeated iteration and refinement of models and field management.

Combining these elements forms a cyber-physical system (CPS) that has the ability to fill data gaps and enhance resilience against air pollution effects in rice farming. The following subsection will discuss concrete examples of environmental monitoring DTs and their applicability to Indian agriculture.

14.3.2 Environmental Monitoring Using Digital Twins: Sector Examples

Digital twins have gained increasing usage across several global environmental and agricultural sectors, and there are useful lessons to be learnt from them for use in air pollution-affected rice cultivation.

In precision agriculture (Awais et al. 2025), DTs have been developed to monitor soil moisture, nutrient status, and crop health in real time. For example, the "Virtual Farm" initiative in Europe brings together satellite images, weather, and IoT sensors to maximize irrigation and fertilizer applications, minimizing wastage and environmental runoff (Escribà-Gelonch et al. 2024). Likewise, "Digital Vineyard" initiatives in various countries make use of DTs to help manage microclimate and risk of disease in a proactive manner (Maesano et al. 2021; Faluomi 2023).

Beyond agriculture, DTs are more and more applied in urban environmental surveillance. The "Virtual Singapore" DT of the City of Singapore combines air quality, traffic, and meteorological data to model pollution dispersion and help in urban planning (SLA 2025). The Smart Forest DT in China tracks forest well-being and fire danger using satellite data combined with ground sensors (Qiu et al. 2023). These initiatives demonstrate the value of real-time environmental data fusion and predictive modeling in complex systems.

In the context of air pollution and crop systems, pioneering efforts have started integrating DTs with pollutant dispersion models and crop response simulations. A recent study by Ta Bui and Hoang Nguyen (2024) focuses on assessing the impacts of $PM_{2.5}$ pollution on rice production and quantifying economic cost losses in the Long Xuyen Quadrangle (LXQ) of the Vietnam Mekong Delta, but there are limited studies that developed a DT prototype combining real-time $PM_{2.5}$ sensing with rice phenology models to predict the yield impacts. Although still experimental, such frameworks illustrate how DTs can address the spatial and temporal variability of pollution exposure relevant to crop health.

These examples point to important benefits transferable to the Indo-Gangetic Plains: enhanced environmental intelligence, timely decision-making support, and multi-scale integration. However, implementing DTs into Indian rice systems involves overcoming challenges involving sensor deployment costs, complexity of data integration, and farmers' engagement.

The following subsection will discuss how these insights can be tailored to the unique conditions of Punjab and Haryana rice agriculture.

14.3.3 Digital Twin Adaptation Potential in Indian Rice Systems

Adapting digital twin technology to the rice-producing states of Punjab and Haryana holds promising avenues for mitigation of air pollution effects. These states are marked by intensive double-cropping regimes, extensive mechanization, and periodic stubble burning cycles that cause seasonal $PM_{2.5}$ peaks during risky crop phases. A digital twin custom-designed for these agro-ecosystems can consolidate varied data streams such as localized air quality observations, crop phenology, meteorological conditions, and management operations to yield detailed, actionable insights.

First, the widespread availability of satellite and UAV imagery over Indian fields supports frequent spatial updates on crop health and residue management. Coupled with an expanding network of low-cost air pollution sensors, this creates an infrastructure backbone for real-time environmental monitoring. Incorporating soil and irrigation data, which are increasingly digitized by agricultural extension services, would enhance the model accuracy and relevance.

Second, crop growth models used in India, such as DSSAT (Decision Support System for Agrotechnology Transfer) or CERES-Rice (specifically simulate rice growth), can be extended to factor in pollution-induced stress mechanisms and calibrated with local experimental data. Machine learning approaches may complement these by identifying complex nonlinear patterns in multi-source data, helping to predict yield variations attributable to $PM_{2.5}$ exposure.

Third, farmer-centric interfaces embedded within the DT system could facilitate adoption by delivering timely alerts on pollution episodes, suggesting adaptive practices like modified irrigation or nutrient application, and providing residue management guidance to minimize burning risks.

Lastly, blending the digital twin with government air quality and agricultural policy frameworks can improve compliance monitoring and facilitate evidence-based policymaking. Digital twins can offer simulated "what-if" scenarios that facilitate interventions to enable productivity objectives and environmental sustainability.

Although the technical potential is considerable, several challenges persist, including rural connectivity, data standardization, and the need for user capacity building. Still, the urgent air pollution situation and economic significance of rice in Punjab and Haryana make digital twin adoption a necessity for research and practice.

14.4 Application Context: Punjab and Haryana

Punjab and Haryana are among the most important places in India's rice- and wheat-producing landscape, which is called the country's "food granary." Both the states embody agricultural intensification and mechanization characteristic of the Indo-Gangetic Plains. Although this intensification has driven high productivity, it has also resulted in acute environmental issues, most of which are seasonal air pollution fueled by extensive stubble burning. Understanding the particular agricultural, environmental, and socio-economic conditions of Punjab and Haryana is crucial to developing effective digital twin solutions to counteract the effects of air pollution on rice farming.

14.4.1 Patterns of Rice Production and Stubble Burning

Punjab and Haryana are India's leading rice producers, together accounting for approximately 15% of the national rice output, despite cultivating only about 10% of the country's rice area. The intensive rice–wheat cropping system dominates in the agrarian countryside, with almost 95% of cultivated land being brought under this cycle (Downing et al. 2022; Korav et al. 2022). Crop intensity is as high as almost 200% in Punjab, which mirrors double-cropping patterns supported by intensive groundwater irrigation (Abdurrahman et al. 2020; Chaudhary et al. 2023; Smolenaars et al. 2023).

Mechanization is widespread, with most farmers employing harvesters for paddy harvesting. Mechanization increases productivity but leaves large rice stubble (~25–30 cm) on the land, which needs to be removed within a short period prior to the next sowing of wheat. This limited crop-to-crop duration leads the farmers to opt for open-field burning of stubble, releasing vast amounts of $PM_{2.5}$ and other emissions into the environment (Singh et al. 2021).

Only about 20% of paddy residue in Punjab is used for alternate uses such as fodder or for biomass power, while almost 15 million tonnes get burnt every year. This occurs even with the bans put in place in 2013 and enforcement by the National Green Tribunal in 2015; stubble burning continues to be widespread. Such continuance is primarily because of the lack of affordable alternatives and socio-economic limitations faced by farmers. The practice has been viewed as the cheapest and simplest way of getting land ready for the following winter crop, so it becomes challenging for farmers to go for more environment-friendly approaches despite the government measures to discourage the burning of stubble (Pandey 2022).

Recent years saw efforts toward reducing burning incidents by encouraging "happy seeders," rotavators, and residue briquetting technologies, as well as crop diversification strategies. Satellite observations of fire occurrence indicate a decline in stubble burning events from around 33,000 toward the end of 2023 to approximately 8400 during the equivalent period of 2024 in Punjab (Priyangi Agarwal 2025) However, ambient air monitoring shows that total $PM_{2.5}$ pollution has not dropped proportionally, which indicates incomplete compliance or hidden burning sources.

These long-standing practices of rice cultivation and residue disposal are the backdrop in which digital twin technologies can be designed and applied to cushion against the effects of pollution and promote sustainable behaviors.

14.4.2 $PM_{2.5}$ Spikes and Seasonal Patterns of Air Quality

Punjab and Haryana's agriculture calendar is directly correlated with unique seasonal air pollution patterns, specifically for fine particulate matter ($PM_{2.5}$). The maximum pollution events occur during the post-harvest season of the Kharif rice crop (October–November) when farmers carry out massive stubble burning to quickly prepare fields for sowing wheat. This results in acute, short-duration spikes in $PM_{2.5}$ concentrations that far exceed the national and World Health Organization (WHO) air quality standards (CPCB 2023).

Satellite observations and ground-based monitoring stations reveal that the PM2.5 levels during these (CPCB 2023) episodes can reach concentrations exceeding 500 μg/m^3, often persisting for several days (Kulkarni et al. 2020). The Indo-Gangetic Plains, over Delhi, Punjab, and Haryana, are never among the cleanest spots on the planet during this season. These peaks are not only responsible for regional haze and reduced visibility but also influence the atmospheric chemistry and radiative forcing at the regional level.

Meteorological conditions such as low wind speed, temperature inversions, and high humidity in late autumn enhance near-surface pollution accumulation, which entrains particulate matter and increases the risk of exposure. Spatial $PM_{2.5}$ variation is heterogeneous with hotspots corresponding to locations of intense burning activities and meteorological conditions such as valley topography.

Seasonal emissions from agriculture are coupled with those of urban and industrial emission sources, further deteriorating air quality. The concurrent existence of pollution sources complicates attribution studies but demarcates the extreme need for integrated monitoring systems.

For rice crop production, the timing of $PM_{2.5}$ spikes overlaps with susceptible developmental stages like grain filling and maturation, which can exacerbate physiological stress and yield losses. For developing adaptive management practices supported by digital twin technologies, understanding these temporal dynamics is important.

14.4.3 Risk Mapping: Overlapping Crop Cycles with Pollution Episodes

One of the major difficulties in managing air pollution impacts for rice cultivation in Punjab and Haryana is the time overlap between periods of pollution and sensitive phases of crop growth. The Kharif season for rice, from June to October, peaks in harvesting in late October and early November, which is when the burning of stubble reaches its peak. Therefore, mature rice crops are frequently exposed to heavy $PM_{2.5}$ concentrations at grain filling and ripening stages, which are critical for yield and grain quality (Table 14.2).

Using satellite-derived fire counts, aerosol optical depth (AOD), and crop calendar data, multiple studies have identified high-risk windows when pollution stress coincides with key phenological stages. For example, analysis by recent studies demonstrated that $PM_{2.5}$ concentrations during the last 30 days before harvest were strongly correlated with reduced yields in districts with high fire activity. Places such as Sangrur, Ludhiana, and Karnal show both high stubble-burning intensity and yield suppression consistently, making them priority areas for intervention.

Risk mapping can be augmented further by digital twin platforms that integrate pollution predictions with real-time monitoring of crop stages. These platforms can then provide spatially explicit alerts and scenario simulation to help farmers and

Table 14.2 $PM_{2.5}$ exposure vs. rice growth stages in Punjab–Haryana (India)

Growth stage	Typical calendar window	$PM_{2.5}$ levels (avg.)	Sensitivity to pollution	Potential Impact
Nursery Establishment	Late June—Early July	Moderate	Moderate	Slowed seedling vigor
Tillering	Mid-July—August	High	High	Reduced tiller number
Panicle Initiation	August—September	Very High	Very High	Grain number reduction
Anthesis (Flowering)	Late Sept—Mid Oct	Severe	Critical	Sterility, spikelet loss
Grain Filling	October—Early Nov	Very High	High	Poor grain weight
Maturity and Harvest	Mid—Late November	Moderate	Low	Residue management burden

policymakers identify areas and periods when exposure to pollution is most likely to cause the most agronomic risk.

Such spatial–temporal risk maps would play a critical role in informing the deployment of mitigation strategies, either through incentive subsidies on residue management machinery, advisories on changed sowing dates, or burn bans during pivotal crop stages. By capturing the intersection of environmental hazard and crop vulnerability, digital twins can translate complex atmospheric data into actionable insights at the farm and district scale.

14.5 Conceptual Framework for a Rice-Agriculture Digital Twin

To create a digital twin (DT) model of the rice–pollution relationship in Punjab and Haryana, the distinctive characteristics of the crop system as well as local air quality dynamics need to be taken into consideration. In this context, a DT is a data-driven, real-time simulation of rice crop development superimposed with environmental stressors like $PM_{2.5}$. It combines data collection, computational modeling, and decision-support output to create a cyber-physical system (CPS) that dynamically adjusts to evolving field conditions.

14.5.1 Data Requirements and Technology Stack

An effective DT begins with comprehensive and continuous data acquisition. This includes remote sensing imagery (such as NDVI from MODIS), ground-based air quality monitoring ($PM_{2.5}$ levels from low-cost sensors and CPCB stations), micro-climatic data (temperature, humidity, and solar radiation), and crop management records (sowing dates, irrigation, and fertilizer inputs). UAVs and IoT-enabled devices enhance spatial and temporal resolution, allowing the system to localize

Table 14.3 Digital twin system architecture for $PM_{2.5}$-resilient rice farming, detailing functional layers, technologies, and their roles in pollution monitoring and crop management

Layer	Function	Example tools/ technologies	Relevance to PM2.5 Monitoring
Perception layer	Capture real-time environmental and crop-level data	IoT sensors ($PM_{2.5}$, humidity), UAVs, satellite (MODIS, Sentinel)	Tracks pollution exposure and microclimate variations
Network and integration	Transmit and unify data across sources	5G, LPWAN, cloud platforms, data fusion APIs (OpenAQ, IMD)	Harmonizes pollution and crop datasets; supports spatio-temporal accuracy
Modeling and simulation	Simulate crop–environment interactions	DSSAT, APSIM, WRF-Chem, AERMOD, ML models	Projects $PM_{2.5}$ impact on yield, phenology, photosynthesis
Analytics & Decision	Derive insights and generate advisories	AI/ML tools, digital dashboards, anomaly detection	Enables yield risk alerts and pollution-based advisories
Feedback and actuation	Inform farmers and systems in real time	Mobile apps, smart irrigation, control actuators	Supports timely interventions and adaptive crop management
User interface	Visualize insights for decision-makers	Web dashboards, mobile alerts, AR-based visualizations	Communicates pollution impact and options to farmers, planners

changes across farms. Data infrastructure needs to accommodate high-frequency ingestion and cloud storage with standard protocols to allow for interoperability (Table 14.3).

The computational stack involves coupling process-based crop models such as DSSAT (Decision Support System for Agro-technology Transfer) or APSIM (Agricultural Production Systems sIMulator) with pollution dispersion and impact models. These may be complemented by machine learning elements trained on past experience to predict yield under pollution stress. Real-time simulation engines update the crop growth model constantly with new environmental data, generating forecasts, stress warnings, and decision suggestions.

14.5.2 Integration of Pollution Data with Crop Growth Models

The major innovation lies in the ability of the DT to simulate the effect of pollution exposure at different growth stages of a crop. This involves linking $PM_{2.5}$ time-series observations to physiological processes such as stomatal conductance, efficiency in photosynthesis, and grain filling rates. For instance, prolonged exposure to $PM_{2.5}$ in the reproductive phase can be simulated to reduce effective leaf area and carbohydrate translocation to simulate the impact on yield. Coupling with meteorological inputs enables compensation for synergism effects of radiation and temperature stress.

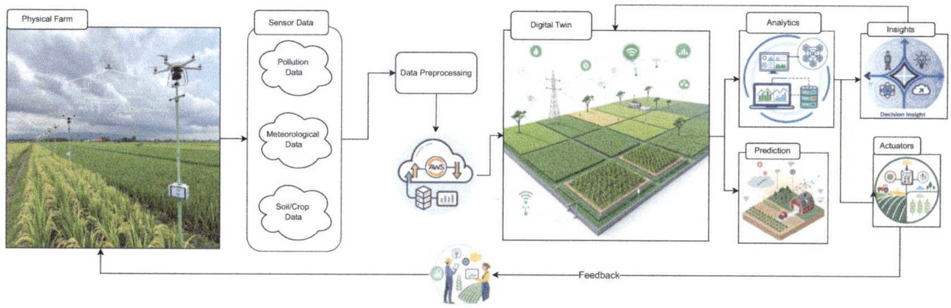

Fig. 14.2 Digital Twin conceptual framework for rice farming integrating $PM_{2.5}$, weather, and crop data to enable real-time prediction, analytics, and decision support

Calibration and validation are essential. Field experiments and past yield data from high-burn and low-burn places provide ground truth for model coefficient refinement. Sensitivity analyses establish which parameters most impact output under pollution conditions. The addition of adaptive thresholds such as $PM_{2.5}$ exceedance duration associated with physiological deterioration adds real-world relevance (Fig. 14.2).

14.5.3 Real-Time Simulation and Decision-Support Potential

Once operational, the DT system can deliver several decision-support functions. These include real-time dashboards visualizing pollution stress, forecasts of likely yield outcomes under current pollution trends, and scenario analyses for alternative management practices (such as shifting sowing dates and applying anti-stress treatments). Farmers can be alerted when $PM_{2.5}$ levels pass over vital thresholds at critical crop phases, along with advisories.

At the institutional level, DT outputs can facilitate district-level planning, such as area prioritization of residue management subsidy, burn ban enforcement, or policy simulations evaluating the effectiveness of mitigation interventions. As scaling up takes place, these DTs have the potential to act as an empirical foundation for agri-environmental governance, linking sustainability objectives with on-ground realities within India's most productive yet pollution-prone rice systems.

14.6 Policy and Practice Implications

The integration of digital twin systems in rice farming has important policy and practice implications at several levels. At the farm level, DTs are able to provide farmers with timely location-specific advice that improves decision-making and reduces risk. For example, a DT system can inform farmers about extreme pollution episodes concurrent with grain-filling phases so that farmers will be able to optimize irrigation or fertilizer application to reduce yield loss. This directly adds to resilience, especially for smallholders who are exposed to environmental stress more than others.

At the institutional level, DTs provide new instruments of regulatory compliance and enforcement. Government agencies can use real-time data and simulation to track stubble-burning hotspots, assess exposure to pollution in crop zones, and determine areas of priority intervention. Simulations of policy interventions like subsidizing in situ residue management equipment or shifting sowing windows enable evidence-based policymaking based on spatial and temporal realities.

In terms of precision agriculture, DTs enhance the efficiency of input use. By identifying stress conditions early, they can reduce over-application of fertilizers or unnecessary irrigation, improving both productivity and environmental outcomes. This aligns well with national goals on climate-smart agriculture and sustainable intensification.

Yet, various issues need to be resolved to achieve such benefits. Rural infrastructure constraints like poor connectivity, sparse sensor density, and limited technical support can constrain DT efficacy. In addition, large-scale farmer adoption relies on evident value demonstration, localized support tools and sufficient training. Institutional capacity to read and respond to DT insights also needs to be enhanced through interagency coordination and investment in digital literacy.

The success of DT integration into agri-environmental systems is therefore not merely a matter of technological readiness but also of institutional commitment, stakeholder engagement, and enabling policy frameworks.

14.7 Future Directions and Research Opportunities

The future for digital twin technology in air pollution-resilient rice production includes focused innovation, field-level verification, and institutional partnership. One key opportunity is increasing the density and richness of sensor networks. Affordable $PM_{2.5}$ sensors, soil sensors, and meteorological stations must be extensively deployed in rural Punjab and Haryana to enable the fine-resolution data needed for DT modeling. Combining satellite and drone imagery with ground sensors improves spatial accuracy and dynamic updating of the digital model.

Field validation is another priority. Long-term field trials under different $PM_{2.5}$ exposure levels can measure the rice physiological and yield responses and optimize DT parameters. Establishing experimental fields in high-stubble-burning-density districts (such as Sangrur and Fatehabad) could provide empirical data to validate and calibrate model assumptions.

Research also needs to be focused on the interoperability of DT platforms. Establishing open standards for data format, APIs, and model sharing will ensure collaborative innovation. Moreover, the inclusion of socio-economic modules like farmer behavior, input costs, and constraints to adoption within the DT ecosystem can improve real-world usability and policy significance.

Collaboration across multiple stakeholders is necessary. Public–private partnerships between agri-tech companies, research institutions, and government departments may drive scalable deployment. Farmer cooperatives and the local extension system need to be engaged in co-design efforts to enhance usability as well as ensure that solutions are adjusted to fit the realities of users.

Ultimately, investment in training, outreach, and feedback mechanisms will support the evolution of DTs from research prototypes to practical tools embedded in India's agricultural resilience strategy.

14.8 Conclusion

This chapter has discussed the complex relationship between rice crop production and air pollution in India's northern plain region focusing particularly on the effects of $PM_{2.5}$ exposure at important phases of plant growth. Punjab and Haryana, being highly productive regions, are becoming more environmentally stressed due to activities such as stubble burning and high-input cultivation. These stresses not only undermine public health but also threaten the long-term sustainability of India's rice–wheat systems.

The chapter suggested a conceptual digital twin (DT) approach specially designed for $PM_{2.5}$ monitoring and impact assessment in rice farming to address the same. The framework integrates sensor-based data collection, modeling and simulation, analytics, and feedback mechanisms to provide real-time data and decision-making support. This framework is the basis for dynamic, responsive, and scalable DT applications in agro-environmental scenarios.

Digital twins, as shown, possess transformative potential in reducing the negative effects of air pollution on crop systems. By leveraging real-time data, crop and pollution models, and predictive analytics, DTs can help optimize resource use, reduce yield loss, and issue early warnings benefiting farmers, policymakers, and researchers alike.

Moving forward, the development of interoperable, open-access DT platforms, expanded empirical studies on pollution–crop dynamics, and close collaboration among stakeholders will be crucial. Bringing digital solutions in sync with local realities and institutional environments will finally hold the key to their success at scale.

In the broader push toward climate-resilient and sustainable agriculture, digital twins may serve as a cornerstone of next-generation farming systems where environmental intelligence and precision decision-making converge.

References

Abdurrahman MI, Chaki S, Saini G (2020) Stubble burning: effects on health & environment, regulations and management practices. Environ Adv 2:100011. Available at: https://doi.org/10.1016/j.envadv.2020.100011

Abiri R et al (2023) Application of digital technologies for ensuring agricultural productivity. Heliyon 9(12):e22601. Available at: https://doi.org/10.1016/j.heliyon.2023.e22601

Agarwal P (2025) "TOI," Delhi Pollution: are Punjab-Haryana farmers hiding fires from satellites?, 29 June. Available at: https://timesofindia.indiatimes.com/city/delhi/farmers-allegedly-out smart-satellites-to-conceal-stubble-burning-in-india/articleshow/115392826.cms. Accessed 29 June 2025

Anand S, Kaur H (2024) Challenges and opportunities in sustainable stubble management in Punjab: a review. Int J Environ Clim Change 14(3):274–297. Available at: https://doi.org/10.9734/ijecc/2024/v14i34040

Awais M et al (2025) Advancing precision agriculture through digital twins and smart farming technologies: a review. AgriEngineering 7(5):137. Available at: https://doi.org/10.3390/agriengineering7050137

Ayala-Cortés M et al (2023) Impact of particulate matter (PM10 and PM2.5) from a thermoelectric power plant on morpho-functional traits of Rhizophora mangle L. leaves. Forests 14(5):976. Available at: https://doi.org/10.3390/f14050976

Bhatia S, Singh SP (2024) Assessing groundwater use efficiency and productivity across Punjab agriculture: district and farm size perspectives. Agriculture 14(8):1299. Available at: https://doi.org/10.3390/agriculture14081299

Bhatt R et al (2021) Rice–wheat system in the northwest Indo-Gangetic plains of South Asia: issues and technological interventions for increasing productivity and sustainability. Paddy Water Environ 19(3):345–365. Available at: https://doi.org/10.1007/s10333-021-00846-7

Burney J, Ramanathan V (2014) Recent climate and air pollution impacts on Indian agriculture. Proc Natl Acad Sci 111(46):16319–16324. Available at: https://doi.org/10.1073/pnas.1317275111

Chakrabarti S et al (2019) Risk of acute respiratory infection from crop burning in India: estimating disease burden and economic welfare from satellite and national health survey data for 250 000 persons. Int J Epidemiol 48(4):1113–1124. Available at: https://doi.org/10.1093/ije/dyz022

Chaudhary BU, Sood A, Bal SK (2023) Multi-temporal AWiFS and SAR data for cropping system analysis in Indian Punjab. Indian J Soil Conserv 51(2). Available at: https://doi.org/10.59797/ijsc.v51.i2.123

Cheabu S et al (2018) Effects of heat stress at vegetative and reproductive stages on spikelet fertility. Rice Sci 25(4):218–226. Available at: https://doi.org/10.1016/j.rsci.2018.06.005

Chetna et al (2023) Trends and variability of PM2.5 at different time scales over Delhi: long-term analysis 2007–2021. Aerosol Air Qual Res 23(5):220191. Available at: https://doi.org/10.4209/aaqr.220191

Cowell NH et al (2025) Moving from monitoring to real-time interventions for air quality: are low-cost sensor networks ready to support urban digital twins? Front Sustain Cities 6:1500516. Available at: https://doi.org/10.3389/frsc.2024.1500516

CPCB (2023) Annual Report 2023. Available at: https://cpcb.nic.in/openpdffile.php?id=UmVwb3J0RmlsZXMvMTY2OV8xNzI3NDE0NTc1X21lZGlhcGhvdG8yOTAyNy5wZGY. Accessed 29 June 2025

Das KR et al (2024) Physiological responses and yield performance of selected rice (Oryza sativa L.) genotypes under deficit moisture stress. Saudi J Biol Sci 31(4):103961. Available at: https://doi.org/10.1016/j.sjbs.2024.103961

Digital Twin Consortium (2025) Digital consortium. Available at: https://www.digitaltwinconsortium.org/initiatives/the-definition-of-a-digital-twin/. Accessed 27 June 2025

Dong D, Wang J (2023) Air pollution as a substantial threat to the improvement of agricultural total factor productivity: global evidence. Environ Int 173:107842. Available at: https://doi.org/10.1016/j.envint.2023.107842

Downing AS et al (2022) Unlocking the unsustainable rice-wheat system of Indian Punjab: assessing alternatives to crop-residue burning from a systems perspective. Ecol Econ 195:107364. Available at: https://doi.org/10.1016/j.ecolecon.2022.107364

Dutta A et al (2022) A state of the art review in crop residue burning in India: previous knowledge, present circumstances and future strategies. Environ Chall 8:100581. Available at: https://doi.org/10.1016/j.envc.2022.100581
Economic survey of Haryana (2025) Economic Survey of Haryana, Economic Survey, 18 June. Available at: https://cdnbbsr.s3waas.gov.in/s32b0f658cbffd284984fb11d90254081f/uploads/2025/03/20250317378593833.pdf. Accessed 18 June 2025
EPA (2025) Health and environmental effects of particulate matter (pM), 24 June. Available at: https://www.epa.gov/pm-pollution/health-and-environmental-effects-particulate-matter-pm. Accessed 24 June 2025
Escribà-Gelonch M et al (2024) Digital twins in agriculture: orchestration and applications. J Agric Food Chem 72(19):10737–10752. Available at: https://doi.org/10.1021/acs.jafc.4c01934
Faluomi V (2023) A digital twin application for vineyards sustainable management. BIO Web Conf Edited by P. Roca, 68:01038. Available at: https://doi.org/10.1051/bioconf/20236801038
Groundwater governance in the Indo-Gangetic and Yellow river basins: realities and challenges (2009) CRC press, Boca Raton (International association of hydrogeologists selected papers, vol. 15)
Grover D, Chaudhry S (2019) Ambient air quality changes after stubble burning in rice–wheat system in an agricultural state of India. Environ Sci Pollut Res 26(20):20550–20559. Available at: https://doi.org/10.1007/s11356-019-05395-5
Gupta V (2025) Punjab paddy dilemma, Punjab sticks to paddy amid depleting groundwater, 18 June. Available at: https://india.mongabay.com/2023/01/punjabs-paddy-dilemma-amid-depleting-groundwater/. Accessed 18 June 2025
Huang X, Xiao J, Ma M (2019) Evaluating the performance of satellite-derived vegetation indices for estimating gross primary productivity using FLUXNET observations across the globe. Remote Sens 11(15):1823. Available at: https://doi.org/10.3390/rs11151823
Jarchow CJ et al (2018) Application and comparison of the MODIS-derived enhanced vegetation index to VIIRS, Landsat 5 TM and Landsat 8 OLI platforms: a case study in the arid Colorado River Delta, Mexico. Sensors (Basel) 18(5):1546. Available at: https://doi.org/10.3390/s18051546
Jethva H (2022) Assessing predictability of post-monsoon crop residue fires in Northwestern India. Front Earth Sci 10:1047278. Available at: https://doi.org/10.3389/feart.2022.1047278
Jethva H et al (2018) Agricultural burning and air quality over Northern India: a synergistic analysis using NASA's A-train satellite data and ground measurements. Aerosol Air Qual Res 18(7): 1756–1773. Available at: https://doi.org/10.4209/aaqr.2017.12.0583
Jones JW et al (2017) Toward a new generation of agricultural system data, models, and knowledge products: state of agricultural systems science. Agric Syst 155:269–288. Available at: https://doi.org/10.1016/j.agsy.2016.09.021
Joshi DC et al (2025) Fine particulate matter (PM2.5, PM10): a silent catalyst for chronic lung diseases in India; a comprehensive review. Environ Challenges:101215. Available at: https://doi.org/10.1016/j.envc.2025.101215
Kaur A et al. (2025) Ground Water Yearbook, Ground Water Year Book, 20 June. Available at: https://www.cgwb.gov.in/old_website/Regions/NWR/Reports/Ground%20Water%20Year%20Book%20of%20Punjab%20State%202020-2021.pdf. Accessed 20 June 2025
Kaushal LA (2020) Examining the policy-practice gap- the issue of crop burning induced particulate matter pollution in Northwest India. Ecosyst Health Sustain 6(1):1846460. Available at: https://doi.org/10.1080/20964129.2020.1846460
Khan AA et al (2023) Effects of stubble burning and firecrackers on the air quality of Delhi. Environ Monit Assess 195(10):1170. Available at: https://doi.org/10.1007/s10661-023-11635-6
Korav S et al (2022) Impacts of mechanized crop residue management on rice-wheat cropping system—a review. Sustainability 14(23):15641. Available at: https://doi.org/10.3390/su142315641

Kulkarni SH et al (2020) How much does large-scale crop residue burning affect the air quality in Delhi? Environ Sci Technol 54(8):4790–4799. Available at: https://doi.org/10.1021/acs.est.0c00329

Kumar P et al (2019) Monitoring of rice crop in Punjab and Haryana with respect to residue burning. Int Arch Photogramm Remote Sens Spat Inf Sci XLII-3/W6:31–36. Available at: https://doi.org/10.5194/isprs-archives-XLII-3-W6-31-2019

Lan R et al (2022) Air quality impacts of crop residue burning in India and mitigation alternatives. Nat Commun 13(1):6537. Available at: https://doi.org/10.1038/s41467-022-34093-z

Liu T et al (2020) Crop residue burning practices across North India inferred from household survey data: bridging gaps in satellite observations. Atmospheric Environ: X 8:100091. Available at: https://doi.org/10.1016/j.aeaoa.2020.100091

Maesano M et al (2021) Smart monitoring technologies for defining variability in vineyard microclimate, and vinegrape performances. In: 2021 IEEE international workshop on metrology for agriculture and forestry (MetroAgriFor). IEEE, Trento-Bolzano, pp 17–21. Available at: https://doi.org/10.1109/MetroAgriFor52389.2021.9628416

Mansoor S et al (2023) Heavy metal induced oxidative stress mitigation and ROS scavenging in plants. Plants 12(16):3003. Available at: https://doi.org/10.3390/plants12163003

Mourtzis D (2023) Digital twin inception in the era of industrial metaverse. Front Manufact Technol 3:1155735. Available at:. https://doi.org/10.3389/fmtec.2023.1155735

Nair M et al (2020) Assessment of contribution of agricultural residue burning on air quality of Delhi using remote sensing and modelling tools. Atmos Environ 230:117504. Available at: https://doi.org/10.1016/j.atmosenv.2020.117504

Nayak HS et al (2023) Pathways and determinants of sustainable energy use for rice farms in India. Energy 272:126986. Available at: https://doi.org/10.1016/j.energy.2023.126986

Neelam N, Rathee RK, Mishra SK (2025) Unraveling the nexus between crop residue burning and air quality in Haryana state, India. Paddy Water Environ 23(1):111–128. Available at: https://doi.org/10.1007/s10333-024-01002-7

Pandey R (2016) Groundwater irrigation in Punjab: some issues and a way forward. In: Singh L, Singh N (eds) Economic transformation of a developing economy, India Studies in Business and Economics. Springer Singapore, Singapore, pp 97–117. Available at: https://doi.org/10.1007/978-981-10-0197-0_5

Pandey SL (2022) Stubble burning in the north-west of India: problems and potential solutions. Int J Adv Chem Res 4(1):110–113. Available at: https://doi.org/10.33545/26646781.2022.v4.i1a.134

Panwar AS et al (2018) Enhancement in productivity, nutrients use efficiency, and economics of rice-wheat cropping systems in India through farmer's participatory approach. Sustainability 11(1):122. Available at: https://doi.org/10.3390/su11010122

Prabhjyot-Kaur et al (2021) Rice yield variability in Punjab: an overview of five decades. Paddy Water Environ 19(4):673–681. Available at: https://doi.org/10.1007/s10333-021-00866-3

Qiu H et al (2023) Forest digital twin: a new tool for forest management practices based on spatio-temporal data, 3D simulation engine, and intelligent interactive environment. Comput Electron Agric 215:108416. Available at: https://doi.org/10.1016/j.compag.2023.108416

Rai PK (2016) Impacts of particulate matter pollution on plants: implications for environmental biomonitoring. Ecotoxicol Environ Saf 129:120–136. Available at: https://doi.org/10.1016/j.ecoenv.2016.03.012

Rathore J et al (2025) Characteristics of haze pollution events during biomass burning period at an upwind site of Delhi. J Geophys Res Atmos 130(7):e2024JD042347. Available at: https://doi.org/10.1029/2024JD042347

Sankhla TS et al. (2024) Observational evidence of the impact of aerosol-radiation interaction on wheat productivity. SSRN. Available at: https://doi.org/10.2139/ssrn.4681127

Saxena P et al (2021) Impact of crop residue burning in Haryana on the air quality of Delhi, India. Heliyon 7(5):e06973. Available at: https://doi.org/10.1016/j.heliyon.2021.e06973

Sharma A et al (2022) Digital Twins: State of the art theory and practice, challenges, and open research questions. J Ind Inf Integr 30:100383. Available at: https://doi.org/10.1016/j.jii.2022.100383

Singh G et al (2021) Stubble burning and its impact on air quality in Delhi NCT: a case study. J Univ Shanghai Sci Technol 23(06):58–68. Available at: https://doi.org/10.51201/JUSST/21/05223

Singh B, Gahlaut V, Rani N (2024) Seasonal and annual emission trends of PM2.5 and PM10 over the national capital Delhi from 2015 to 2020. J Air Pollut Health [Preprint]. Available at: https://doi.org/10.18502/japh.v9i2.15923

Sinha B et al (2015) Assessment of crop yield losses in Punjab and Haryana using 2 years of continuous in situ ozone measurements. Atmos Chem Phys 15(16):9555–9576. Available at: https://doi.org/10.5194/acp-15-9555-2015

SLA (2025) Virtual Singapore - a 3D city model platform for knowledge sharing and community collaboration, 27 June. Available at: https://www.sla.gov.sg/articles/press-releases/2014/virtual-singapore-a-3d-city-model-platform-for-knowledge-sharing-and-community-collaboration. Accessed 27 June 2025

Smolenaars WJ et al (2023) Exploring the potential of agricultural system change as an integrated adaptation strategy for water and food security in the Indus basin. Environ Dev Sustain 26(6):15177–15212. Available at: https://doi.org/10.1007/s10668-023-03245-6

Ta Bui L, Hoang Nguyen P (2024) Assessing impacts on rice production and quantifying economic cost losses with PM2.5–based damage—a case of the Long Xuyen Quadrangle, Vietnam Mekong Delta. Comput Electron Agric 217:108570. Available at: https://doi.org/10.1016/j.compag.2023.108570

Tagarakis AC et al (2024) Digital twins in agriculture and forestry: a review. Sensors (Basel) 24(10):3117. Available at: https://doi.org/10.3390/s24103117

Teng Z et al (2023) Environmental stimuli: a major challenge during grain filling in cereals. Int J Mol Sci 24(3):2255. Available at: https://doi.org/10.3390/ijms24032255

Thepbandit W, Athinuwat D (2024) Rhizosphere microorganisms supply availability of soil nutrients and induce plant defense. Microorganisms 12(3):558. Available at: https://doi.org/10.3390/microorganisms12030558

Vadrevu K, Lasko K (2018) Intercomparison of MODIS AQUA and VIIRS I-band fires and emissions in an agricultural landscape—implications for air pollution research. Remote Sens 10(7):978. Available at: https://doi.org/10.3390/rs10070978

Von Schneidemesser E et al (2020) How will air quality effects on human health, crops and ecosystems change in the future? Philos Trans R Soc Lond A Math Phys Eng Sci 378(2183):20190330. Available at: https://doi.org/10.1098/rsta.2019.0330

Wu H et al (2024) How does PM2.5 impact the urban vertical temperature structure? A case study in Nanjing. Aerosol Air Qual Res 24(4):230214. Available at: https://doi.org/10.4209/aaqr.230214

Yadav P, Usha K, Singh B (2022) Air pollution mitigation and global dimming: a challenge to agriculture under changing climate. In: Climate change and crop stress. Elsevier, pp 271–298. Available at: https://doi.org/10.1016/B978-0-12-816091-6.00015-8

Digital Twins in Sustainable Agriculture in Precision Irrigation and Water Management: Challenges and Future Perspectives

15

Braj Kishor Prasad, Vivek Raj, Kanchan Kumari, Sunny Kumar, Shivansh Singh, Santosh Kumar Karn, and Arun Lal Srivastav

Abstract

The combination of digital twins (DTs) and Artificial Intelligence (AI) in water resources management for agriculture is a revolutionary development that brings about a new era of precision and sustainability in the agriculture of the future. This section of the text visualizes the DTs' virtual copies of the material world in agriculture, which utilize AI-enabled analytics as a type of green technology in a CPS that performs both offline and online activities of system operation and decision support. Using such kinds of technologies for adaptive irrigation is a great example of the many benefits of uninterrupted data flows. The primary benefit of this technological method is that it enables the AI system in this environment to undergo continuous learning and upgrading. The idea of combining AI with DT is an AI machine that can carry out complicated pattern

B. K. Prasad
Department of Computer Applications (MCA), Patna Women's College, Patna, Bihar, India

V. Raj (✉) · K. Kumari
Department of Biotechnology, Patna Women's College (Autonomous), Patna University, Patna, Bihar, India

S. Kumar
Department of Electrical Engineering, IIT Delhi, New Delhi, India

S. Singh
School of Energy Systems, LUT University, Lappeenranta, Finland

S. K. Karn
Department of Biochemistry & Biotechnology, Sardar Bhagwan Singh University, Dehradun, Uttarakhand, India

A. L. Srivastav
Department of Applied Sciences, Chitkara University, Solan, Himachal Pradesh, India

© The Author(s), under exclusive license to Springer Nature Singapore Pte Ltd. 2026

A. L. Srivastav et al. (eds.), *Digital Twin Technology for Sustainable Agriculture*,
https://doi.org/10.1007/978-981-95-5915-2_15

recognition and optimization tasks in the digital twin environment, thus achieving higher precision and scalability than almost any of the traditional methods. Issues such as data fragmentation, lack of interoperability, and shortage of digital infrastructure still exist in rural farming areas. These issues have now become the biggest stumbling blocks for the uptake of DT–AI solutions. It is necessary for people from different fields to collaborate, create standards, and invest in digital literacy as well as infrastructure to overcome these challenges. The main theme of this chapter is the crucial part played by the combination of digital technologies and artificial intelligence in water management that is efficient, which is the very heart of the sustainable agricultural practices and the environmental protection that is necessary, therefore resulting in the fulfillment of the need for food and the sustainability of the earth.

Keywords

Digital twins · Artificial intelligence · Agricultural water management · Precision agriculture · Sustainability

15.1 Introduction

Global agriculture is facing an unimaginable crisis in water scarcity that is becoming increasingly severe due to various factors, including rapid population growth, climate change, and wasteful water usage. Water shortage is the main cause of this crisis. It is no secret that agriculture accounts for approximately 70% of the total freshwater resources withdrawn from the earth. However, losses and inefficiencies in the irrigation systems continue. Traditional irrigation is generally based on fixed schedules or manual estimations that typically do not take into account the real-time changes in weather, soil moisture, and crop water requirements (Mishra et al. 2021). The inefficient practices not only waste water but also bring about a bad harvest, pollute the environment, and contribute to climate change; thus, they are at the same time inefficient and overproduce and are the main culprits of environmental degradation. It clearly shows that today, global agriculture is in an unimaginable crisis of water shortage that is becoming increasingly severe due to various factors such as population explosion, the changing climate, and water wastage. Water scarcity is the root cause of this disastrous situation. The fact that the agricultural sector is responsible for around 70% of the total freshwater resources withdrawn from the earth is well known. However, losses and inefficiencies in the irrigation systems continue. Traditional irrigation methods typically operate on preset routines or through manual estimations, often without considering real-time changes in weather, soil moisture, and crop water requirements (Mishra et al. 2021). Hence, the farmers have gone beyond the limit of water use and have also got a lower crop yield, pollute the environment, and, at the same time, they are the reason for climate change that negatively impacts the ecosystem. This clearly shows that there is a need for data-driven, adaptive water management strategies in agriculture.

15.1.1 Emerging Technologies in Agriculture

The transition to Industry 4.0 technologies is the primary driving force behind a revolution in the use of natural resources. It is a revolution that enables more precise, automated, and efficient use of natural resources. DTs belong to the new products of the market that have been talked about for their potential huge effect on resource efficiency and sustainability, however, only if they are used correctly. A DT is a very interactive virtual character that constantly reflects the state of a real object. The AI-enabled device and analytical systems just search for solutions and are updated. The car-making and aerospace sectors were the pioneers in implementing the concept of the digital twin principle. However, the technology has also been very helpful in many different areas, including agriculture, which is a perfect place for DTs to monitor, model, and control operations that are complicated and energy-intensive, like irrigation and water management (Fasciolo et al. 2024).

Decision-making in water management for agriculture has been revolutionized by the implementation of digital twins. These digital twins are in a good relationship with the merged forces of live sensor data for soil moisture, crop health, weather changes, and irrigation infrastructure. To be more precise, the digital twins can execute several simulations; thus, not only do they forecast the results, but also, they virtually live them; additionally, they can also make irrigation as accurate as they want. It is the case that water is being saved; simultaneously, a plant is developing quite well, and it is still capable of adapting to changing weather, which may occur during its growth phase (Awais et al. 2025). In addition, employing AI and IoT is a significant enabler of the QT's capability enhancement since they have the possibility to absorb information as well as energy from the environment in the same way as unaided human senses or activities.

15.2 Objective

The chapter is about digital twins being employed in environmentally friendly agriculture that additionally involves utilizing precision irrigation and intelligent water management. It aims to establish the conceptual basis of DTs, identify supporting technologies such as IoT and AI, and examine their convergence in real-time irrigation systems. The aim is to supply the last pointers to a composite picture of the capabilities of DTs to economize with water, enhance crop yield, and induce climate-friendly farming practices in case studies, benefits, drawbacks, and potentials that further draw attention to the study of lacunae and the requirements of policy for the extensive usage of DT-powered solutions.

15.3 Conceptual Framework of Digital Twins in Agriculture

DTs are virtual replicas of physical systems that are incessantly data-fed to simulate, predict, and improve the system's performance. DTs were initially introduced in the manufacturing and aerospace sectors to enhance predictive maintenance and system design through simulation. The idea of DTs was also applied to the intricate and ever-changing agricultural sector. In much the same way, DTs have changed into algorithms that take data to make decisions and keep on going over the state of crops, soil, and water supply for better and greener agricultural practices (Mchirgui et al. 2024). This kind of forward movement in the farm business is their method of addressing the skyrocketing need for energy, which can help to relieve the environmental and climatic uncertainties through the use of innovative technologies.

15.4 Components of a Digital Twin in Agriculture

When farmers refer to a DT in general, they usually mean three main elements which are interconnected: the tangible object, the computer-generated model, and the network. A real object, therefore, is a representation of the parts of the farm that are indeed the soil, water supply, devices, and obviously the plants. The virtual entity is merely a computer model that reflects the state and the performance of the physical system through alterations. That model consists of parameters of a biophysical nature, local environmental variables, and the growth of the plant model. The data interface operates as a communication layer, thereby enabling a seamless two-way data flow between the physical and the virtual environments (Tagarakis et al. 2024). This nonstop data transfer not only enables the visualization of the DT of various scenarios and events and the prediction of the outcomes, but it also becomes the foundation on which decisions in irrigation and water management are based.

15.5 Enabling Technologies

The essential nature of digital twins, which are specifically designed for agricultural purposes, is yet largely a very intricate technological network. IoT sensors collect highly detailed information about soil (like moisture and temperature), air (humidity), and plants' condition. The work of providing the strength of the decision, searching for the directions of the development, and determining the irrigation schedules has been carried out by AIs and MLs through the analysis of this data.

Cloud computing acts as the main source for storage and computational power, and GIS and remote sensing are the tools for carrying out special analysis and monitoring of the environment. Moreover, blockchain provides a very safe space that allows the exchange of agricultural information in a decentralized manner. The combination of these technologies creates a smart ecosystem that is not only able to adapt to the changes of the environment but also capable of performing the function of water conservation (Nsoh et al. 2024). The smart ecosystem that comes from the bundling of these technologies can certainly be of help in water conservation, as well as being part of the environmental change response.

15.6 Digital Twins vs. Other Digital Tools

Compared to simulation models or Decision Support Systems (DSS), which are based on data that are either static or updated only from time to time, digital twins (DTs) are created to be live, and therefore, they are still being fueled. Forecasting is the attribute of a simulation, but it cannot give a response immediately. DSS, in the main, is a rule-based system that employs assistance to reach a decision. Nevertheless, digital twins incorporate real-time sensor data, prediction, and reaction parts, thus enabling the system to function without interruption and to be able to adjust if the solution is altered, as in the case of agriculture (Zhao et al. 2025). Thus digital twins are ideal partners for going on the journey of dealing with the complexity and invisibility of the issues that are common to the precision irrigation and water management problems.

15.7 Current Challenges in Irrigation Practices

First of all, the traditional irrigation method has definitely been water-wasteful due to inefficient water use. The inefficient aspect of the irrigation system was the key reason for water misuse, which became the main factor in the decrease in crop productivity. Actually, all irrigation systems are hugely based on being either fixed-schedule-operated or manual-calculation-driven, which, however, mostly ignore the influence of weather, soil properties, and plant water requirements. The "one-size-fits-all" approach is still a common way in farming irrigation, which, however, usually results in the unnecessary use of water during times of water shortage. Therefore, the use of this approach will not only injure the plant health but also diminish the yield. It is worth noting that water management through irrigation systems is still a major point that causes the inefficient use of water in agriculture, and there are several reasons for that. A lot of areas are still heavily dependent on old and traditional infrastructure, which does not have features like automation and real-time monitoring. The absence of solar data at the local level and the lack of real-time information make adaptive management inefficient. Variability in climate affects these issues negatively by changing the distribution of rainfall and evapotranspiration, thus causing uncertainties (Nikolaou et al. 2020). Thus, the farmers are most probably going to have to deal with huge uncertainties about the amount of water that they will get and when the irrigation will take place, in particular, in those areas where water is very limited or that which is available is only through rain.

15.8 Need for Data-Driven Irrigation Solutions

Precision irrigation using sensors and IoT (Internet of Things) for water management and instant decision-making in agriculture makes it possible to apply the exact amount of water to each unit of the field. Sensors, weather forecasts, and AI algorithms contribute to supplying only water that the plant requires, thus not only preventing water deficit but also overwatering (Roy et al. 2024). Implementing such a management system is necessary, as the conjunctive use of groundwater with surface water may lead to overexploitation of groundwater, which is becoming a serious issue in many parts of the world. Hence, the affected farmers should be encouraged and supported to adopt water-saving techniques such as deficit irrigation (DI) and regulated deficit irrigation (RDI), which have been proven to be sustainable and water-conserving practices in agriculture. Water-saving irrigation strategies, such as regulated deficit irrigation (RDI) and partial root-zone drying (PRD), have been shown in several studies to reduce water use while improving water productivity, maintaining or even increasing yield and quality. We chose RDI for our study because it is a relatively basic irrigation technique that requires only adjustments in frequency and quantity of irrigation, without the use of an advanced automation system for scheduling control. Additionally, these water-saving techniques can be mutually synergized to embody a higher level of water-saving efficiency, hence, amplifying the amount of saved water utilized in the PRO and RDI. In the first place, while deciding the mode of irrigation, it is ideal to select those that can easily adapt to the change in conditions with minimum human intervention, for instance, drip irrigation, sprinkler irrigation, and automatic irrigation. Normally, drip irrigation can save 40–60% water as compared to flood irrigation, and sprinkler irrigation can save 20–50% (Fig. 15.1).

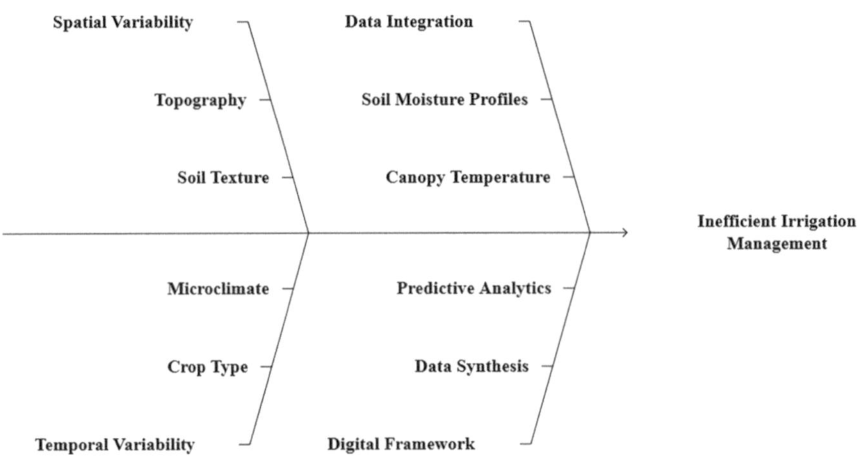

Fig. 15.1 Challenges in implementing data-driven irrigation

15.9 Benefits of Precision Irrigation Using DTs

Digital twins (DTs) are the perfect match for a smart irrigation system, as they provide an opportunity to establish a continuous collaboration between the real field and the intelligent irrigation decision-making system. By continually extracting data and executing simulations, DTs can delve much deeper into the soil–plant–atmosphere process, thereby becoming more precise in determining water requirements in small areas. This, in effect, allows the real execution of irrigation scheduling optimization, which is a water-saving measure, thus leading to lower input costs and also to a rise in crop production. Besides that, the DTs offer the farmer the possibility of trying out different scenarios, such as the effect of an unexpected heat wave or a water supply shortage, thus guaranteeing efficient irrigation management without any incidents along the way.

However, by the substitution of the human factor with the digital twin system's intuition and input of the manual work, this will be not only more sustainable but also less vulnerable to different kinds of dangers and more efficient in the use of resources (Peladarinos et al. 2023). Also, these systems represent the enormous potential of smart agriculture in the fight against climate change, but they are also the ones that prove they increase the efficiency of water use and manage the uncertainty of water supplies, caused by climate change, as depicted in the future in the simulations of various time intervals.

15.10 Application of Digital Twins in Precision Irrigation and Water Management

The employment of DTs in exact irrigation is a computer network that is the heart of the process of data transfer from the physical world to the digital one. Actually, the sensor networks strongly implanted in agricultural fields are the brain of this system, which gather inputs on soil moisture, temperature, humidity, plant health, irrigation flow rates, and other parameters with high resolution. In this way, sensors send real-time data to cloud-based platforms without any gaps, where the data is converted and used in the virtual twin's setup—an empowered, digital copy of the irrigation environment. The twins, besides, continue their simulations both in time and space even more with the help of GIS, remote sensing, and meteorological information. The data is deeply analyzed by the advanced analytics and ML algorithms, which make the DT the one that can simulate various irrigation scenarios and give the last water distribution schedules. So, the constant feedback loop, thus, makes it possible for the irrigation choices to be updated immediately, not only for the current situation but also for the future ones.

15.11 Use Cases and Case Studies

Highly impactful digital images of human activities, such as water irrigation operations, have been the result of the spread of the same idea to vast regions of the earth, and as a result, the twins are the radiating examples of new opportunities and good effects. As an instance from Israel, a country that is very much known for its scarcity of water resources, it can be mentioned that the irrigation systems, which are configured with the help of digital twins, have been deployed for high-value crops such as tomatoes and citrus, and the case has been extremely successful. The supervision relied on the records of evapotranspiration and soil moisture from the sensors. Along the same linen, in the Netherlands, indoor crop digital twin systems—where IoT sensors interface with environmental models—enable the application of the exact quantity of water and nutrients to horticultural crops. The method mentioned above will not only save resources but also improve the quality of the crops. Bringing India into the picture, merely to illustrate, there were first attempts toward precise farming that traditionally require the use of DTs for managing the paddy plots, using not only the satellite data but also sensors in the field to decide on the water supply change according to the new crop development phases and climatic variation (Kalyani et al. 2024).

The potential benefits of the twin to the agriculture sector have been evidenced through the sign-tests that were carried out in areas that are affected by drought, where the grape and vineyard are the main agricultural portfolio. The DTs have been employed to keep track of the water situation of plants and also to decide the most efficient drip irrigation times. On the other hand, the wheat systems also make use of DTs that carry out the simulation-based output for each irrigation strategy, hence, forecasting the yield response.

15.12 Predictive Analytics and Decision Support

One of the incredible characteristics that allows digital twins (DTs) in irrigation to be very efficient is their power, which enables them not only to access predictive analytics but also to be a good decision support system. A digital twin executes a weather forecast simulation together with the archival crop and evapotranspiration data that are suitable for determining the water consumption of the plant in the next period with very little error. Therefore, these models provide not only farmers and water managers the opportunity to devise irrigation strategies before the occurrence of the problem instead of responding but also facilitate the efficient use of water even under the situation of climate stress. On the one hand, the water management of the digital twins is still being used for those that are capable of yield prediction, alert systems for irrigation anomalies, and long-term planning of scenarios (El Sakka et al. 2025). This ability to predict can result in sustainable and climate-smart farming by wisely applying irrigation that is in agreement with the forecast of the crop water demand and the environmental sustainability objective (Fig. 15.2).

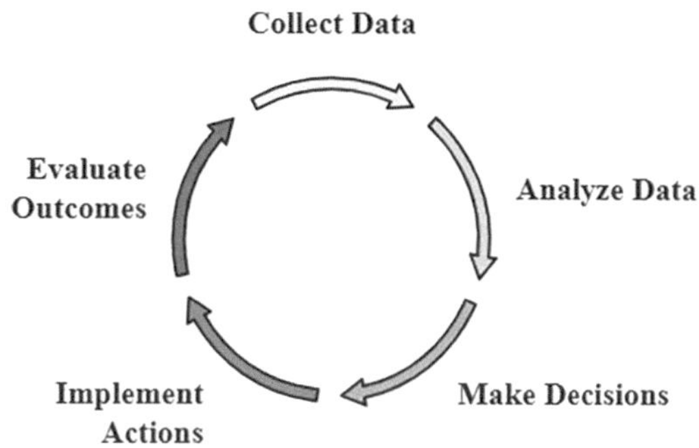

Fig. 15.2 Predictive irrigation cycle

15.13 Integration with AI and ML

The use of Artificial Intelligence and Machine Learning with DTs gives a tenfold increase in their ability to provide smart, context-sensitive irrigation systems. Artificial Intelligence techniques, especially deep learning architectures such as ANNs, are the leading approaches to estimate the water needs of plants. The algorithms are designed to detect complex, nonlinear interdependencies with multiple variables—such as soil moisture, temperature, humidity, solar radiation, and stages of crop growth. These machines can change their response based on past and present information to determine the best irrigation quantities with a very high accuracy (Rojek et al. 2024). Besides this, the machine learning technique, called RL, is a quickly developing tool that aims at the ongoing improvement of water supply management. One example of such a system is the water distribution policy that is observed to be changing in this way. In such a situation, the method of trial and error is used to make the decision process more certain, during which the system gets information from the environment (for example, the change in crop conditions or soil saturation) that allows it to improve its later decisions in a sequence of chosen actions.

15.14 Digital Twins + AI for Adaptive Irrigation

Digital twins, when used with AI, are not just static digital models but also become self-regulating and intelligent systems. Such enhanced decision trees can outperform human decision-making; therefore, they can react to changes in the environment without any assistance from humans. For example, if the sensor data shows that there will be high evapotranspiration due to the temperature rise, the DT can not only detect it but also change the irrigation schedule in such a way that the soil moisture is at the optimum level. Besides that, DTs employing AI can spot problems like a leak in the hose, a failure in the pump, or a disease that occurs in the plants, and they can choose the best action instantly. The learning capability of these systems is what allows the gradual improvement of the irrigation schemes, which is very important, due to the climate change history, the appearance of new and changing weather patterns, and the transformation of agricultural practices (Alfaro-Viquez et al. 2025). It is still the rule-based automation that is being converted into intelligent adaptation, and the water management sector of high precision is certain to benefit from this massive step.

15.15 Benefits for Sustainability and Resilience

The integration of a virtual world along with AI and ML results in a substantial improvement in the sustainability of the agriculture sector. One example here is a drastic decrease in the consumption of water by customers who are totally informed that the irrigation is no different from the water requirement of the crop. Besides being the most compatible with the prevention of over-irrigation and the minimization of runoff, they not only keep the soil health at a high level but also act as a source of reduction of nutrient leaching. With the forecasting feature, the farmers will be able to forecast and, thus, if any climatic events such as droughts or heat waves occur, they will be prepared; therefore, the agricultural systems become more potent in solving the climate change issue. On the other hand, consumption of energy is also being optimized by irrigation time, which occurs during that time period of the day when the energy demand is at its lowest or the energy is used most efficiently. Another point is that these intelligent systems cover huge areas, and at the same time, they are of great help in making climate-smart agriculture, which is fully compatible with Sustainable Development Goals (SDGs). The latter primarily addresses the issues of water efficiency, food security, and environmental protection (Nie et al. 2022).

15.16 Barriers, Risks, and Ethical Considerations

15.16.1 Technical Barriers

DTs are seen as a revolution in water-saving irrigation and water management; however, they are still in the experimental phase and facing many technical challenges. One of the main issues is data incompatibility, which is caused by different agricultural systems. Usually, agrarian operations rely on a lot of sensors, platforms, and data formats, which make it very difficult to integrate them into one uninterrupted DT architecture. Besides that, the main issues of sensor calibration and sensor maintenance have to be solved—sensors must be very accurate and reliable in the changing field environment if they are to provide good data; however, they can still be influenced by drift, malfunction, or they can become noisy, which is undesirable. The major technical problem mentioned is the complexity of the model. The authors point out that it is very difficult to generate real-time, biophysical models, which can absolutely represent soil–plant–atmosphere processes, because it needs to bring together interdisciplinary knowledge and also a lot of computing power (Purcell et al. 2023). Such problems are like the stones lying in the way for the small farmers or those areas with limited resources who plan to set up and continue their DT systems running sustainably.

15.16.2 Economic and Social Challenges

Economic limitations are certainly the primary cause that restricts the use of digital technology in agriculture. Besides, the initial expenses that were astronomically high for purchasing sensors, building communication networks, utilizing cloud computing, and integrating AI cloud have made farmers opt out of using such technologies, most especially in developing countries. Usually, farmers with small and marginal land pieces do not have the money they need, subsidies, or financing options that can help them use digital technology in an efficient way. What is more, digital literacy and farmer awareness are very important parts of the problem. Most farmers cannot relate to the technology of DT and may not have the skills required for understanding its outputs or using it in their daily decision-making. The digital divide is a threat to the equality of the socio-economic situation because those with large and tech-savvy businesses will win, while those with small businesses will lose (Senoo et al. 2024). There is a shortage of local demonstrations, training, and services, which results in increased trust issues and difficulties in adoption.

15.16.3 Ethical and Environmental Risks

Data privacy and ownership are by far the most significant ethical issues and are closely related because of an exponential growth in decision tree systems that consume real-time data and hence make decisions. The farmers might be very careful and decide not to disclose any information about land use, crop performance, or irrigation schedules to anyone, unless they are sure that the data is theirs and that it will be used only in accordance with the agreed conditions. In this case, they may find themselves not willing to share the data. Overusing technology, along with many other things, is the most likely danger, which, as a consequence, may lead to human beings' absence in the process of decision-making. Such a situation is still very dangerous if a technical failure or a cyber-attack occurs. If DTs are used wrongly or designed in the wrong way, they might cause only bad consequences for the environment despite every attempt to use them for good, for example, irrigation that is more efficient for yield production can definitely cause biodiversity disappear, or soil salinity may even become higher if it goes a long time (Gardezi et al. 2023). Ethical guidelines, laws, and participatory governance models are the first steps toward the right direction of responsible and just development of the deployment of DT technology in farming. Table 15.1 reflects the main technical, economic, social, ethical, and environmental challenges associated with digital technology (DT) utilization in agricultural water management. The table points out that there are a number of major problems, such as data compatibility, high charges,

Table 15.1 Summary of key barriers, risks, and ethical considerations in digital twin adoption for agricultural water management

Category	Key issues	Implications
Technical barriers	Interoperability, calibration, complexity	Integration, accuracy, accessibility
Economic and social challenges	Cost, literacy, training	Adoption, inequality, capacity
Ethical and environmental risks	Privacy, automation, vulnerability, impact	Trust, resilience, sustainability

casual digital skills, privacy issues, and the necessity for responsible management, which, if not dealt with, will result in none of the implementations to be sustainable and green.

15.17 Future Trends and Research Directions

The future of exact irrigation and water management is set to be dependent on robots in electronic farms operated by highly advanced DTs very quickly. Robotics and Unmanned Aerial Vehicles' (UAVs) partnership with DT will enable not only the execution of irrigation commands in real time based on digital simulations but also the predictions. Robots that have precision irrigation nozzles may, on their own, provide water only to the plants that need it and agree with the data-driven input from AI models relying on DT data. The 5G rollout will certainly play a very important role in the energy transition as it provides very low latency and high capacity, which perfectly match the requirements for real-time sensor data, edge computing, and cloud analytics. Edge computing here plays a major role by allowing the data to be processed directly at the farm, which definitely reduces the need for a cloud, and at the same time, it makes faster, more localized reactions possible—this is very important in places where the connection is not good (Sibona et al. 2025). Along with these new technologies, the picture is a future where not only the automation of decision-making, resource allocation, and field work but also the use of digital intelligence for those activities is done smoothly.

15.18 Policy and Governance Implications

Even though the use of digital twin (DT) technologies in the agricultural industry is widespread in order to improve crop yields, there is still a significant lack of solid and good governance systems that are needed for the safe and ethical use of these technologies. It is necessary to develop such policy frameworks, which have inter-disciplinary, bright, inclusive, and flexible characteristics. Furthermore, these protocols need to be not only consistent for efficient data exchange between different platforms but also reliable in terms of farmers' privacy as well as their rights of ownership. Apart from that, policies have to become more inclusive by providing incentives, subsidies, or making public–private partnerships that will support the smallholder farmers in technology adoption, thus overcoming the high cost of adoption. Furthermore, utilizing agricultural extension services digitally may allow disseminating training, stockpiling capacity, and providing decision support suitable for DT-enabled systems. Certainly, a regulatory mechanism will provide the necessary support in assessing whether the implementation of the efficient water program in the lower part of the river has an impact, such as reduced water flow or other effects on the ground, which may, for instance, lead to soil degradation. Therefore, sustainability assessments and adaptive governance models must be included as a part of the solution to this problem (Adibi et al. 2024).

15.19 Open Research Challenges

Several open research challenges become visible in spite of the very noticeable progress. A key challenge here is the standardization of DT architectures to make them modular, interoperable, and scalable not only locally but also globally across different agroecosystems. The inability to use the same modeling frameworks for different initiatives makes it difficult to replicate and adapt them. Real-time checking of the DT forecast by high-frequency field data is vitally important for the system to create trust and become reliable. Existing DTs are frequently working with partial or old datasets; thus, there is a possibility that the forecast will not be accurate. Besides that, scalability still remains a very important issue—technologies that have been tested as effective in proof-of-concept projects or research farms can fail in their attempt to be implemented widely in agricultural systems that are larger or more diverse (Sigler et al. 2024). Taking care of such issues will definitely demand that scientists in different fields, such as agronomy, data science, engineering, and even policy-makers, come together and agree on what kind of agile, adaptive, and ethically aligned DT solutions should be developed for the agriculture of the future.

15.20 Government Initiatives in Digital Agriculture and Water Management

Data.gov.in and the National Data Sharing and Accessibility Policy in the domain of agriculture data promote clarity and innovation based on data. Further, AgriHubs for innovation and insights based on data, with systems like IIT-Indore's AgriHub, along with the state-level AgriHubs in Andhra Pradesh, like APAIMS 2.0, both drive integrated data collection and AI-based insights. State-based schemes, like the Rs. 500 crore MahaAgri-AI scheme launched by Maharashtra, combine blockchain, AI, and satellite/remote sensing data into advisory services for farmers in real time. Other pilot projects include public–private partnerships (PPP) with "Smart Intelligent Village" in Satnavri, where the goal is to modernize the postharvest process and focus on AG growth value chain in UP, which involves real-time monitoring.

EU initiative Horizon 2020 is working to drive IoT and data interoperability across a number of projects, including IoF2020, DEMETER, ATLAS, and SmartAgriHubs. The SWAMP project demonstrates the impact of precision irrigation. The Common European Agricultural Data Space (CEADS) is being developed by the EU through the Data Act and will use secure and standardized means to share agricultural data in Europe, including satellite data integration with environmental data. Open data portals and datasets like EuroCrops provide access to important geospatial data and crop data that are important for any digital twin (DT) applications.

15.21 Quantitative Impact of Digital Twins on Water Savings and Yield in Precision Irrigation

A number of studies and pilot implementations have quantitatively indicated the positive impact of digital twin (DT) systems in value-based precision irrigation. For example, DT-enabled irrigation systems have provided water savings between 25% and 40% in areas of agricultural masses in the U.S. and resulted in higher crop yields of up to 5–15%. With DT, an irrigation scheduler in Canada based on predictive control and machine learning uses DT irrigation systems, resulting in a reduction of 6.4–22.8% water usage, together with an increase in yield of 2.3–4.3% in trials encompassing a 26.4-hectare experimental field. Some additional smart irrigation case studies have reported potential water savings of up to 59% through DT irrigation simulation of interconnected tomato irrigation; 29.2% reduction in basin systems based on design by DT at 40% decrease of water use; and a 46% reduction in paddy rice fields of the interconnected systems. These also indicate greater potential implications for the integration of DTs intersecting with IoTs and AI technologies. Drip irrigation can reduce consumption (in and of itself) by 40–60%, while sprinklers can reduce by 20–50%, as DT systems optimize improved water reduction goals, through a system/framework in terms of agriculture, which could vastly enhance overall efficiency and sustainability outcomes.

Table 15.2 provides material evidence of the effectiveness of different geographical and technological DTs and smart irrigation systems on decreasing water use and increasing crop yield. For reduced water use, studies reported reduced water use ranging from 6.4% and 72%, with DT resulting in consistent water use reductions of 25–40%, and AI-enabled precision agriculture studies conducted in the USA showed yield increases generally ranged from 2.3% to 15%. The results suggest the evidence of AI, machine learning, and sensor technology with respect to water sustainability, while the performance values associated with the accuracy of predictions and monitoring only represent the missed opportunity of optimizing decision-making from accurate predictions and monitoring and associated sustainable farming methods.

Table 15.2 Comparative performance values of digital twin and smart irrigation systems in agriculture

System/study type	Water savings	Yield increase/error improvement
AI precision Ag (general)	25–40%	5–15% yield; 30–50% monitoring accuracy
DT precision irrigation (U.S. farms)	25–40%	5–15% yield; ~92% prediction accuracy
Smart irrigation systems (SIS)	15–59% (varies by crop)	Yield improvements (coffee, rice, etc.)
ML-controlled DT scheduler (Canada)	6.4–22.8%	2.3–4.3% yield
Deep RL irrigation (almond case)	~9.5%	–
Leaf sensor control	~25%	–
DT estimate (NumberAnalytics)	20–30%	Maintain/improve yield
Precision sensor irrigation (Reddit)	20–72%	–

15.22 Conclusion

The implementation of DTs and AI in the water management for agricultural products has revolutionized the precision farming sector, which is a perfect example of the synergy between technology and environmental issues. This chapter of the book has focused on DTs and AI collaboration as a way of radically transforming the water utilization in agriculture and hence creating a novel and data-guided instrument for the accommodation of natural variations and the plant requirements on-the-go. Digital twins' usage as the representation of the physical factories and domains here, therefore, very much aligns with the concept of decision-making accuracy and operational efficiency achieved by the system's behavior being continuously observed, simulated, and predicted. In addition, the use of AI in such virtual worlds of synthetic nature not only provides them with more processing power, but it also enables them to carry out operations like unearthing secret connections, forecasting future incidents, and selecting the optimum solutions, which are unimaginably far ahead of human capabilities. Here, it is not only guaranteed that water is distributed and consumed most efficiently, but also that energy is conserved by the system being resilient to climate change, and pollution is reduced due to lower input costs and an environmentally friendly system. In spite of the potential being very high, the distribution of the DT-AI framework in the agricultural sector is, nevertheless, limited due to data silos, interoperability problems, lack of digital infrastructure in rural areas, and the need for cross-disciplinary work. Only the collaboration of academic institutions, industries, and policymakers who agree on the development of standardized platforms, the promotion of open data ecosystems, and the education and training of digital skills and capacity building can fill these gaps.

References

Adibi S, Rajabifard A, Shojaei D, Wickramasinghe N (2024) Enhancing healthcare through sensor-enabled digital twins in smart environments: a comprehensive analysis. Sensors 24(9):2793. https://doi.org/10.3390/s24092793

Alfaro-Viquez D, Zamora-Hernandez M, Fernandez-Vega M, Garcia-Rodriguez J, Azorin-Lopez J (2025) A comprehensive review of AI-based digital twin applications in manufacturing: integration across operator, product, and process dimensions. Electronics 14(4):646. https://doi.org/10.3390/electronics14040646

Awais M, Wang X, Hussain S, Aziz F, Mahmood M (2025) Advancing precision agriculture through digital twins and smart farming technologies: a review. AgriEngineering 7(5):137. https://doi.org/10.3390/agriengineering7050137

El Sakka M, Ivanovici M, Chaari L, Mothe J (2025) A review of CNN applications in smart agriculture using multimodal data. Sensors (Basel) 25(2):472. https://doi.org/10.3390/s25020472

Fasciolo B, Panza L, Lombardi F (2024) Exploring the integration of industry 4.0 technologies in agriculture: a comprehensive bibliometric review. Sustainability 16(20):8948. https://doi.org/10.3390/su16208948

Gardezi M, Joshi B, Rizzo D, Ryan M, Prutzer E, Brugler S, Dadkhah A (2023) Artificial intelligence in farming: challenges and opportunities for building trust. Agron J 116. https://doi.org/10.1002/agj2.21353

Kalyani Y, Vorster L, Whetton R, Collier R (2024) Application scenarios of digital twins for smart crop farming through cloud–fog–edge infrastructure. Futur Internet 16(3):100. https://doi.org/10.3390/fi16030100

Mchirgui N, Quadar N, Kraiem H, Lakhssassi A (2024) The applications and challenges of digital twin technology in smart grids: a comprehensive review. Appl Sci 14(23):10933. https://doi.org/10.3390/app142310933

Mishra BK, Kumar P, Saraswat C, Chakraborty S, Gautam A (2021) Water security in a changing environment: concept, challenges and solutions. Water 13(4):490. https://doi.org/10.3390/w13040490

Nie J, Wang Y, Li Y, Chao X (2022) Artificial intelligence and digital twins in sustainable agriculture and forestry: a survey. Turk J Agric For 46:642–661. https://doi.org/10.55730/1300-011X.3033

Nikolaou G, Neocleous D, Christou A, Kitta E, Katsoulas N (2020) Implementing sustainable irrigation in water-scarce regions under the impact of climate change. Agronomy 10(8):1120. https://doi.org/10.3390/agronomy10081120

Nsoh B, Katimbo A, Guo H, Heeren DM, Nakabuye HN, Qiao X, Ge Y, Rudnick DR, Wanyama J, Bwambale E, Kiraga S (2024) Internet of things-based automated solutions utilizing machine learning for smart and real-time irrigation management: a review. Sensors (Basel) 24(23):7480. https://doi.org/10.3390/s24237480

Peladarinos, N., Piromalis, D., Cheimaras, V., Tserepas, E., Munteanu, R.A. and Papageorgas, P., 2023. Enhancing smart agriculture by implementing digital twins: a comprehensive review. Sensors (Basel) (Basel), 23(16), p.7128. doi:https://doi.org/10.3390/s23167128

Purcell W, Neubauer T, Mallinger K (2023) Digital twins in agriculture: challenges and opportunities for environmental sustainability. Curr Opin Environ Sustain 61. https://doi.org/10.1016/j.cosust.2022.101252

Rojek I, Marciniak T, Mikołajewski D (2024) Digital twins in 3D printing processes using artificial intelligence. Electronics 13(17):3550. https://doi.org/10.3390/electronics1317355

Roy S, Rathour S, Mehta A, Dwivedi R, Pandey A (2024) Precision water management for resource conservation in India's dryland agriculture: strategies and technologies. Int J Environ Clim Chang 14:464–480. https://doi.org/10.9734/ijecc/2024/v14i84367

Senoo EEK, Anggraini L, Kumi JA, Karolina LB, Akansah E, Sulyman HA, Mendonça I, Aritsugi M (2024) IoT solutions with artificial intelligence technologies for precision agriculture:

definitions, applications, challenges, and opportunities. Electronics 13(10):1894. https://doi.org/10.3390/electronics13101894

Sibona F, Harrison N, Wallace N, Stevens C, Sukkarieh S, Raubenheimer D, Simpson S, González L (2025) A framework to enable eco-cyber-physical systems for robotics-focused digital twins in smart farming. J Field Robotics. https://doi.org/10.1002/rob.22562

Sigler L, Ubach de Fuentes P-A, Mora J, Oñate E (2024) A review of technologies and challenges for integrated modeling analysis. Arch Comput Methods Eng 32. https://doi.org/10.1007/s11831-024-10187-3

Tagarakis AC, Benos L, Kyriakarakos G, Pearson S, Sørensen CG, Bochtis D (2024) Digital twins in agriculture and forestry: a review. Sensors (Basel) 24(10):3117. https://doi.org/10.3390/s24103117

Zhao T, Song C, Yu J, Xing L, Xu F, Li W, Wang Z (2025) Leveraging immersive digital twins and AI-driven decision support systems for sustainable water reserves management: a conceptual framework. Sustainability 17(8):3754. https://doi.org/10.3390/su17083754

Applications of Digital Twins in Agriculture: Enhancing Crop Monitoring, Livestock Health, Soil Fertility, and Water Management

16

Gurwinder Singh and M. R. M. Veeramanickam

Abstract

This chapter explores the transformative role of Digital Twin (DT) technology in revolutionizing the agricultural sector by enhancing the efficiency and sustainability of crop management, livestock health, soil fertility, and precision irrigation. The implementation of DTs integrates real-time monitoring systems with predictive analytics, fostering smarter agricultural practices, and enabling data-driven decision-making. By creating virtual replicas of physical assets and processes, DTs allow for continuous optimization and monitoring, which can lead to significant improvements in productivity, sustainability, and the management of resources. This chapter will delve into the specific applications of DTs across key agricultural domains, supported by case studies, and an analysis of technological advancements and best practices for their successful integration.

Keywords

Digital Twins · Smart agriculture · Crop Monitoring · Livestock HEALTH · Soil fertility · Precision irrigation · Sustainable farming

G. Singh (✉)
Department of AIT-CSE, Chandigarh University, Mohali, Punjab, India

M. R. M. Veeramanickam
Department of Computer Science & Engineering, Chandigarh University, Chandigarh, Punjab, India

© The Author(s), under exclusive license to Springer Nature Singapore Pte Ltd. 2026

A. L. Srivastav et al. (eds.), *Digital Twin Technology for Sustainable Agriculture*,
https://doi.org/10.1007/978-981-95-5915-2_16

16.1 Introduction

Agriculture is undergoing a paradigm shift driven by the integration of digital technologies, data analytics, and automation into farming practices. Among these innovations, Digital Twin (DT) technology has emerged as a transformative approach capable of revolutionizing agricultural systems through the creation of virtual replicas of physical entities, processes, and environments. These replicas are continuously synchronized with real-time data from sensors, Internet of Things (IoT) devices, and remote monitoring systems, enabling dynamic simulation, prediction, and optimization of agricultural activities. By providing a digital counterpart of crops, livestock, soil, and water resources, DTs offer an interactive platform for monitoring performance, detecting anomalies, and optimizing resource utilization. Climate in the agricultural sector is struggling with ongoing issues such as climate change, water shortages, soil falls, insect pressure, and increasing global demand for food. Traditional agricultural methods, although useful in some scenarios, are limited by their reactive approaches and dependence on manual monitoring. However, digital twins (DTs) provide an active, data channel solution by taking advantage of future indication analysis, artificial intelligence, and simulation models to enable real-time decisions. This allows problems, accurate intervention, and initial detection of permanent use of resources.

DTs have different uses in agriculture. For crop monitoring, they enable real-time tracking of plant growth, early detection of stress, and accurate dividend forecasts. In livestock farming, DT's digital copies of farms and health models are made to monitor animal welfare, adapt food, and diseases. Soil-state models and soil fertility are improved through the future model and sensor networks, which enables targeted fertilization and long-term stability. Water management must benefit from DT-driven precise watering, weather data, soil moisture, and crop integration and reduce the plan and reduce the waste. Despite their ability, the adoption of broad DT in agriculture is still newborn. The challenges include high costs, limited rural connections, data integration problems, and complications of biological systems. However, progress in IoT, AI, Edge Computing, and Blockchain continuously overcome these obstacles, making DTs more viable and scalable. With supportive drivers, peasant training, and modular implementation, DT-enabled agriculture is ready for Agriculture 5.0, and strengthening productivity, stability, and global food security.

This chapter offers a detailed exploration of digital twin applications in agriculture, focusing on crop monitoring, livestock health, Earth's fertility, and water management. It includes practical examples, case studies, recent technological progress and opportunities and challenges to integrate DTs into modern agricultural systems.

16.2 Related Works

Digital Twin (DT) technology has aroused considerable interest in different fields; its use in agriculture appears as a high prominent field of research. Several studies have offered ideological and practical approaches to the development, adoption, and integration of DTs to improve agricultural efficiency and stability.

The main concept with DTs and their defined features were completely discovered by Jones et al. (2020), which outlined their features as a synchronization of digital copies of physical systems. Within agriculture, DT is quickly in line with the goals of Agriculture 4.0 and 5.0. Studies by Wang (2024) and Zhang et al. (2025) have undergone the recent progress of agriculture–DTs, which indicates both the ongoing challenges related to technological progress and differences, data integration, and scalability. Similarly, the role of artificial intelligence and DT in permanent agriculture and forestry emphasizes the importance of future analysis for resource adaptation. Also comprehensive reviews of machine-learning and deep-learning methods for crop yield prediction are provided by Jabed and Murad (2024).

Applications in crop monitoring have received particular attention. Skobelev et al. (2021) developed a DT of rice for decision-making in precise farming, while Singh and Saini (2023) explored crop growth modeling through virtual reality-based DTs. Banerjee et al. (2025) proposed DT-enabled crop recommendation systems for yield optimization, and Soundankar (2024) made available smart farming sensor datasets that support DT-based yield prediction. The use of UAV and 3D visualization for crop management was examined by Feldmane and Rubauskis (2023) and Catala-Roman et al. (2024), demonstrating how advanced data acquisition enhances the accuracy of DT models.

Livestock management has also benefited from DT applications. Jeong et al. (2023) presented a virtual pig house as a DT platform for digital livestock farming, while Raba et al. (2022) proposed a DT framework for optimizing livestock feeding strategies. Herlin et al. (2021) investigated the welfare implications of digital tools for monitoring cattle and sheep, emphasizing the balance between productivity and ethical considerations.

Soil health and fertility represent another critical domain where DTs have been applied. Silva et al. (2023) conducted a systematic review of DTs for soil quality assessment, outlining methodologies for real-time soil monitoring. Research by Eremina et al. (2022) highlighted the use of digital technologies to strengthen agricultural enterprise information systems, while predictive approaches to soil fertility management were further advanced by Escribà-Gelonch et al. (2024) and Zake and Majore (2022). These studies underscore the importance of accurate simulation of soil dynamics for sustainable nutrient management.

In the area of water management, DT-based precision irrigation systems have emerged as highly effective tools. Purcell et al. (2023) analyzed the opportunities and challenges of DTs for environmental sustainability, particularly in irrigation and resource efficiency. Sadowski and Spachos (2020) highlighted wireless IoT technologies with energy harvesting for agricultural monitoring, supporting water resource optimization. Weather-driven irrigation scheduling, with DT simulation, which shows the integration of cyber-physical systems with drone applications in agriculture, was discussed by Tsolakis et al. (2021).

Integration of digital twins (DTs) with competent technologies has been an important focus for recent research. Cakir et al. (2023) proposed intermediate products for Smart Farm IoT network, while Liu et al. (2023) examined the combination of DTs with Generative AI. DT architecture that includes age, fog, and cloud computing was detected by highlighting the scalability and real-time treatment properties of Hananto et al. (2024) and Yogeswaranathan and Collier (2022). In addition, Verdouw and Kruize (2017) provided early examples of DTs in farm management through European smart agricultural projects.

From a comprehensive perspective, the adoption of digital twins (DTs) in agriculture raises concerns related to security, policy, and stability. Agnusdei et al. (2021) examined the role of DTs in safety management, while Wilberforce and Mwebaze (2025) introduced an IoT-enabled framework for smart agriculture. The ability of.

DTs to enhance decision-making and stability was emphasized by Elijah et al. (2021) in the context of Nigerian agriculture. Zhao et al. (2023) extended DT applications beyond traditional agriculture by demonstrating their role in autonomous agricultural vehicles. Finally, Talkhestani et al. (2018) highlight the synchronization techniques between physical and digital systems, which remain central to DT applications in agriculture.

Together, these studies suggest that DTs provide a significant ability to meet agricultural challenges through real-time and data-driven approaches. However, questions such as interoperability, cost-effectiveness, and complexity when modeling biological systems require further research and development of practical structure to ensure efficient DT integration in agricultural systems.

16.3 Crop Monitoring and Management

Crop monitoring goes into a new era and goes from reactive estimates to data-driven control. There are the digital twins in the hearts of the shift-three models located in the regions of the world that are constantly updated with live data. These powerful models give farmers a unique window in their crops so that they can follow different scenarios, predict the results, and make smart decisions to promote health and dividends. It is made possible by a network of technologies working on concerts, from earth sensors and drone images to sophisticated data analysis, feeding information into digital twins to create a dynamic, real-time model (Jones et al. 2020; Wang 2024; Zhang et al. 2025).

16.3.1 Real-Time Monitoring of Crop Health and Growth

Gone are the days to rely on a trip to the periodic field. Digital twins enable surveillance of real time by synthesizing a continuous stream of information.

Drone overheads equipped with multisacral cameras fly, while the ground-based sensors track the conditions at the root level. Together, they measure important signs such as plant height, chlorophyll material, and canopy density and paint a comprehensive picture of crop growth in the virtual model. For example, scientists have created a digital twin for rice fields that directly provide practical insight to farmers, which helps them trace growth stages with accuracy. Other studies have used advanced 3D imaging to create incredibly detailed visualization of crop structures, making real-time monitoring more accurately than ever. This allows continuous vigilance for timely adjustment, to ensure that crops are provided exactly what they need to bloom.

16.3.2 Early Detection of Stress, Pests, and Diseases

Perhaps the main advantage of this technique is its forecasting power. Digital twins can act as an initial warning system and identify hidden dangers such as nutrients, insect invasions, or outbreaks of the disease before they become visible to the naked eye. By comparing live sensor data against the baseline of the "healthy" development model, the system can mark micro-discrepancies that indicate problems. Researchers have done this using virtual reality-based digital twins to simulate crop growth and successfully discover the early signs of plant stress. This concept is expanded to complete recommendation systems that not only identify problems but also suggest the best crops or tasks to reduce losses and adapt crops. By integrating artificial intelligence, these digital twins are becoming increasingly effective, ultimately helping to protect the asset's value and reduce the need for extensive intervention.

16.3.3 Precision Farming and Resource Optimization

The purpose of accurate agriculture aims to optimize the use of resources such as water, fertilizers, and pesticides, and DTs play an important role in achieving this goal. By following different management strategies, DT's recommendations for input allocation reduce waste and environmental impact. Feldmane and Rubauskis (2023) illustrated the application of UAV-assisted DTs for orchard management, supporting precision spraying and resource targeting. Escribà-Gelonch et al. (2024) emphasized the orchestration of DTs for sustainable agricultural practices, particularly in resource-intensive environments. These applications demonstrate how DTs can reduce input costs while maintaining or increasing crop productivity.

16.3.4 Case Study: Yield Prediction with Sensor Data

To illustrate the role of digital twins in crop management, we analyzed the Smart Farming Sensor Data for Yield Prediction dataset (Soundankar 2024). The dataset

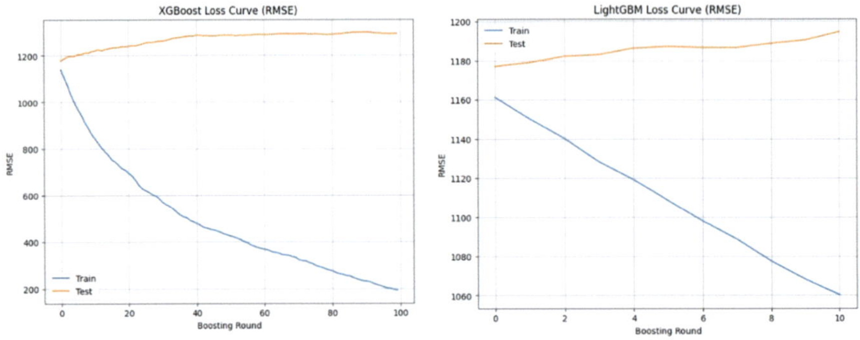

(a) XGBoost Loss Curve (RMSE).

(b) LightGBM Loss Curve (RMSE).

Fig. 16.1 Comparison of loss curves for XGBoost (left) and LightGBM (right) during model training. (**a**) XGBoost loss curve (RMSE). (**b**) LightGBM loss curve (RMSE)

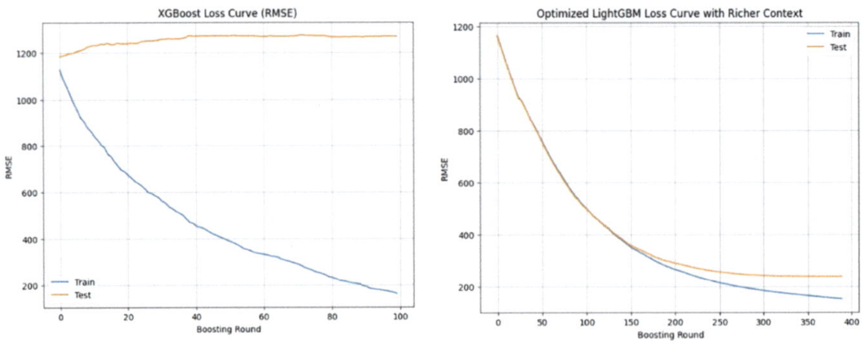

(a)XGBoostLossCurve(RMSE)-Extended Training.

(b) Optimized LightGBM Loss Curve with Richer Context.

Fig. 16.2 Comparison of extended XGBoost loss curve (left) and optimized LightGBM loss curve with richer context (right). (**a**) XGBoost loss curve (RMSE)-extended. (**b**) Optimized LightGBM loss curve with training richer context

covers 500 farms across diverse agro-climatic zones and includes soil moisture, pH, temperature, rainfall, humidity, sunlight, pesticide usage, and NDVI values, alongside crop yield outcomes. The variety of conditions makes it suitable for testing DT-based predictive models.

Figure 16.1 compares the training and test RMSE curves for XGBoost and LightGBM models. The XGBoost curve (left) shows consistent reduction in training error with a relatively stable test error, highlighting model generalization and potential overfitting trends. In contrast, the LightGBM curve (right) demonstrates a balanced evolution of train and test RMSE, indicating good generalization and suitability for crop monitoring applications in the digital twin framework.

Figure 16.2 presents a comparison of advanced model training curves. The XGBoost extended training loss curve (left) illustrates progressive learning and

Table 16.1 Performance of machine learning models for crop yield prediction

Model	MSE R^2 Score	
Linear regression	1,493,551	0.05
Random forest	1,475,050	0.38
XGBoost	1,613,467	0.46
LightGBM (basic)	1,373,314	0.66
LightGBM (optimized)	**56,874**	**0.959**

model convergence over a larger number of boosting rounds, guiding optimal training duration.

In contrast, the optimized LightGBM loss curve (right), enhanced with richer contextual features such as genotype, historical trends, management practices, and soil attributes, demonstrates a significant reduction in both train and test RMSE. This highlights the effectiveness of digital twin-driven feature engineering for robust and accurate yield modeling.

To benchmark predictive performance, multiple machine learning models were applied to the dataset. As shown in Table 16.1, linear regression performed poorly due to the nonlinear nature of crop growth. Random forests and XGBoost improved predictions, but LightGBM models offered the best accuracy. The optimized LightGBM model, enhanced with contextual features such as genotype and management practices, achieved an R^2 score above 0.95, demonstrating the potential of DTs in yield forecasting.

This case study highlights how DTs can combine environmental sensing with advanced machine learning to guide agricultural decision-making. By modeling real-world variability, they enable precise forecasting and optimization that would not be possible with traditional tools. Importantly, the results show that model accuracy depends heavily on the richness of contextual features, underscoring the value of integrated data ecosystems in agricultural digital twins.

16.4 Livestock Health and Productivity

Livestock farming is a critical component of global agriculture, contributing significantly to food security, income generation, and rural livelihoods. However, challenges such as disease outbreaks, feeding inefficiencies, welfare concerns, and the need for sustainable production demand innovative solutions. Digital twins(DTs) are increasingly being applied in livestock systems to provide real-time monitoring, predictive modeling, and decision-support capabilities that enhance both animal health and farm productivity (Nie et al. 2022; Zhang et al. 2025) (Table 16.2).

DTs for livestock replicate animal physiology, behavior, and farm environment through continuous integration of data from video surveillance and Smart Farm IoT network. These digital representations allow farmers to evaluate health parameters, predict disease risk, and optimize feeding strategies. By combining organic data with simulation models, DT enables an active approach to livestock management, where

Table 16.2 Representative sensors for livestock health monitoring

Sensor type	Parameter(s)	Use-case	Frequency
Heart rate monitor	bpm	Stress, fever, disease detection	Seconds–minutes
Temperature probe	Skin/core °C	Detect fever, heat stress	Minutes
Accelerometer/IMU	Activity, posture	Behavior, estrus, lameness	High frequency
Rumination tracker	Jaw movement	Digestive health	Minutes–hours
GPS collar	Position, movement	Grazing, isolation, estrus	Minutes
Environmental node	Temp., humidity, gases	Barn/pasture conditions	10–60 min

intervention can be used before worsening health problems or declining productivity.

16.4.1 Health Monitoring and Welfare Assessment

Continuous health monitoring is one of the most important uses of DT in livestock farming. Portable equipment and environmental sensors collect data by animal movement, feeding behavior, heart rate, and temperature, synchronized with DT models. Jeong et al. (2023) used of a virtual pig house as a DT platform, which enabled simulation-based livestock farming. Similarly, Herlin et al. (2021) examined the digital surveillance tool for cattle and sheep and emphasized welfare implications and ability to track animals on pasture. These applications not only improve health monitoring but also support welfare -operated decision -making.

16.4.2 Feeding Optimization and Productivity Enhancement

DTs also help to improve control and productivity. Combined nutritional models with real-time consumption data can simulate different feeding events to determine the optimal strategies for milk or meat production. Raba et al. (2022) presented a DT-based decision support equipment to feed livestock, highlighting its role in balanced nutrition, reduced waste, and improved efficiency. Built-in predictive models in DTs help identify the least animals, so that goal intervention can maximize productivity.

16.4.3 Disease Prevention and Risk Management

Early detection of disease risk is important for livestock productivity and public health. DTs can integrate veterinary records and real-time surveillance data to predict disease spread and outbreaks. This means that the affecting capacity of the future can distinguish the affected animals, adapt the treatment plans, and prevent mass damage. Research of Elijah et al. (2021) highlighted the wider capacity of DTs in resource-limiting references, where such systems can act as initial warning platforms against epidemics.

16.4.4 Sustainability and Ethical Farming Practices

In addition to enhancing productivity, the application of digital twins (DTs) supports the stability and ethical practices of livestock farming. DTs can simulate environmental impacts such as greenhouse gas emissions from feeding strategies or fertilizer management, providing valuable insights into sustainable production options. Purcell et al. (2023) discussed DTs in relation to environmental sustainability, while Agnusdei et al. (2021) emphasized their relevance to safety and management aspects in agricultural operations. This double focus on productivity and stability is consistent with the goals for Agriculture 5.0, where animal welfare, resource efficiency, and environmental management are equally preferred.

So, DT in livestock farming improves health monitoring, adjusts feeding systems, improves the prevention of the disease, and supports permanent practice. By combining real-time data with future modeling, these systems establish an active and moral structure for livestock management, which contributes to high productivity and well-being standards in modern agriculture.

16.5 Precision Irrigation and Water Management

Water shortages is one of the most pressing challenges in modern agriculture, affecting the crop yield, the profitability of the farm, and long-lasting stability. Traditional irrigation practices are often disabled, which use water resources and soils. Digital twins (DTs) are revolutionizing irrigation by helping us monitor, simulate, and adjust systems with incredible accuracy. By pulling in real-time data on soil, crops, and weather, these digital replicas make it easy to distribute water precisely and manage resources effectively (Nie et al. 2022; Zhang et al. 2025).

16.5.1 Optimizing Water Use with Digital Twins

When it comes to water management, digital twins focus on using each drop wisely to help plants grow when they cut waste. They do this by integrating data from soil moisture sensors, remote measurement, and other smart units, enabling movement of water through the soil, to build their real-time model. This lets farmers see where the water is moving, test different irrigation programs, and see the results of their decisions before they turn on the sprinkler. Using digital twins to guide watering not only saves water but also helps the environment by preventing excess watering and protecting our precious groundwater. They mainly become the control centers, capable of orchestrating accurate watering even in the most complex farm layouts.

16.5.2 Integration of Weather Data and Irrigation Scheduling

In addition to soil conditions, weather is an important factor for water management. Digital twins also make the irrigation plan smart by including rainfall, temperature, humidity, and evaporation rates such as meteorological data. This allows them to estimate future needs and adjust irrigation strategies on the fly, ensuring that plants are never over- or under-watered. For example, some systems use drones and digital twins to sync irrigation perfectly with changing weather patterns. The ability to simulate how irrigation will perform under different climate scenarios makes these systems especially valuable in areas vulnerable to climate change.

16.5.3 Water Resource Management and Waste Reduction

Beyond the field-level applications, DTs also provide value in extensive water resource management. From modeling irrigation networks, aquifers, and reservoirs, DTs can support regional plans and optimize water distribution in many stakeholders. Sadowski and Spachos (2020) discussed how wireless IoT equipment with energy-harvesting skills can expand the DT functionality for large-scale agricultural monitoring, which can improve the efficiency of water distribution. Studies of Elijah et al. (2021) further explain how DT frameworks can be used in line with resource-limiting areas and provide a cost-effective solution for water conservation and crop flexibility.

16.6 Integration and Future Trends

Various applications of digital twins (DTs) in crop monitoring, livestock health, and earth's fertility and water management emphasize their ability to serve as the backbone of future ecosystems for smart agriculture. However, the true value of DTs lies in their integration across platforms that span multiple agricultural domains. By combining real-time sensor data, remote sensing technologies, weather forecasts, and predictive analytics within an interoperable DT framework, continuous insights can be generated throughout the entire production cycle (Escribà-Gelonch et al. 2024; Cakir et al. 2023; Yogeswaranathan and Collier 2022). Such integration transforms fragmented digital solutions into comprehensive decision-support systems, enhancing productivity, resilience, and adaptability in agriculture.

Future trends point toward increasing convergence of DTs with other frontier technologies. For example, integration of artificial intelligence (AI) and machine learning enables self-learning DTs which continuously improves predictions and recommendations as more data are collected (Nie et al. 2022; Liu et al. 2023). The use of generative AI further improves simulation capacity, which allows climate change effects, detection of various landscapes such as market outbreaks or market-driven adjustments in the crop scheme. Similarly, cloud–fog–edge architectures are expected to play a critical role in ensuring scalability and responsiveness of agricultural DT platforms (Zake and Majore 2022). Emerging 5G and IoT connectivity will support real-time synchronization between physical and digital entities, even in remote rural environments.

Beyond technological innovation, the future of DTs in agriculture must also address challenges of interoperability, data privacy, and cost-effective adoption. Standardization of DT architectures and protocols will be necessary to ensure that diverse systems can exchange and utilize data seamlessly (Wang 2024). Farmer-centric adoption strategies, including training programs and low-cost deployment models, will be essential to bridge the digital divide and encourage widespread implementation. In addition, DTs are likely to become integral to policy frameworks for sustainable farming, supporting compliance with environmental regulations and carbon footprint management.

16.7 Conclusion

Digital twin (DT) technology is transforming agriculture by providing real-time monitoring, future insights, and adaptive crop control, livestock, soil, and water resources. Through integration of IoT, artificial intelligence, and simulation models, DTs are proactive and capable of data-driven decisions that improve productivity and stability. Applications in crop monitoring, livestock health, soil fertility, and accurate watering demonstrate the versatility of DT in tackling major agricultural challenges. These systems increase efficiency, reduce resource dissemination, and support flexibility against climate change. However, high implementation costs, connectivity gaps, interoperability issues, and the complexity of biological modeling continue to hinder large-scale adoption. With ongoing advancements in digital technologies and supportive policy frameworks, DTs in agriculture are expected to play a central role in Agriculture 5.0. Their capacity to integrate resilience with productivity positions them as a cornerstone of future-ready agricultural systems.

References

Agnusdei GP, Elia V, Gnoni MG (2021) Is digital twin technology supporting safety management? A bibliometric and systematic review. Appl Sci 11(6):2767

Banerjee S, Mukherjee A, Kamboj S (2025) Precision agriculture revolution: integrating digital twins and advanced crop recommendation for optimal yield. J Agric Technol 21(3):45–62

Cakir LV, Bilen T, Özdem T, Canberk B (2023) Digital twin middleware for smart farm IoT networks. In: Proceedings of the 2023 international Balkan Conference on Communications and Networking (BalkanCom), pp 1–5

Catala-Roman P, Navarro EA, Segura-Garcia J, Garcia-Pineda M (2024) Harnessing digital twins for agriculture 5.0: a comparative analysis of 3D point cloud tools. Appl Sci 14(5):1709

Elijah O, Rahim SKA, Emmanuel AA, Salihu YO, Usman ZG, Jimoh AM (2021) Enabling smart agriculture in Nigeria: application of digital-twin technology. In: International Conference on Multidisciplinary Engineering and Applied Science (ICMEAS)

Eremina I, Yudin A, Tarabukina T, Oblizov A (2022) The use of digital technologies to improve the information support of agricultural enterprises. Int J Technol 13(7):1393–1402

Escribà-Gelonch M, Liang S, van Schalkwyk P, Fisk I, Long NVD, Hessel V (2024) Digital twins in agriculture: orchestration and applications. J Agric Food Chem 72(19):10737–10752

Feldmane D, Rubauskis E (2023) Digital twin: orchard management using UAV. Environ Technol Resour 1:247–251

Hananto AL et al (2024) Digital twin and 3D digital twin: concepts, applications, and challenges in industry 4.0. Computers 13(4):100

Herlin A, Brunberg E, Hultgren J, Högberg N, Rydberg A, Skarin A (2021) Animal welfare implications of digital tools for monitoring and management of cattle and Sheep on pasture. Animals 11:829

Jabed MA, Murad MAA (2024) Crop yield prediction in agriculture: a comprehensive review of machine learning and deep learning approaches. Heliyon 10:e40836

Jeong D-Y, Jo S-K, Lee I-B, Shin H, Kim J-G (2023) Digital twin application: making a virtual pig house toward digital livestock farming. IEEE Access 11:1–11

Jones D, Snider C, Nassehi A, Yon J, Hicks B (2020) Characterising the digital twin: a systematic literature review. CIRP J Manuf Sci Technol 29:36–52

Liu J et al (2023) Exploring the integration of digital twin and generative AI in agriculture. In: Proceedings of the 2023 International Conference on Intelligent Human-Machine Systems and Cybernetics (IHMSC), pp 223–228

Nie J, Wang Y, Li Y, Chao X (2022) Artificial intelligence and digital twins in sustainable agriculture and forestry: a survey. Turk J Agric For 46:642–661

Purcell W, Neubauer T, Mallinger K (2023) Digital twins in agriculture: challenges and opportunities for environmental sustainability. Curr Opin Environ Sustain 61:101252

Raba D, Tordecilla RD, Copado P, Juan AA, Mount D (2022) A digital twin for decision making on livestock feeding. INFORMS J Appl Anal 52(2):267–282

Sadowski S, Spachos P (2020) Wireless Technologies for Agricultural Monitoring Using Internet of things devices with energy harvesting capabilities. Comput Electron Agric 176:105–117

Silva L, Rodríguez-Sedano F, Baptista P, Coelho JP (2023) The digital twin paradigm applied to soil quality assessment: a systematic literature review. Sensors 23(2):1007

Singh K, Saini M (2023) Towards digital twin of crops for growth modelling using virtual reality. In: Proceedings of the 5th ACM International Conference on Multimedia in Asia (MMAsia)

Skobelev P et al (2021) Digital twin of Rice as a decision-making Service for Precise Farming. In: Proceedings of the 2021 7th IEEE International Conference on Information Technology and Nanotechnology (ITNT), pp 1–5

Soundankar A (2024) Smart farming sensor data for yield prediction. Kaggle. Available at: https://www.kaggle.com/datasets/atharvasoundankar/smartfarming-sensor-data-for-yield-prediction/data. Accessed 10 June 2025

Talkhestani BA, Jazdi N, Schloegl W, Weyrich M (2018) Consistency check to synchronize the digital twin of manufacturing automation based on anchor points. Procedia CIRP 72:159–164

Tsolakis N, Bechtsis D, Vasileiadis G, Menexes I, Bochtis DD (2021) Sustainability in the digital farming era: a cyber-physical analysis approach for drone applications in agriculture 4.0. In: Springer optimization and its applications information and communication technologies for agriculture—theme IV: actions, vol 185, pp 29–53

Verdouw C, Kruize JW (2017) Digital twins in farm management: illustrations from the FIWARE accelerators SmartAgriFood and fractals. In: Proceedings of the 7th Asian-Australasian conference on precision agriculture held in Hamilton, New Zealand, October, 1–5

Wang L (2024) Digital twins in agriculture: a review of recent Progress and open issues. Electronics 13:2209

Wilberforce N, Mwebaze J (2025) A framework for IoT-enabled smart agriculture. Internet of Things J 10(1):123–135

Yogeswaranathan K, Collier RW (2022) Towards a new architecture: MultiAgent based cloud-fog-edge computing and digital twin for smart agriculture. Springer, Cham

Zake M, Majore G (2022) Application of multi-perspective modelling approach for building digital twin in smart agriculture. In: Proceedings of the 2022 63rd international scientific conference Information Technology and Management Science (ITMS), pp 1–7

Zhang R, Zhu H, Chang Q, Mao Q (2025) A comprehensive review of digital twins technology in agriculture. Agriculture-Basel 15(9):903

Zhao X et al (2023) Digital twins in smart farming: an autoware-based simulator for autonomous agricultural vehicles. Int J Agric Biol Eng 16(04):184–189

Digital Twin-Driven Crop Monitoring and Management: A Machine Learning Approach for Crop Yield Optimization

17

Kanaka Raju Kalla, Budumuru Giri Teja, D. Vijaya Kumar, Ashok Ganga, and Jaganmohan Rao Tarra

Abstract

Predicting crop yields with high accuracy is essential for precision agriculture, especially in countries like India where variable climate and diverse soil conditions strongly influence productivity. In this chapter, we develop a machine learning framework to forecast crop yield using a dataset that integrates soil characteristics, nutrient levels (N, P, and K), weather variables such as temperature, humidity, and wind speed, along with the categorical details of crop and soil types. Three ensemble-based models, Random Forest, Gradient Boosting, and CatBoost, were trained and assessed using regression metrics including RMSE, MAE, MSE, and R^2. Among these, Random Forest achieved the best performance ($R^2 = 0.9746$), followed by CatBoost ($R^2 = 0.9704$), while Gradient Boosting showed lower accuracy ($R^2 = 0.8988$). Analysis of feature importance highlighted temperature, humidity, and soil quality as the most influential factors affecting yield. These findings demonstrate the effectiveness of ensemble learning methods for reliable and efficient agricultural forecasting. The study also points to their potential integration into decision-support systems for farmers. Finally, the chapter discusses practical implications for agri-technology adoption and policy and suggests future directions, including the use of temporal models and hybrid AI–physical approaches.

K. R. Kalla (✉) · D. Vijaya Kumar · A. Ganga · J. R. Tarra
Department of Electrical and Electronics Engineering, Aditya Institute of Technology and Management, Tekkali, Srikakulam, Andhra Pradesh, India
e-mail: kallakanakaraju@adityatekkali.edu.in; dr.vijaykumar@adityatekkali.edu.in; ashokganga@adityatekkali.edu.in; jagan@adityatekkali.edu.in

B. Giri Teja
Department of CSE (AI & ML), Aditya Institute of Technology and Management, Tekkali, Srikakulam, Andhra Pradesh, India
e-mail: bgiriteja.tpc@adityatekkali.edu.in

© The Author(s), under exclusive license to Springer Nature Singapore Pte Ltd. 2026

A. L. Srivastav et al. (eds.), *Digital Twin Technology for Sustainable Agriculture*,
https://doi.org/10.1007/978-981-95-5915-2_17

Keywords

Crop yield prediction · Random forest · Gradient boosting · CatBoost · Machine learning in agriculture · Feature importance

17.1 Introduction

For centuries, agriculture had been the foundation of India's social and economic edifice, enriching not only millions of livelihoods but also the culture and strategy of national development. It is reported in the Economic Survey of India that the industry still engages approximately 47% of workers in the country, and the contribution to the GDP is almost 18% (Biswal and Bahinipati 2025; Barman 2025). Despite the fast growth of industry and services, agriculture is deeply crosscut with India's growth story from which rural areas, where crop productivity directly bears on food security and employment, have not been entirely removed. This high dependency makes it even more urgent to have accurate and scalable crop yield forecasting, particularly in light of increasing climatic uncertainties (Sarkar et al. 2024).

The challenges to agricultural productivity in India are multifaceted. With over 60% of India's net sown area dependent on monsoon rainfall, interannual variability in precipitation and temperature often disrupts both sowing and harvesting schedules. Additionally, soil degradation, nutrient imbalances, changing pest patterns, and water stress continue to impede optimal yield realization. These challenges are compounded by regional disparities in landholding patterns, infrastructure, and market access, which collectively contribute to uneven agricultural outcomes across states. In such a complex landscape, traditional statistical models rooted in linear assumptions and narrow variable sets often prove inadequate in capturing the nuances of yield variability (Basantaray et al. 2024; Maurya et al. 2015).

More recently, digital markers like machine learning (ML) and artificial intelligence (AI) have been used to address these issues. Thereby, the ensemble machine learning (ML) methods, especially those capable of learning complex and nonlinear relationships in the data without making any stringent statistical assumptions, are very appropriate. Data-driven by nature, they are particularly useful in agriculture where vast amounts of data on soil, climate, and crop conditions are readily available but hard to analyze using traditional methods.

Among the array of machine learning methods, decision-tree-based ensemble algorithms have shown considerable promise. Random Forest (RF), introduced by Breiman (2001), is a robust ensemble technique that aggregates predictions from multiple decision trees to reduce overfitting and improve generalization. RF's inherent ability to handle multicollinearity, outliers, and missing values makes it ideal for agricultural datasets that are often noisy or incomplete. Likewise, Gradient Boosting Machines (GBM), originally conceptualized by Friedman (2000), offer a more refined approach by sequentially training weak learners to minimize loss functions. GBM models can capture subtle patterns and are highly customizable through hyperparameter tuning.

A more recent development in this family is CatBoost, a gradient boosting framework developed by Dorogush et al. (2018) at Yandex (Dorogush et al. 2018). Unlike conventional gradient boosting models, CatBoost is explicitly designed to handle categorical features without extensive preprocessing. This capability is particularly valuable in the Indian context, where variables such as "Crop Type" or "Soil Type" are inherently categorical and regionally diverse. CatBoost uses a novel technique called ordered boosting, which avoids target leakage and significantly enhances generalization, even on small datasets.

This chapter benchmarks the learning performance of three ensemble learners, Random Forest, Gradient Boost, and CatBoost on real data collected from various agricultural regions of India. The dataset includes environmental conditions (e.g., temperature, humidity, and wind speed) and soil parameters (e.g., pH, nitrogen, phosphorus, potassium, and soil type). It also includes such categorical information for major crops, like rice, wheat, and soybean. The dependent variable, crop yield (in quintals ha^{-1}), has the effect of numerous temporal, spatial, and agronomic factors. The purpose of this chapter was to evaluate the performance of these models in terms of prediction accuracy, interpretability, and the ability for adaptation to different agro-climatic environments.

Figure 17.1 shows the structured workflow for predicting crop yield based on machine learning models using a set of environmental, soil, and crop-sensitive features. The key part of this framework is a data-driven pipeline that starts by

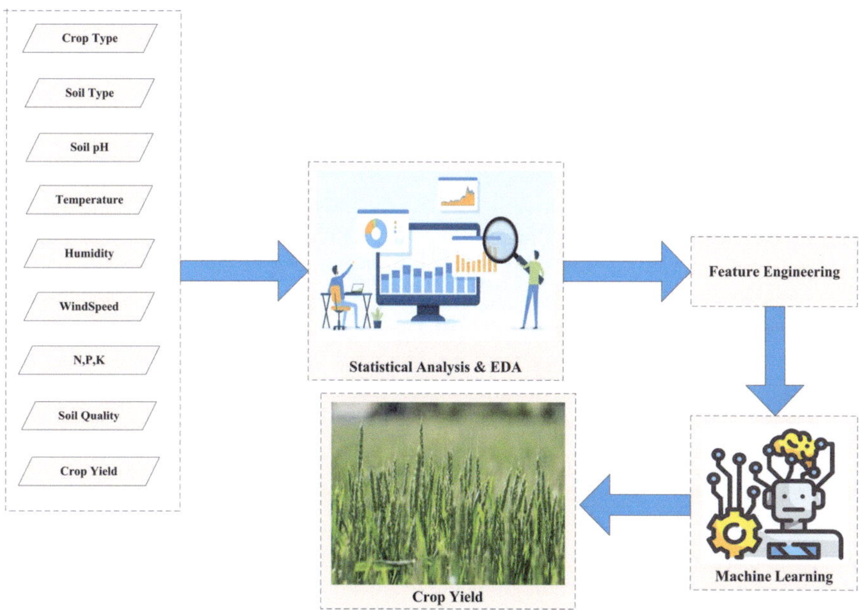

Fig. 17.1 Workflow architecture for crop yield prediction using machine learning

collecting the pertinent agronomic data, proceeds with exploratory and statistical analytics, and terminates in the applications of predictive algorithms in order to produce an estimate of crop output. The input layer contains primary features such as crop type and soil type, which are categorical variables reflecting agricultural variability. It also involves soil pH, soil quality, and concentrations of two most essential macronutrients, nitrogen (N) and potassium (K), as well as phosphorus (P) that are directly related to plant growth and productivity. Additionally, environmental conditions such as temperature, humidity, and wind speed are integrated into the dataset, as these climatic parameters have a significant impact on crop growth and yield variability.

Following the collection of raw data, the system performs statistical analysis and Exploratory Data Analysis (EDA) to better understand the distribution, relationships, and underlying patterns in the dataset. This step typically includes visualizing correlations, detecting outliers, and assessing feature variability to guide the model design. Subsequently, the feature engineering stage refines the data through normalization, encoding of categorical variables, and construction of derived features to enhance model learning. This preprocessing ensures that the machine learning models receive well-prepared, high-quality input data.

The final stage uses different machine learning techniques such as Random Forest, Gradient Boosting, and CatBoost to predict the crop yield. They are particularly appropriate for capturing nonlinear relationship and mixed data types and thus are effective to work within agronomic datasets. The learned model communicates from the historical behavior and intervariable relationships to predict crop production or some unit in which the crops are produced, such as kilograms per hectare. This leads to a predictive tool which may aid the farmers, agronomists, and policymakers in optimizing agricultural practice, enhancing resource allocation and food security strategy.

A key advantage of ensemble ML models is their interpretability, particularly through techniques such as feature importance ranking and SHAP (SHapley Additive exPlanations) values (Koushik et al. 2024). For instance, feature importance metrics in RF and CatBoost can help identify the most influential variables, be it soil pH, temperature, or nitrogen concentration, thereby enabling more targeted agronomic interventions. For policymakers and extension agencies, such insights can inform fertilizer subsidy distribution, cropping pattern recommendations and climate risk assessments (Coello et al. 2025).

Beyond prediction accuracy, we also examine issues of model fairness, data imbalance, and regional bias. In India, yield data often reflects regional disparities, and ML models trained without appropriate stratification may inadvertently favor data-rich states while underperforming in underrepresented regions. By evaluating performance across multiple strata and using cross-validation strategies, we aim to ensure that the models developed are not only accurate but also equitable and generalizable.

Furthermore, we highlight the practical implications of deploying such models in the field. With increasing penetration of digital agriculture platforms and satellite-based monitoring systems (e.g., ISRO's Bhuvan and Mahalanobis National Crop

Forecast Centre), integrating machine learning pipelines into existing decision-support tools is both feasible and impactful. Mobile-based agro-advisory services can leverage these predictive insights to issue timely alerts and optimize input application, thereby enhancing resilience and profitability for smallholder farmers.

It is also important to acknowledge the challenges in this domain. Data quality, granularity, and availability remain persistent issues. Much of the agricultural data in India is collected manually and is prone to inconsistencies. In this study, preprocessing steps including outlier handling, feature encoding, and normalization were critical in preparing the dataset for effective model training. Additionally, balancing model complexity and computational efficiency was necessary to ensure that the solutions proposed remain accessible even in resource-constrained settings.

Ultimately, the motivation behind this work is to bridge the gap between technological potential and practical utility. By benchmarking the performance of these three machine learning models under real-world agronomic conditions in India, we seek to contribute to a growing body of evidence that AI can play a transformative role in agriculture. This chapter not only advances methodological knowledge but also provides actionable insights for agronomists, data scientists, and policymakers alike.

In the sections that follow, we describe the dataset in detail, outline the methodology for model training and evaluation, present comparative results, and discuss key findings with an emphasis on practical deployment. Through this structured approach, we aim to highlight how intelligent data-driven systems can support India's journey toward sustainable and resilient agriculture in the twenty-first century.

17.2 Literature Review

Agricultural yield forecasting in India is crucial for food security and economic planning, given the country's reliance on staple crops (rice, wheat, maize, and soybean) and monsoon-sensitive climate (Nikhil et al. 2024; Clercq and Mahdi 2024). Traditionally, crop yields have been estimated using statistical regression or process-based models. For example, multivariate regression on meteorological and remote-sensing indices has long been applied to crops such as paddy (Pazhanivelan et al. 2022). Crop growth models (e.g., DSSAT-CERES simulations) integrate soil, genotype, and weather processes to simulate yields, often after local calibration. However, these conventional methods assume linear or mechanistic relationships that may not capture complex climate crop interactions, and they require extensive ground data for calibration. In many Indian studies, linear models and DSSAT have achieved moderate accuracy (e.g. $R^2 \approx 0.80$, NRMSE $<10\%$ for rice in Tamil Nadu) (Pazhanivelan et al. 2022). Yet their applicability is limited by heterogeneous conditions. DSSAT tends to excel at site-specific scales but struggles to generalize across diverse regions (Gumma et al. 2024).

Remote sensing data (satellite and aerial imagery) have become important complements to ground observations. Vegetation indices (NDVI, EVI, LAI, etc.)

and SAR-derived measures of canopy moisture provide timely, large-scale insights on crop health (Sah et al. 2024). Studies have shown that integrating satellite indices with crop models or regressions can substantially improve estimates: for instance, Sentinel-1 SAR data fed into regression or semiphysical models produced rice yield maps with $R^2 > 0.80$ and NRMSE <10% (Pazhanivelan et al. 2022). In one Uttarakhand study, combining optical (Sentinel-2) and SAR indices with in-field biophysical measures (LAI, biomass) and machine learning yielded early rice forecasts with high R^2 ($\approx$0.80–0.93) depending on the growth stage (Sah et al. 2024). Remote sensing thus adds spatial and temporal coverage, enabling yield monitoring over large tracts, while ground weather/soil data supply local context. The best approaches fuse both: e.g., one analysis for Tamil Nadu rice used regression on SAR indices and a DSSAT model driven by satellite LAI data, both achieving $R^2 > 0.8$ (Pazhanivelan et al. 2022).

17.2.1 Traditional Statistical and Process-Based Models

Linear and nonlinear regression methods remain baseline tools. Early Indian studies often applied multiple linear regression (MLR) between yields and climatic or vegetation predictors (e.g., rainfall, growing degree days, and NDVI) (Pazhanivelan et al. 2022). These models are simple and interpretable but assume additive or linear effects. For example, classical crop area surveys in India relied on regression of yield on predictors like NDVI or phenology dates. Autoregressive time-series (ARIMA) models have also been used regionally, but they require stationary trends and struggle with outliers (Nidhi Parashar et al. 2024). Crop growth simulations like DSSAT (Decision Support System for Agrotechnology Transfer) represent a major traditional approach. DSSAT's CERES subsystems for rice, wheat, soybean, maize, etc. use detailed cultivar parameters and soil/water models to simulate phenology and yield. Such models can be accurate when well-calibrated: e.g., in a recent Indian case, DSSAT-simulated rice followed observed trends over 17 years, with normalized RMSE $\approx$13–17% (Kudikala et al. 2025). Likewise, DSSAT has shown good wheat and cotton results in other tropical contexts, typically with R^2 in the range 0.7–0.9 under controlled conditions (Ghimire et al. 2025). However, DSSAT demands intensive data (soil profiles, cultivar coefficients, and management details) and local expertise. It often performs best at the experimental field or small-region scale, while accuracy can decline when extended to heterogeneous areas. In summary, conventional models are interpretable and grounded in crop science, but they can be inflexible and time-consuming to deploy across all of India's agro-climatic zones.

17.2.2 Machine Learning Approaches

In the last decade, data-driven machine learning (ML) methods have surged in crop yield prediction due to their ability to model complex, nonlinear relationships.

Unlike fixed-form regressions, ML algorithms can ingest many features (weather, soil, remote indices, and management) and automatically learn interactions. A systematic review notes that artificial neural networks (ANNs) were historically most common, with Random Forest (RF) and Support Vector Machines (SVM) also widely used (Klompenburg et al. 2020). In practice, ensemble tree methods (RF and Gradient-Boosted Trees) dominate recent literature for Indian crops. For example, an ML survey found that temperature, rainfall, historical yields, and NDVI are the top predictors across studies and that RF, ANNs, SVM, and XGBoost are the "mostly used ML algorithms" (Sharma et al. 2025). Deep learning (convolutional and recurrent neural nets) is emerging but remains less common in India due to data requirements.

17.2.3 Random Forest and Tree Ensembles

Random Forest regression is often a go-to model for yield because it handles nonlinearity and interactions automatically. In Indian case studies, RF frequently outperforms linear models. For instance, in a multi-crop dataset from South India (including rice, sorghum, cotton, sugarcane, and oilseeds), Extra Trees (a variant of RF) achieved the highest $R^2 \approx 0.96$, with RF at ≈ 0.94 (versus ~0.86 for linear regression). Similarly, studies in Punjab and Uttar Pradesh have found RF yields (with weather/remote inputs) surpass linear baselines. A large Indian rice study using 20 years of weather and satellite data found tree-ensemble models (extremely randomized trees, CatBoost, and LightGBM) could predict district yields with out-of-sample R^2 up to ~0.82 (Clercq and Mahdi 2024). Globally, RF's popularity in yield prediction is reflected by repeated comparison papers: e.g., RF scored among the top methods for US maize and Pakistani wheat in multi-model tests (Ashfaq et al. 2025; Joshi et al. 2025). RF is robust to overfitting on high-dimensional data and provides variable importance metrics, aiding interpretability. However, it can be computationally heavy and slower to train than linear models, especially on very large datasets. In practice, RF models trained on Indian crop data (ground plus remote features) have yielded low RMSE values and high R^2 in cross-validation, demonstrating substantial gains over statistical models.

17.2.4 Gradient Boosting Machines

Gradient-boosted tree methods (such as XGBoost, LightGBM, and CatBoost) are now commonly used to push performance further. These "boosting" algorithms build trees sequentially to reduce errors and often outperform RF. In the Uttarakhand rice example, XGBoost gave the best yield predictions at all growth stages, exceeding RF by a margin; overall, XGB, Cubist (another tree ensemble), and a neural network were ranked top (sRPI index) (Sah et al. 2024). CatBoost (which handles categorical variables efficiently) and LightGBM have likewise shown strong results. The Oxford/Princeton team's recent rice study explicitly included CatBoost and

LightGBM among 19 models, finding tree-based ensembles achieving R^2 up to 0.82 (Clercq and Mahdi 2024). Outside India, research on crops like sugarcane, potato, and maize consistently reports XGBoost or LightGBM giving lower RMSE than older methods (Sah et al. 2024). Gradient boosting tends to be even more prone to overfitting than RF, so careful tuning (cross-validation and regularization) is needed. Still, these algorithms scale relatively well with data size. CatBoost in particular has been applied in at least one recent Indian study where a tuned CatBoost regression achieved outstanding accuracy ($R^2 \approx 0.99$ on 21 years of mixed-crop data) outperforming a complex CNN + BiLSTM hybrid. In summary, modern GBMs often yield the highest predictive accuracy in Indian crop datasets, at the cost of additional hyperparameter tuning.

17.2.5 Comparative Performance and Interpretability

Overall, machine learning methods consistently achieve higher accuracy than traditional statistical models in empirical tests but with trade-offs in interpretability and data needs. For example, in a side-by-side comparison, an Indian dataset yielded $R^2 \approx 0.94$ for RF versus $\approx$0.86 for linear regression. Likewise, Deep Learning and boosting models often surpass both linear and RF baselines if enough data are available (Joshi et al. 2025; Sah et al. 2024). However, each approach has strengths: linear models are simplest and highly interpretable (coefficients directly show the effect of each variable), while crop models like DSSAT embed domain knowledge (soil–water–crop processes) and can simulate what-if scenarios. Tree-based ML (RF, GB, and CatBoost) handles interactions automatically and shows robustness across heterogeneous conditions, as seen by high sRPI scores in rice predictions (Sah et al. 2024). Importantly, recent Indian studies use explainable AI tools to open the “black box”: SHAP analysis on ML ensembles has identified key drivers (e.g., temperature, soil moisture, and Leaf Area Index) and their effect directions (Joshi et al. 2025; Clercq and Mahdi 2024). These help translate ML forecasts into agronomic insights. In terms of real-world deployment, scalability and data availability are key considerations. ML models can rapidly ingest new meteorological and remote-sensing data to update forecasts mid-season, which is appealing for dynamic decision support. For example, the Oxford rice study built an interactive dashboard to generate district forecasts monthly, a workflow difficult to achieve with conventional survey methods (Clercq and Mahdi 2024). By contrast, running DSSAT over large regions (many soil grids and cultivar variants) is computationally intensive and requires local calibration data, making it less nimble. A reviewer noted that ML “exhibits effectiveness in homogeneous areas with similar cultivars,” whereas DSSAT “faces difficulties” when extrapolating beyond experimental plots (Gumma et al. 2024). Thus, ML tends to be favored for large-scale, in-season prediction, while DSSAT or hybrid approaches may be used for research-grade assessments of yield gaps or climate impacts.

17.2.6 Applications to Major Indian Crops

17.2.6.1 Rice (Paddy)

Rice is India's largest crop by area, and many yield studies are focused here. Traditional approaches (SAR or optical index regression, DSSAT-CERES-Rice) have yielded R^2 ~ 0.80–0.90 under ideal conditions (Pazhanivelan et al. 2022). Recent ML studies have advanced this: for example, district-level rice yields in 247 districts were predicted with R^2 up to ~0.82 using CatBoost, LightGBM, and extra-trees models trained on 20 years of climate reanalysis and satellite data (Clercq and Mahdi 2024). Field-scale work in Uttarakhand found that XGBoost models combining NDVI/EVI and water indices could predict rice yield 2–3 weeks ahead of harvest with R^2 ~ 0.84 (mid-season prediction) (Sah et al. 2024). Overall, ensemble ML methods consistently outperform linear regression on Indian rice data. They have captured subtle drivers, e.g., one analysis showed higher August temperature *combined with* high Leaf Area Index boosts yield (Clercq and Mahdi 2024). Limitations remain: cloud cover in monsoon can delay optical data, and fine-resolution ground yield data for all districts are scarce, so many models are validated at a coarse (district/state) scale rather than at the farm level.

17.2.6.2 Wheat

Wheat yield in India (Rabi season) has also been addressed by ML. Deep sequential models (LSTM) have been explored using weekly weather and phenological inputs, yielding R^2 around 0.8–0.9 in test datasets (Joshi et al. 2025; Ashfaq et al. 2025). The Bi-LSTM study mentioned above (though on Pakistan wheat) suggests similar performance is achievable in the region; a comparable Indian study on wheat found convolutional nets and tree ensembles outperformed SVR and linear models for Punjab harvest forecasts (Joshi et al. 2025). Traditional DSSAT applications on Indian wheat generally require careful calibration but can capture genotype x environment effects. No single ML winner emerges universally: in one meta-analysis, RF and XGB outperformed others, but SVM or LASSO did best in some contexts (Ashfaq et al. 2025; Sah et al. 2024). Integration of remote sensing (e.g., Sentinel-derived LAI or SAR) is less common for wheat in India, partly due to the shorter season. Overall, ML models for wheat have shown promise but are still maturing relative to rice.

17.2.6.3 Soybean

Soybean (a major oilseed) yield prediction in India has been less studied than cereals, but some multi-model efforts exist. For instance, in Madhya Pradesh—India's top soybean state—researchers combined the DSSAT CROPGRO-Soybean model with ML approaches (ANN and SVM) to estimate yields across 16 districts (Nargund et al. 2024). They reported that both ANN and SVM performed comparably to DSSAT, suggesting a multi-model ensemble can capture climate impacts. Global reviews note that moisture, solar radiation, and NDVI are critical soybean predictors and that RF and boosting give good results in other countries. Indian work is still emerging, but indications are that ML can match or exceed DSSAT for soybean if given adequate weather–soil inputs.

17.2.6.4 Maize (Corn)

Maize (sometimes called corn) is an important Kharif cereal in parts of India. Yield forecasting for maize has used similar methods: remote-sensing vegetation indices (NDVI) combined with temperature/precipitation inputs. A few studies (mostly global) show RF and gradient boosting outperform linear regression for maize yield, with error reductions of 10–30%. In India, published studies are sparse, but one multi-crop analysis included maize with others (rice and sorghum) and found tree ensembles again outperformed linear baselines. Given maize's intermediate scale (planted area and data quality between rice and oilseeds), its prediction accuracy tends to be in line with rice forecasts.

17.2.7 Data, Interpretability, and Limitations

Despite their promise, ML approaches face challenges in Indian applications. A recurring issue is data quality and availability. Historical yield data at fine scale (farm or village) are often limited or aggregated, and missing values abound. Many ML studies report that "insufficient data" and short time series hamper model training (Klompenburg et al. 2020). Combining multiple data sources (ground stations, remote sensing, and soil maps) can mitigate this but increases complexity. The heterogeneity of India's environments (soil types, hybrid varieties, and management) means models trained in one state may not generalize to another. Some works use transfer learning or region-specific models, but a national-scale ML system must incorporate location as a feature or build localized models. Another concern is interpretability. Traditional statistics and crop models are inherently understandable (regression coefficients and growth equations), whereas ML ensembles are "black boxes." This raises trust issues for policymakers and agronomists. To address this, researchers have begun using Explainable AI tools. For example, SHAP value analysis identified that in India's districts, temperature and soil moisture indices were the strongest drivers of rice yield, providing actionable insights (Clercq and Mahdi 2024). Similarly, the Bi-LSTM wheat study employed LIME and integrated gradients to show that late-season EVI and rainfall were key to predictions (Joshi et al. 2025). While these tools help, they add complexity to the modeling pipeline.

In terms of scalability and timeliness, ML models are generally advantageous: once trained, they can ingest new weather and satellite data throughout the season to update forecasts. This enables in-season yield updates (important under erratic monsoons) that traditional survey methods cannot match. Several studies demonstrate this dynamic prediction: e.g., one workflow issued monthly rice forecasts at the district scale using Boosted Trees, outperforming analogous approaches in Brazil and China (Clercq and Mahdi 2024; Sah et al. 2024). However, ML models must be retrained when underlying conditions change (e.g., new hybrid introduction or climate shifts), whereas process models can adapt by modifying inputs.

Gaps in the literature include a paucity of comprehensive regional studies for soybean and maize and limited integration of farm management data (fertilizer rates and planting dates) into ML models. Many Indian ML studies focus on meteorology and satellite features, but factors like irrigation or fertilizer use (harder to measure) are often omitted. Most papers also concentrate on a few states or districts; pan-India analyses remain rare. Furthermore, few works compare ML models head-to-head with DSSAT or other crop models on the same dataset. One Indian review noted that when such comparisons are made, DSSAT may simulate potential yields well, but ML captures actual yield variation more flexibly (Gumma et al. 2024). Finally, few studies have probed uncertainty: yield forecasts ideally should quantify confidence intervals, which classic regressions can do, but most ML studies report only point-prediction metrics (R^2 and RMSE).

In sum, post-2010 literature shows a clear shift toward machine learning for Indian crop yield forecasting. Ensemble tree methods (Random Forest, Gradient Boost, and CatBoost) consistently achieve higher predictive accuracy than simple regressions or standalone crop models across rice, wheat, and other crops. Remote-sensing data play a key role in these gains, enabling spatially explicit models. ML models are generally more scalable and adaptive, although they require careful data curation and interpretation. Conventional approaches (linear regression and DSSAT) remain valuable for their transparency and grounding in agronomy and are sometimes combined with ML (e.g., calibrating ML inputs via DSSAT outputs). The best current systems often fuse process knowledge, statistical insight, and machine learning: for example, assimilating satellite-derived LAI into DSSAT or using hybrid ML/DSSAT ensembles.

Despite many successes, major research gaps persist in integrating fine-scale management data, expanding studies to all major crop zones, and validating models under climate extremes. Future work will likely expand deep learning (e.g., transformer models for multi-spectral time series) and leverage large ensembles of multiple algorithms to hedge against model biases. The Indian context with its climatic diversity and data challenges will continue to be a proving ground for yield forecasting methods. Overall, the trend is clear: machine learning models, especially boosted decision trees, have become essential tools in India's agricultural prediction arsenal, complementing rather than completely replacing traditional agronomic models in the quest for reliable, scalable yield forecasts.

17.3 Methodology

The methodological design of this study is structured around a data-driven machine learning pipeline aimed at accurately predicting crop yield using agronomic, environmental, and soil-based features. The approach combines statistical analysis, data preprocessing, feature engineering, and ensemble-based supervised machine learning models, specifically Random Forest (RF), Gradient Boosting (GB), and CatBoost. Each component of the methodology is detailed below, from dataset characteristics to model evaluation strategies.

17.3.1 Dataset Description and Feature Space

The dataset used in this study includes a multi-dimensional matrix of both numerical and categorical variables, capturing a variety of soil-, crop-, and weather-related conditions. The features include:

- *Crop Type* (categorical)
- *Soil Type* (categorical)
- *Soil pH* (continuous)
- *Soil Quality* (continuous)
- *N, P, K* values: nitrogen, phosphorus, and potassium concentrations (continuous)
- *Temperature, Humidity, Wind Speed* (continuous weather indicators)
- *Crop Yield* (target variable)

Initial exploratory data analysis (EDA) was conducted to understand the structure and distribution of these variables. As shown in Fig. 17.2, most continuous features such as temperature, humidity, wind speed, and soil pH exhibit approximately normal or unimodal distributions, though some features like Wind Speed and Crop Yield show moderate skewness.

Boxplots were used to detect potential outliers, particularly in nutrient content variables (N, P, K), where several instances of high kurtosis were identified. The correlation matrix, as shown in Fig. 17.3, provides insights into multicollinearity among features. Notably, P and K exhibit strong mutual correlation ($r \approx 0.91$), and

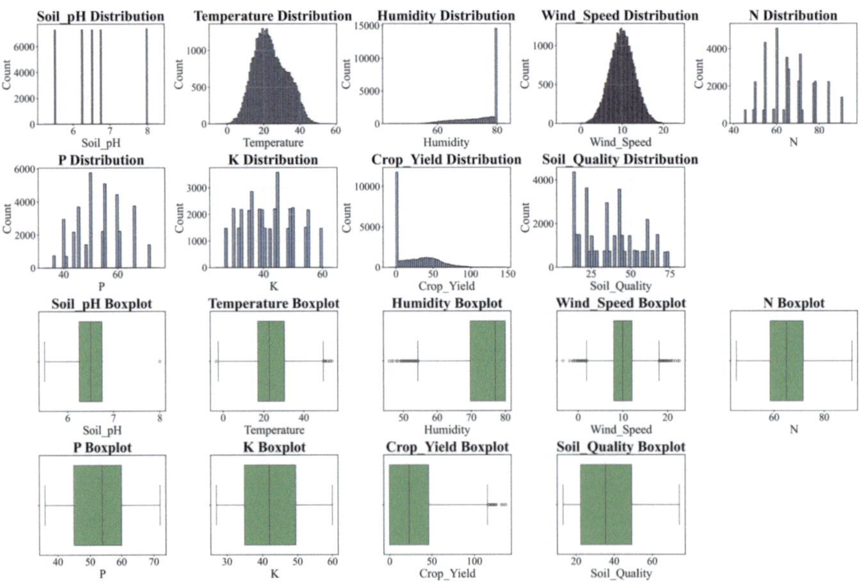

Fig. 17.2 Distribution and boxplots of continuous features: soil pH, temperature, humidity, wind speed, N, P, K, soil quality, and crop yield

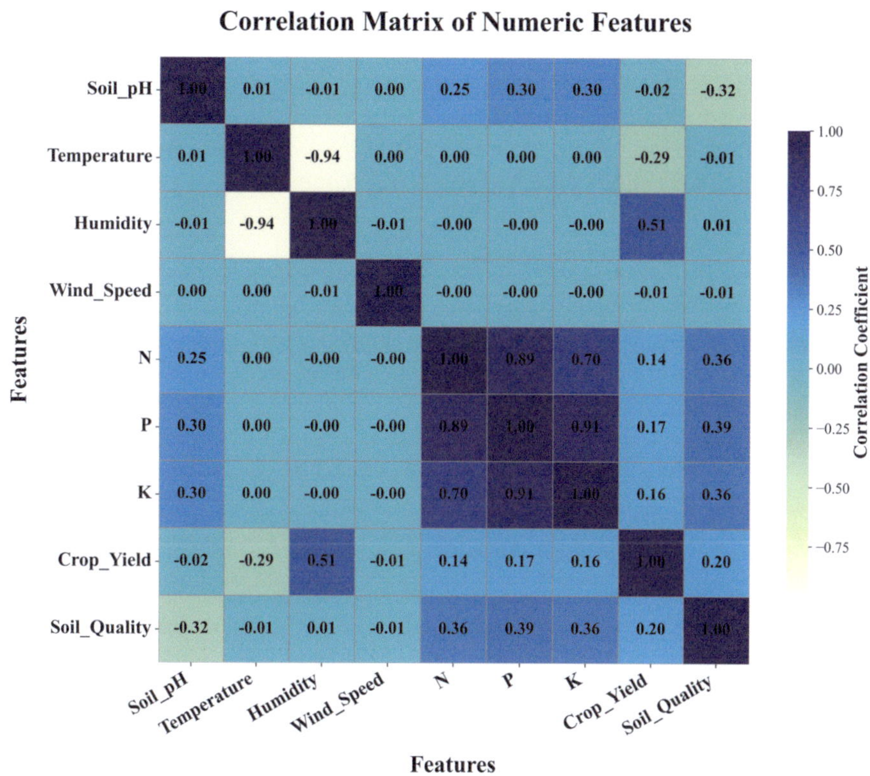

Fig. 17.3 Pearson correlation matrix of numerical variables

both are also strongly correlated with N ($r \approx 0.89$). These high correlations suggest a shared influence of macro-nutrient content on crop development. Meanwhile, Humidity and Temperature show a strong negative correlation ($r \approx -0.94$), aligning with known atmospheric interactions. Importantly, Crop_Yield shows moderate positive correlation with Humidity ($r \approx 0.51$) and weak-to-moderate correlation with P, K, and Soil_Quality. These findings inform model feature prioritization.

17.3.2 Data Preprocessing and Feature Engineering

To ensure the reliability and robustness of machine learning models, a systematic preprocessing pipeline was implemented.

(a) *Handling Missing Values*
Although the dataset was generally clean, any missing values were handled using appropriate strategies:

- For continuous variables, mean or median imputation was applied based on the distribution shape.
- For categorical variables, the mode imputation strategy was used.

(b) *Categorical Encoding*
Categorical features Crop_Type and Soil_Type were transformed using label encoding for Random Forest and Gradient Boost models. For CatBoost, which handles categorical variables natively, raw labels were retained.

(c) *Normalization and Scaling*
Although tree-based models are not sensitive to feature scaling, normalization was applied in parallel tests to compare performance consistency with scale-sensitive models. Min–max scaling was applied where relevant, primarily for visualization and exploratory regression benchmarking.

(d) *Feature Construction*
A new feature Soil Fertility Index was created as a weighted average of N, P, and K concentrations to simplify nutrient modeling. In addition, categorical encoding of seasonal patterns (e.g., mapping specific Crop Types to Kharif or Rabi seasons) was tested to capture temporal variability.

(e) *Outlier Handling*
Extreme outliers (e.g., abnormally high NPK levels) were managed through IQR-based filtering, ensuring that models were not biased by improbable agronomic scenarios.

17.3.3 Model Selection and Training Strategy

The study evaluated three state-of-the-art machine learning algorithms for predicting Crop Yield:

(a) *Random Forest Regressor*
RF, introduced by Breiman (2001), is an ensemble of decision trees trained via bootstrap aggregation. It is known for its robustness to overfitting and strong performance in high-dimensional, nonlinear spaces. The model was tuned using a grid search over hyperparameters such as:
- n_estimators (number of trees)
- max_depth (tree depth)
- min_samples_split and min_samples_leaf
- bootstrap (sampling strategy)

(b) *Gradient Boosting Regressor*
GBM constructs additive trees sequentially, each correcting the residuals of the previous one. While more sensitive to noise than RF, GBMs can achieve better generalization when properly regularized. Hyperparameter tuning included:
- learning_rate
- n_estimators
- max_depth

- subsample
- loss function selection (Huber, least squares)

(c) *CatBoost Regressor*
CatBoost, developed by Yandex, is a high-performance boosting algorithm particularly well-suited for datasets with categorical features. It uses ordered boosting to mitigate target leakage and gradient-based one-hot encoding to efficiently encode categorical data. The model was trained with:
- iterations, depth, and learning_rate as key tunables
- loss_function set to RMSE
- cat_features explicitly defined (Crop_Type, Soil_Type)

All models were implemented using Scikit-learn and CatBoost Python libraries.

17.3.4 Model Evaluation and Validation

To ensure robust performance evaluation, a 70-15-15 train–validation–test split was used, maintaining proportional representation across different crop types. The following metrics were computed to assess model quality:

- *R^2 (Coefficient of Determination)*: Measures variance explained
- *RMSE (Root-Mean-Squared Error)*: Penalizes large errors
- *MAE (Mean Absolute Error)*: Provides an intuitive error measure
- *RMSLE (Root-Mean-Squared Log Error)*: Useful for skewed yield data
- *MAPE (Mean Absolute Percentage Error)*: Highlights relative error magnitude

K-fold cross-validation ($k = 5$) was conducted on the training set to avoid overfitting and test for generalizability. Model hyperparameters were optimized using GridSearchCV where applicable.

17.3.5 Feature Importance and Model Interpretation

A critical part of the methodology involves not just prediction but also interpretation. Feature importance scores were extracted from all models to assess which variables most influenced the output. Across all three algorithms, Humidity, Phosphorus, and Soil_Quality emerged as the most influential predictors, consistent with the correlation matrix insights. Additionally:

- CatBoost's SHAP integration was used to visualize local and global feature impacts.
- Permutation importance was used in Random Forest to confirm feature rankings.

These interpretability tools provide valuable agronomic insights, suggesting that interventions in soil fertility and humidity control may yield significant benefits.

17.3.6 Computational Environment

All experiments were conducted on a system with the following specifications:

- Python 3.11
- Scikit-learn 1.4.2, CatBoost 1.2
- Pandas, NumPy, Matplotlib, Seaborn
- Intel i7 CPU with 32GB RAM (no GPU acceleration was used)

The codebase is modular, enabling reproducibility and scalability for deployment in real-world agro-advisory systems.

17.4 Results and Discussion

This section presents a comprehensive analysis of the performance and interpretability of the three machine learning models Random Forest (RF), Gradient Boosting (GB), and CatBoost applied to the task of crop yield prediction. Evaluation metrics, feature importance analysis, and actual vs predicted yield plots are systematically discussed to provide both quantitative and qualitative insights into model efficacy.

17.4.1 Random Forest (RF)

Random Forest delivered the most robust performance among all three models. As shown in Table 17.1, the RF model achieved:

- R^2: 0.9746
- *RMSE*: 4.12
- *MAE*: 2.39
- *MSE*: 16.94

An R^2 value of 0.9746 indicates that the model can explain approximately 97.5% of the variance in crop yield based on the input features. This high level of explanatory power, combined with low RMSE and MAE, reflects the RF model's ability to capture complex, nonlinear relationships without significant overfitting. The slight difference between RMSE and MAE also suggests a relatively consistent error distribution, with few extreme outliers.

Table 17.1 Model performance comparison

Model	R^2 Score	RMSE	MAE	MSE
Random Forest	0.9746	4.1154	2.3866	16.9365
Gradient Boost	0.8988	8.2145	6.0825	67.4783
CatBoost	0.9704	4.4391	3.1182	19.706

17.4.2 Gradient Boosting (GB)

The GB model yielded considerably lower performance than RF and CatBoost:

- R^2: 0.8988
- *RMSE*: 8.21
- *MAE*: 6.08
- *MSE*: 67.48

While an R^2 of 0.8988 still signifies acceptable predictive quality, the comparatively high error metrics reveal that GB struggled to approximate the target values with precision. The difference between RMSE and MAE suggests that this model was more sensitive to outliers and may have underfit or overfit certain samples in the validation set.

17.4.3 CatBoost

CatBoost demonstrated excellent predictive capability, performing slightly below RF but significantly better than GB:

- R^2: 0.9704
- *RMSE*: 4.44
- *MAE*: 3.12
- *MSE*: 19.71

The CatBoost model was especially strong in capturing categorical interactions (e.g., Crop Type and Soil Type), likely due to its native handling of categorical features via ordered boosting. The MAE value of 3.12 suggests a relatively low average error per prediction, although the slightly higher RMSE compared to RF implies a few larger residuals.

17.4.4 Comparative Visualization of Predictions

To further validate the model outputs, actual versus predicted yield plots were generated for each algorithm (Fig. 17.4). These plots offer intuitive visual assessments of each model's fidelity in reproducing ground truth values.

17.4.4.1 Random Forest Actual vs Predicted Yield

The RF model exhibits a near-linear distribution along the diagonal line, indicating strong agreement between actual and predicted values. Most data points cluster tightly around the ideal line ($y = x$), with minimal dispersion, especially in the mid-to-high yield ranges. The consistency of prediction across all ranges reflects the model's robustness.

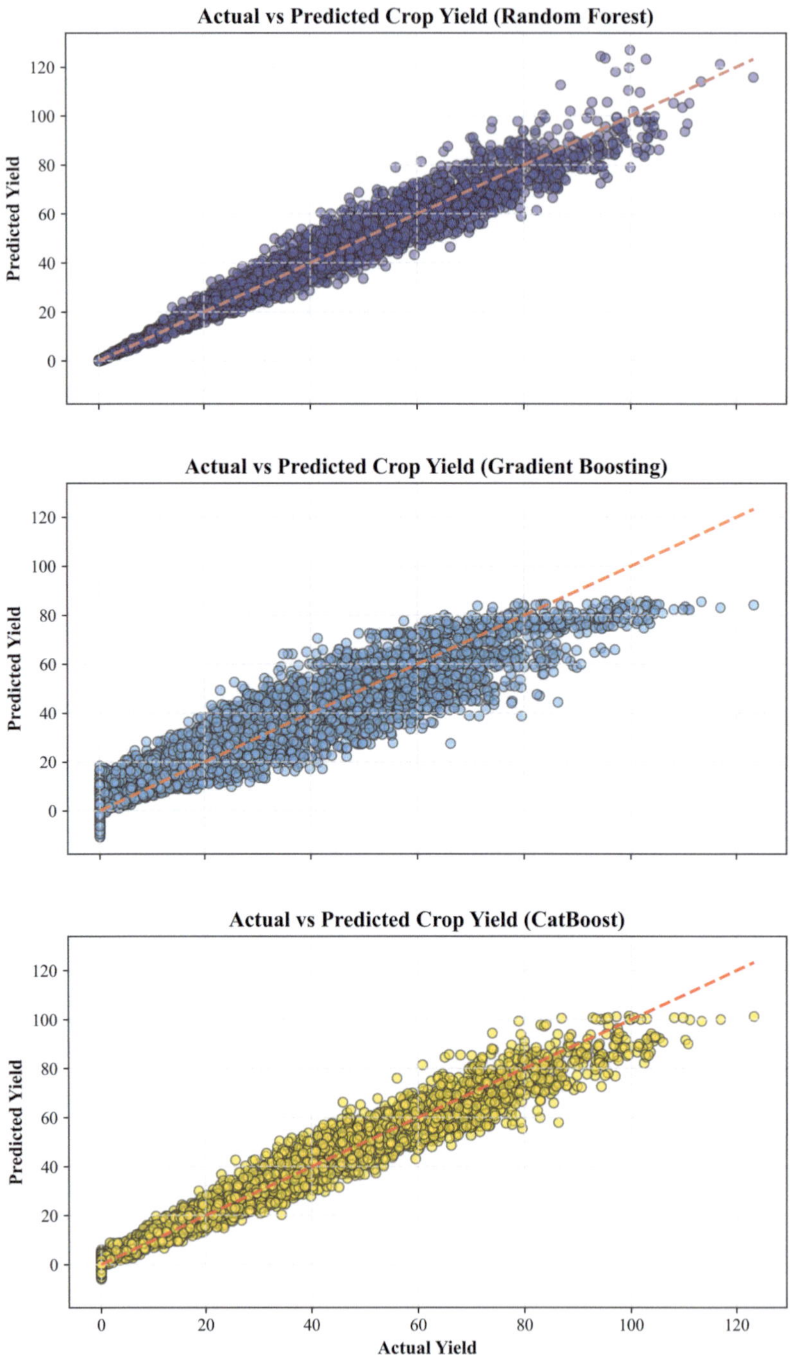

Fig. 17.4 Scatter plots of three models

17.4.5 Feature Importance Analysis (Random Forest)

To understand the decision logic of the RF model, a feature importance analysis was performed based on the Gini impurity reduction criterion. As shown in Table 17.2, Temperature emerged as the most influential predictor by a wide margin, contributing nearly 58% to the model's decision process. This finding aligns with known agro-climatic principles, where temperature directly governs photosynthesis, flowering, and crop maturation rates. The second most important feature was Humidity (15.5%), followed by Soil Quality (9.5%), Crop Type (encoded) (7.1%), and Nitrogen (N) (5.7%). Interestingly, Soil pH, Soil Type, and Potassium (K) were found to be the least influential, with combined importance below 1%.

This importance ranking not only aids interpretability but also guides practical recommendations. For instance, investment in temperature- and humidity-monitoring sensors may yield higher returns in predictive precision than extensive soil pH testing in certain regions.

17.4.6 Comparative Model Assessment

17.4.6.1 Accuracy and Generalization

Based on evaluation metrics and visual plots, Random Forest clearly stands out as the most accurate and consistent model. CatBoost follows closely, with minimal trade-offs. Gradient Boosting, while theoretically powerful, underperformed in this case, possibly due to sensitivity to noisy data or suboptimal hyperparameter tuning.

17.4.6.2 Model Robustness

RF and CatBoost displayed robust performance across a wide range of crop yields and feature values. Their ensemble nature, coupled with bagging or ordered boosting strategies, likely contributed to this stability. GB's relatively poorer performance may stem from its sequential training mechanism, which can amplify early modeling errors and suffer under high-variance data unless properly tuned.

Table 17.2 Feature importance

Feature	Importance
Temperature	0.579056
Humidity	0.154562
Soil_Quality	0.094684
Crop_Type_enc	0.070633
N	0.057004
P	0.027545
Wind_Speed	0.008302
K	0.003572
Soil_Type_enc	0.003535
Soil_pH	0.001108

17.4.6.3 Interpretability

RF offers intuitive variable importance rankings, and its decision trees are inherently interpretable. CatBoost, although slightly more complex, includes built-in tools for SHAP value calculation, providing both local and global explanations. GB models, while interpretable in theory, often suffer from lack of built-in interpretability tools unless manually configured.

17.4.7 Agronomic Interpretation and Real-World Implications

17.4.7.1 Agronomic Drivers of Yield

The prominence of Temperature, Humidity, and Soil Quality as the top predictors is consistent with agroecological theory. Temperature fluctuations significantly influence evapotranspiration and phenological stages, especially in cereals like wheat and rice. Humidity impacts disease susceptibility and water balance. Soil Quality, likely a composite of texture, fertility, and organic matter, governs nutrient availability and root health.

The relatively lower importance of NPK individually, despite their known biological roles, suggests that their effects may be captured better in combination or interactively with climatic variables. This underscores the utility of machine learning in capturing subtle interdependencies.

17.4.7.2 Recommendations for Agricultural Policy and Practice

- *Precision Farming*: Deploy temperature and humidity sensors in fields to enhance localized yield forecasting.
- *Soil Quality Management*: Focus on integrated soil health indicators rather than relying solely on NPK metrics.
- *Model Deployment*: RF and CatBoost models can be integrated into agro-advisory platforms to provide real-time, location-specific yield estimates.

17.4.8 Limitations and Future Scope

Despite promising results, several limitations must be acknowledged:

- *Data Granularity*: The dataset is aggregated and may not fully capture intra-field variability or management practices such as irrigation or pesticide use.
- *Temporal Dynamics*: Static feature inputs limit forecasting for changing weather conditions. Incorporating time-series data (e.g., daily weather and crop stages) could improve performance.
- *Model Scalability*: While RF and CatBoost are scalable, their deployment in resource-constrained rural areas must be optimized for computational efficiency.

Future work may explore hybrid models that combine physical crop models (e.g., DSSAT) with machine learning, deep learning architectures (e.g., BiLSTM), or satellite-derived vegetation indices (NDVI and EVI) to enhance both accuracy and spatial resolution.

17.5 Conclusion

This study presented a comprehensive machine learning-based approach for crop yield prediction using agronomic, soil, and environmental variables. By applying three ensemble models—Random Forest, Gradient Boosting, and CatBoost—on a regionally diverse dataset, we demonstrated that advanced data-driven models can effectively capture the nonlinear interactions and complex dependencies that govern crop productivity.

Among the tested models, Random Forest emerged as the most accurate and robust predictor, achieving an R^2 score of 0.9746 and the lowest error metrics (RMSE: 4.12; MAE: 2.39). CatBoost closely followed with a similarly high performance, validating its strength in handling categorical features without manual encoding. Gradient Boosting, although theoretically powerful, lagged behind due to its higher sensitivity to residual noise and potentially suboptimal hyperparameter tuning.

Feature importance analysis highlighted Temperature, Humidity, and Soil Quality as the most influential predictors, confirming well-established agronomic principles. The relatively lower impact of Soil pH and Potassium suggests that integrated soil–health indices offer better predictive power than individual chemical metrics alone. Visual evaluation of actual vs predicted values further confirmed the generalization capability of the RF and CatBoost models across the entire yield range.

From a practical perspective, these findings support the deployment of ensemble models—particularly Random Forest and CatBoost—in operational agricultural forecasting systems. Their ease of interpretation, scalability, and high predictive accuracy make them ideal candidates for real-time crop monitoring platforms, agro-advisory apps, and policy simulations. As India and other agricultural economies embrace data-driven solutions, integrating such models with satellite data, weather APIs, and farm management records could mark a significant step toward achieving precision agriculture at scale.

Future work may extend this research by incorporating dynamic weather sequences (e.g., LSTM or Transformer-based models), exploring explainability frameworks like SHAP and LIME in greater depth, and integrating spatial data from remote sensing sources. Additionally, developing hybrid systems that combine physical crop growth models (e.g., DSSAT) with machine learning may further enhance interpretability and cross-region generalizability.

References

Ashfaq M, Khan I, Afzal RF et al (2025) Enhanced wheat yield prediction through integrated climate and satellite data using advanced AI techniques. Sci Rep 15:18093. https://doi.org/10.1038/s41598-025-02700-w

Barman D (2025) The agricultural sector of India: trends, performance and problems. Int J Res Appl Sci Eng Technol 13:1081–1088. https://doi.org/10.22214/ijraset.2025.67472

Basantaray AK, Acharya S, Patra T (2024) Crop diversification and income of agricultural households in India: an empirical analysis. Discov Agric 2:8. https://doi.org/10.1007/s44279-024-00019-0

Biswal D, Bahinipati CS (2025) Farmers' preference for crop-diversification in India: does crop-insurance play a part? Mitig Adapt Strateg Glob Chang 30:17. https://doi.org/10.1007/s11027-025-10205-0

Breiman L (2001) Random forests. Mach Learn 45:5–32. https://doi.org/10.1023/A:1010933404324

Clercq D, Mahdi A (2024) Feasibility of machine learning-based rice yield prediction in India at the district level using climate reanalysis and remote sensing data. Agric Syst 220:104099. https://doi.org/10.1016/j.agsy.2024.104099

Coello F, Decorte T, Janssens I et al (2025) Global crop-specific fertilization dataset from 1961–2019. Sci Data 12:40. https://doi.org/10.1038/s41597-024-04215-x

Dorogush A, Ershov V, Gulin A (2018) CatBoost: gradient boosting with categorical features support. https://doi.org/10.48550/arXiv.1810.11363

Friedman J (2000) Greedy function approximation: a gradient boosting machine. Ann Stat 29. https://doi.org/10.1214/aos/1013203451

Ghimire B, Adedeji O, Ritchie GL, Guo W (2025) Simulating crop yields and water productivity for three cotton-based cropping systems in the Texas High Plains. Crop Environ 4(2):83–96, ISSN 2773-126X. https://doi.org/10.1016/j.crope.2025.03.001

Gumma MK, Nukala RM, Panjala P, Bellam PK, Gajjala S, Dubey SK, Sehgal VK, Mohammed I, Deevi KC (2024) Optimizing crop yield estimation through geospatial technology: a comparative analysis of a semi-physical model, crop simulation, and machine learning algorithms. AgriEngineering 6(1):786–802. https://doi.org/10.3390/agriengineering6010045

Joshi A, Pradhan B, Chakraborty S, Varatharajoo R, Alamri A, Gite S, Lee C-W (2025) An explainable Bi-LSTM model for winter wheat yield prediction. Front Plant Sci 15. https://doi.org/10.3389/fpls.2024.1491493

Klompenburg T, Kassahun A, Catal C (2020) Crop yield prediction using machine learning: a systematic literature review. Comput Electron Agric 177:105709. https://doi.org/10.1016/j.compag.2020.105709

Koushik A, Manoj M, Nezamuddin N (2024) SHapley additive exPlanations for explaining artificial neural network based mode choice models. Transp Dev Econ 10:12. https://doi.org/10.1007/s40890-024-00200-6

Kudikala R, Keesara V, Buri E (2025) Climate change impact assessment on rice yield using dssat and geospatial methods. https://doi.org/10.21203/rs.3.rs-6557175/v1

Maurya DK, Rao PVN, Dadhwal VK, Dutt CBS (2015) Large area soil moisture variations in response to cyclone Phailin in eastern India. IEEE Geosci Remote Sens Lett 12(2):264–268. https://doi.org/10.1109/LGRS.2014.2334739

Nargund R, Rautela K, Goyal M, Sinha N, Mohanty M, Bhatia VS (2024) Assessing soybean yield in Madhya Pradesh by using a multi-model approach. Field Crop Res 322:109716. https://doi.org/10.1016/j.fcr.2024.109716

Nikhil UV, Pandiyan AM, Raja SP, Stamenkovic Z (2024) Machine learning-based crop yield prediction in South India: performance analysis of various models. Computers 13(6):137. https://doi.org/10.3390/computers13060137

Parashar N, Johri P, Khan AA, Gaur N, Kadry S (2024) An integrated analysis of yield prediction models: a comprehensive review of advancements and challenges. Comput Mater Contin 80(1): 389–425. http://www.techscience.com/cmc/v80n1/57370

Pazhanivelan S, Geethalakshmi V, Tamilmounika R, Sudarmanian NS, Kaliaperumal R, Ramalingam K, Sivamurugan AP, Mrunalini K, Yadav MK, Quicho ED (2022) Spatial rice yield estimation using multiple linear regression analysis, semi-physical approach and assimilating SAR satellite derived products with DSSAT crop simulation model. Agronomy 12(9):2008. https://doi.org/10.3390/agronomy12092008

Sah S, Haldar D, Singh R et al (2024) Rice yield prediction through integration of biophysical parameters with SAR and optical remote sensing data using machine learning models. Sci Rep 14:21674. https://doi.org/10.1038/s41598-024-72624-4

Sarkar S et al (2024) crop yield prediction using multimodal meta-transformer and temporal graph neural networks. IEEE Trans AgriFood Electron 2(2):545–553. https://doi.org/10.1109/TAFE.2024.3438330

Sharma RK, Kaur J, Feng G et al (2025) Maize and soybean yield prediction using machine learning methods: a systematic literature review. Discov Agric 3:64. https://doi.org/10.1007/s44279-025-00215-6

Future Trends and Innovations in Vertical and Layered Farming with Digital Twin Technology

18

Riddhi Rathva and Uma Hapani

Abstract

As the global population continues to rise and arable land becomes increasingly limited, innovative agricultural strategies are essential to meet food security demands. Vertical and layered farming practices emerge as solutions, enabling crop production within compact urban environments. These methods utilize vertical space efficiently by cultivating plants in stacked layers using controlled environment techniques, including hydroponics, aquaponics, and aeroponics. This eliminates the need for soil and reduces reliance on external climatic conditions, supporting year-round cultivation, optimizing resource use, and minimizing waste. Multi-layered or multi-storied cropping systems, a complementary form of vertical farming, involve growing compatible crops at varying heights to maximize sunlight capture, enhance biodiversity, and reduce pest infestations. The integration of Digital Twin (DT) technologies into vertical and layered farming systems marks advancement in smart agriculture, increasing productivity and sustainability. DTs are dynamic, virtual replicas of physical farming systems that allow real-time monitoring (e.g., temperature, humidity, and CO_2 levels), predictive analysis, and simulation of various scenarios. When integrated with enabling technologies such as the Internet of Things (IoT), Artificial Intelligence (AI), Geographic Information Systems (GIS), Cloud Computing, and Blockchain, digital twins form a comprehensive agri-digital ecosystem. These technologies enable tracking of environmental conditions, optimize

R. Rathva
School of Environment and Sustainable Development, Central University of Gujarat, Gandhinagar, Gujarat, India

U. Hapani (✉)
Department of Environmental Studies, The Maharaja Sayajirao University of Baroda, Vadodara, Gujarat, India

© The Author(s), under exclusive license to Springer Nature Singapore Pte Ltd. 2026
A. L. Srivastav et al. (eds.), *Digital Twin Technology for Sustainable Agriculture*,
https://doi.org/10.1007/978-981-95-5915-2_18

irrigation, predict disease outbreaks, ensure resource efficiency, and enhance supply chain transparency. Components such as sensor networks, data analytics platforms, simulation models, and modular IoT frameworks support scalable and energy-efficient operations. However, challenges remain, including scalability in multi-layer environments, data integration, system interoperability, high initial costs, and operational inefficiencies. Addressing these issues requires collaborative research, incremental implementation strategies (e.g., pilot projects), and the development of standardized data protocols. The chapter emphasizes the convergence of vertical-layered farming and digital twin technology as a promising pathway for the future of smart agriculture.

Keywords

Vertical farming · Multi-layer farming · Digital twin (DT) technology · Internet of Things (IoT) · Smart agriculture

18.1 Introduction

Food is the basic requirement for sustaining life. Food insecurity may cause extreme hunger and malnutrition; it becomes a barrier to sustainable development as it leads to less productive individuals. People become prone to diseases and are unable to earn and improve their standards. Yet, it is estimated that around 2.33 billion people globally faced moderate to severe food insecurity in 2023. Factors like pandemic, climate change, conflict, and inequalities have made hunger and food insecurity the global issues (Goal 2: Zero Hunger—United Nations Sustainable Development). Sustainable Development Goal (SDG) 2 ensures that each person has access to safe, nutritious, and sufficient food. Its primary aim is to end hunger, achieve food security, and promote sustainable development (Olabimpe Banke Akintuyi 2024). It extends beyond food production, distribution, and consumption and also encourages the systems that double the productivity and income of small-scale farmers, safeguard biodiversity, and preserve genetic resources for future generations.

Meeting the rising food demands remains a global challenge despite decades of agricultural advancements. By 2050, the global population is expected to reach 10 billion, demanding enough with quality to gain food security, that too under constrained resources and shifting climatic conditions (Food and Agricultural Organization of the United Nations 2009). Conventional rural agriculture alone struggles to keep pace while facing challenges of soil conditions, unpredictable weather patterns, and expanding urban sprawls that consume the valuable arable land.

Rapid urbanization and industrialization increase challenges and complexities. The population in urban areas is more than 55% which is expected to increase to 70% by 2050 (United Nations Department of Economic and Social Affairs 2022). Cities are observed expanding both upwardly and outwardly. Yet, the citizens are heavily reliant on distant food supplies that are vulnerable to damage, and long transportations leave a major carbon footprint.

This brings us to modern ways of food production and introduces urban agriculture. Architecture, along with novel technologies, is taken into consideration and integrated into urban agriculture to enable the cultivation of crops and edible plants into urban infrastructure (Olabimpe Banke Akintuyi 2024). Among the innovative methods of urban agriculture are vertical and layered farming systems. This reduces the pressure on conventional farming by utilizing and transforming urban spaces (rooftops, warehouses, and purpose-built indoor farms) and sparse areas into productive assets for crop cultivation (Kalantari et al. 2020). This boosts the self-sufficiency in food production, enabling citizens to be fed with fresh produce with a longer shelf life due to a reduction in transit times (Olabimpe Banke Akintuyi 2024).

There are several social, economic, and environmental advantages to vertical and layered farming (Clementi et al. 2024). These productive assets provide opportunities for education, skill-building, recreation, and social interaction (Clementi et al. 2024). In contrast to open field agriculture, which is usually impacted by climate variability and uncertainty, controlled environments, such as that in vertical-layered farming, support promising and stable yields (Olabimpe Banke Akintuyi 2024).

To achieve these benefits, these farms must be equipped to monitor certain variables, including temperature, humidity, moisture, sunlight, indoor lighting, shadow patterns, and dimensions and growth behavior of different plants. This urban approach describes the circulation and transformation of resources, water, nutrients, and energy within cities, mimicking the processes by which natural ecosystems manage microclimates and nutrient flows (Clementi et al. 2024).

Recent technological advances make it possible to manage these complex systems with greater precision than before. The integration of sensors, Internet of Things (IoT) devices, Artificial Intelligence (AI), and even robotics enhances the precision of urban farming operations, automating tasks such as irrigation, nutrient delivery, lighting control, and pest management (Olabimpe Banke Akintuyi 2024). These tools reduce reliance on manual labor, lower operational costs over time, and allow farmers and practitioners to monitor and adjust environmental conditions in real time.

As urban farms scale up and become more complex in enclosed buildings, the need for sophisticated monitoring, predictive modeling, and real-time decision support grows. This is where Digital Twin (DT) technology comes into play. By combining the principles of vertical and layered farming with DT systems, cities can virtualize urban food systems to meet the demands of a growing population. The sections that follow explore the practices, enabling technologies, system components, benefits, challenges, and futurc pathways for implementing DT-supported vertical and layered farming. Figure 18.1 shows vertical-layered farming integrated with DT technology supporting urban agriculture.

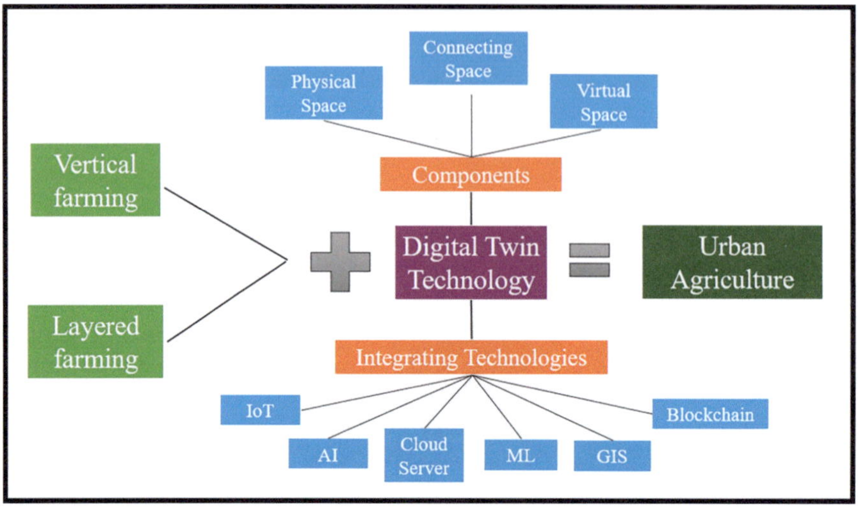

Fig. 18.1 Vertical-layered farming integrated with DT technology supporting urban agriculture

18.2 Vertical Farming Practices

Vertical farming is a transformative solution for cities to boost local food production while efficiently utilizing limited area and urban infrastructure. Vertical farming involves growing plants in multiple stacked layers or along inclined surfaces inside enclosed spaces (Yapp et al. 2025). These can range from converted warehouses and underutilized high-rise buildings for soil-free crop cultivation (Olabimpe Banke Akintuyi 2024). This design flexibility makes vertical farming highly adaptable to dense urban settings, where ground-level space is scarce but vertical space is underutilized. In a vertical farm, irrespective of the external environment and seasons, internal microclimates are precisely managed to provide conditions for optimum growth (Nie et al. 2022).

Hydroponics is the most widely used technique in vertical farming. Instead of using soil, plants are grown hydroponically, that is, plant roots are submerged in a circulating water solution with essential nutrients added to it (Dennison et al. 2025). This eases mineral and pH regulation. Depending on the crop varieties, deep water culture beds or Deep Flow Technique (DFT) or Nutrient Film Technique (NFT) channels are used in the system (Peladarinos et al. 2023). In the NFT system, plants are grown in shallow nutrient solutions (Seyhan and Seyhan 2025). The nutrient solution is pumped through a closed loop of channels where plant roots are submerged in the stream. This system uses less water, prevents waterlogging, and therefore supplies proper oxygen. Although there are a few challenges in this approach, root debris may accumulate in the channels, which may cause blockages in the channels, and continuous power supply is required to maintain the flow of nutrients. Inadequate water supply may cause dehydration and nutrient loss (Seyhan and Seyhan 2025). In the DFT system, plants are grown in deep water systems, and these water cultures are rich in nutrients. The roots are submerged in it, and water is recirculated. Its major advantage is that it is stable and failsafe (Seyhan and Seyhan 2025).

Aquaponics, a combined approach is generated by integrating hydroponics and aquaculture. It appears to be a symbiotic relationship between fish and plants. Fish excrement provides nutrition for the plants; in return, plants purify the water that is circulated back to the fish tank. This closed-loop system of aquaponics reflects the circular economy and lowers the need and utilization of chemical fertilizers (Peladarinos et al. 2023).

Aeroponics is another advanced method used in vertical farming. In aeroponics, plant roots are suspended in the air and regularly misted with a nutrient-rich solution. Because roots have greater access to oxygen compared to hydroponics, aeroponics can accelerate growth rates and use even less water than traditional hydroponic methods (Kumari and Kumar 2019). However, aeroponic systems require high levels of monitoring and precision control, where digital technologies become vital.

Lighting is a critical factor in vertical farming success. Many vertical farms use LED grow lights, which are programmed to deliver specific light wavelengths optimal for photosynthesis at different stages of plant growth. Some systems adjust light intensity and duration automatically in response to sensor data, ensuring plants receive exactly the right amount of energy while minimizing electricity costs (Mengi et al. 2024).

The following examples show the evolution of vertical farming from pilot experiments to food production solutions: Singapore's urban agriculture policy supports high-rise vertical farms to improve the island's food self-sufficiency. In Europe, companies have turned abandoned factories and shipping containers into modular vertical farms, producing pesticide-free leafy greens. New York, Chicago, and Toronto are home to rooftop hydroponic greenhouses and vertical towers that reduce food miles and provide year-round harvests (Olabimpe Banke Akintuyi 2024).

In addition to food production, vertical farms improve building energy efficiency by acting as thermal insulators, reducing urban heat island effects through increased greenery, and contributing to urban biodiversity by providing habitats for beneficial insects when integrated with rooftop gardens.

18.3 Multi-layered or Multi-storied Cropping Systems

Multi-layered or multi-storied farming is an evolution of vertical farming. This system is inspired by nature's stratification, where the flora exists at different heights in the forests, for example, orchids as epiphytes on trees, or mango, neem, and banyan trees at different heights, or canopy layers of the banyan tree at different heights.

A multi-layered cropping system is planned and structured in such a way that light, water, and nutrients are maximized or provided based on the optimum requirement in the same growing space but at different vertical lengths (Solanki et al. 2024). It is about coexistence at different levels or elevations. This principle is adapted into stacked or tiered arrangements, rooftop terraces, indoor racks, and greenhouse layers that function together to maximize resource efficiency.

A three-tier system of three models of vegetable cropping was introduced in two districts of Bihar (Pramanik et al. 2022). Farmers were able to grow bitter gourd, cowpea, and yams in the first model; pointed gourd, okra, and cucumber in the second model; and pointed gourd, okra, and Amaranthus in the third model.

Multi-layered cropping systems capture waste heat from Heating, Ventilation, and Air Conditioning (HVAC) systems, which regulate indoor temperature, humidity, and air quality. Also, they recycle greywater for irrigation, thereby closing resource loops that support stable conditions for controlled agriculture (Clementi et al. 2024).

18.4 Introduction to Digital Twin (DT) Technology in Agriculture

Vertical-layered farming systems face practical constraints, including high installation costs, energy demands for lighting lower layers, and the need for skilled operation and monitoring. By integrating DT frameworks, farms can streamline these complexities, reduce labor needs, and tune production for maximum return on investment (Peladarinos et al. 2023).

A DT can be broadly defined as a dynamic, virtual replica of a physical entity, system, or process that is continuously updated using real-time data (Peladarinos et al. 2023). In agriculture, this means creating a detailed simulation of a farm's physical assets, including crops, soil, nutrient solutions, irrigation systems, climate conditions, and even individual plant traits and integrating these data streams in a virtual platform capable of mirroring, predicting, and optimizing real-world performance.

The concept of DT technology was proposed in the early 2000s by Michael Grieves, who introduced the idea of the "Mirror Space Model"(Nie et al. 2022). The DT model is developing further over time into various sectors, including aerospace, healthcare, smart cities, and agriculture (Shahzad et al. 2022).

With the upgrading of knowledge and advances, a DT for greenhouse tomato crops was developed in 2020 (Nie et al. 2022). In the reported study by Nie et al., 2022 showed interaction between crop qualities and environmental factors that were updated by 3D simulation models. Information were acquired from the sensors in real greenhouses, which supported the crop management.

In 2020, another application was implemented for mango exports (Nie et al. 2022). A DT was created for three mango varieties, Alphonso, Kesar, and Totapuri, to simulate the accurate cooling behavior during shipments with varying airflow and temperatures. This enabled the monitoring of cooling inconsistencies and quality loss during transport and storage while providing logistics for a better shelf life.

A tree-level DT was also developed in 2020 (Nie et al. 2022). It is an automatic crown monitoring system named *AgScan3D+*. It was used for mango, macadamia, avocado trees, and vine orchards. It was installed with 3D cameras to generate digital replicates for 15,000 individual trees. The system tracked health, structure, canopy growth, and crop and fruit quality while supporting disease detection and stress prediction, also supporting real-time decisions.

In 2022, a single-view leaf reconstruction method was proposed (Nie et al. 2022). It was based on ResNet to generate accurate single-view leaf reconstruction within the DT system. It is a computer vision technique that uses a single image with a camera view to recreate a 3D shape. It is used for image classification and recognizing plant parts via images or video recordings. This detects the disease spots, pests, or damage on leaves.

DTs possess the potential to cover the entire agricultural cycle in all its stages. The pre-production stage includes soil or nutrient quality testing, irrigation, and seed selection, followed by mid-production, including growth and microclimate monitoring, and finally post-production including harvest, storage, and supply chain traceability (Nie et al. 2022).

18.5 Integration of Enabling Technologies

DT systems can be utilized with more potential and efficiency when they are incorporated with machine learning models. Geographic Information Systems (GIS) integration extends its value by mapping farm layouts, identifying resource flows in the region, and climate and heat predictability (Clementi et al. 2024). Further, when used with AI, IoT, and blockchain, it becomes a part of an agri-digital ecosystem in which data flows among devices, cloud servers, analytical engines, and dashboards. Altogether, these technologies transform static crop arrangements into smart, intelligent, agricultural cycles.

A DT connects these layers through communication networks that ensure a constant feedback loop between the physical space, the actual farm, and the virtual space, the simulation model that “learns” and “predicts” in real time. The section follows and explores enabling technologies, their integration, the specific system components involved, and the shaping of sustainable urban agriculture.

18.5.1 Internet of Things (IoT)

A DT is an extensive network of IoT devices. These IoT devices transmit data through secure gateways to cloud-based servers, where the information is collected, stored, and fed into the virtual twin model for analysis. In a vertical-layered system, sensors are typically distributed in every rack and tier to monitor the microenvironment. For example, such a farm relies on sensors to track nutrients, pump flow rates, and detect leakages to conserve resources (Peladarinos et al. 2023).

18.5.2 Artificial Intelligence (AI)

Smart farms are not driven solely by raw data, especially when raw data can be collected manually. A decisive value must be added in response to the happenings. AI algorithms process and interpret the incoming sensor data. AI has the potential to forecast plant stresses based on the changes in the environment, predict the optimal harvest period, and detect symptoms for plant diseases (Nie et al. 2022). An AI-based nutrient model can analyze the water chemistry that could include parameters such as pH and electrical conductivity (EC) to optimize the dosing schedules. It also has the potential to fully automate the irrigation systems, considering the crop growth stage.

18.5.3 Machine Learning (ML)

ML models take the predictive capacity of AI a step ahead by providing previous and real-time data. For example, Evapotranspiration (ET) reflects the weather's impact on crop water requirements. It is a key parameter to predict the water needs, adjusting irrigation and balancing indoor climatic conditions. The ML model estimates the past and current ET values, as well as predicts the future values (Nie et al. 2022).

In the context of vertical-layered farming, ML has the potential to manage microclimates for each layer. ML can model light intensity and airflow changes in the stacked vertical-layered farming from the top layer to the bottom layer (Shahzad et al. 2022).

18.5.4 Geographic Information System (GIS)

GIS plays a crucial role in urban metabolism by identifying the production areas in urban (Clementi et al. 2024) as well as rural agriculture. This integration can assist site selection in urban space, resource allocation, and operational planning. If the vertical-layered farm is built indoors, GIS can be used to analyze urban land use, such as available buildings, suitable rooftops, power and water availability, and market proximity. It may also help to optimize delivery routes from farms to consumers.

Water distribution is visually represented by water mapping, which also identifies information on the irrigation sources and vulnerability to floods (Clementi et al. 2024). DT is enhanced by evaluating irradiance, humidity, and temperature across the buildings in urban regions. These parameters produce energy maps to determine the production of suitable crop species (Clementi et al. 2024).

18.5.5 Cloud Computing

A DT system requires a cloud server to store data generated from sensors and IoT devices. Cloud computing aligns the real-time happenings of the physical farm in its virtual twin. This synchronization allows users to access dashboards and data from any location (Peladarinos et al. 2023). The cloud also accommodates AI and ML data, allowing them to give updated control commands back to the actuators of the farm. A secure cloud installation is necessary for storing and protecting large amounts of data while also serving as a backup system.

18.5.6 Blockchain for Traceability

Blockchain technology offers a secure solution applicable in vertical-layered farming. It allows the system to be transparent in records from production to the supply chain movements. This makes the agricultural cycle more secure. When blockchain is combined with DT, it can be expected to track all the environmental conditions from the seeds to harvest (Peladarinos et al. 2023). It simultaneously verifies whether the produce is grown under specified and optimal conditions. Blockchain enables tracking and recording of every transaction step; in general, it makes the journey from farm to table transparent. This transparency strengthens the customers' trust and support, easing certifications. Urban farm produce is sold in the local markets. For high-value crops or export produce, blockchain-integrated DTs can streamline regulatory compliance and traceability requirements (Nie et al. 2022).

This integrated approach reduces resource waste while enabling urban food systems and smart agriculture.

18.6 Components of a Digital Twin System in Vertical-Layered Farming

The DT system was initially made of three main components: a "Physical Space", "Virtual Space", and "Connections between these Spaces" (Nie et al. 2022).

The physical space consists of physical assets such as sensors and actuators, while the virtual space comprises multiscale and probabilistic stimulation models. A connection between these spaces enables continuous data exchange and allows control commands to be transmitted between these two spaces (Nie et al. 2022).

The six-component framework was later proposed by Parrott and Warshaw, comprising five enabling components. In this framework, the "physical space" and "virtual space" are represented as the enabling components "sensors" and "actuators," with the processes "act," "crcatc," and "communicate," belonging to the physical space, and the components such as "data" and "analytics", together with the processes "collect", "analyze", and "predict," are assigned to the virtual space.

The element of "integration" has an important role in connecting the physical and virtual aspects to allow interaction and data flow between them (Nie et al. 2022). The

Table 18.1 Typical components of a digital twin for a vertical-layered farming

Component	Description
Physical assets	Stacked plant trays, nutrient solution tanks, recirculation pumps, LED grow lights, fans, and automated mechanisms for planting & harvesting
Sensors and IoT devices	Integrated devices that monitor environmental and crop growth parameters
Actuators and control interface	Mechanisms that command to adjust pumps, fans, lights, irrigation flow, misting systems, automatically based on decisions made by the virtual twin
Communication layer	A secure network infrastructure ensures real-time transfer of sensor data and operational commands between the physical farm and virtual farm
Data fusion and integration module	Integrates multi-source data collected from sensors, IoT devices to provide the virtual twin with accurate & up-to-date information
Virtual twin core	Continuously mirrors the actual state of the vertical farm and simulates processes such as water flow, plant transpiration, and microclimate conditions across layers
AI/ML analytics engine	Uses advanced machine learning models, such as GBDT to forecast crop growth, predict evapotranspiration rates and nutrient requirements, for farm management
Cloud server and storage	Provides secure storage for large volumes data
User dashboard	A web or mobile application that allows visualising real-time farm status, receive AI-generated insights, approve automated actions, and adjust system settings as needed
Integration layer	Connects the physical and virtual domains for bidirectional data flow and control

ability of DT to present a sound mirror image depends on computing, intelligent data analysis, and VR presentation (Nie et al. 2022). The typical components of a DT for a vertical-layered farming are listed in Table 18.1, representing the typical components present in a DT (Buldakova and Suyatinov 2019; Zhong et al. 2023; Hu et al. 2021).

While a smart system monitors and decides actions using AI based on sensor data, a DT system mirrors the physical environment while monitoring, stimulating, predicting, optimizing, and validating the actions virtually before applying them.

18.7 Benefits of Digital Twin Integration

Resource efficiency becomes a priority in urban agriculture. Vertical-layered farms already use less land than traditional farming. The DT system maximizes water cycles, nutrient uptake, and energy consumption in detail. A well-implemented DT directly contributes to higher yields and better-quality produce. By maintaining stable growing conditions and resources, plants experience less environmental stress and can grow more uniformly (Peladarinos et al. 2023). In commercial urban farms, this results in more harvest cycles.

DTs provide continuous, real-time monitoring of all parameters across every layer of a stacked farm. This reduces the need of frequent manual monitoring and saves labor costs (Peladarinos et al. 2023). For example, if a sensor detects a sudden rise in humidity that may promote fungal growth, the DT automatically adjusts fans or misting systems to restore balance. DTs provide predictive analysis to help growers to make prompt decisions. They also provide a chance to test multiple scenarios virtually, to predict future states with AI, and to automate optimal actions (Nie et al. 2022). This reduces costs on trial-and-error experiments.

As cities meet SDGs, DTs in urban farms demonstrate sustainability performance. For example, real-time carbon footprint tracking, energy usage logs, and water recycling rates can be visualized in the user dashboard. These metrics support green certifications as well. Through GIS and blockchain, a DT system can feed verified data, traceability, and food safety. Consumers and retailers can trace crop's growing conditions and handling.

Finally, DT systems have several social benefits for educational tours or school programs, teaching high-tech farming for urban communities.

18.8 Challenges and Limitations

However, successful farming is not without its challenges. High energy demands, particularly for artificial lighting and climate control, can pass the sustainability gains if not paired with renewable energy. Retrofitting high-rise buildings or constructing dedicated vertical farms demands proper planning, zoning approvals, and structural adaptations. Layered systems must be carefully designed to balance light distribution. Lower tiers often receive less natural light, increasing dependency on energy-intensive LED systems. Poor design can lead to uneven crop growth and inefficiencies (Shahzad et al. 2022).

Upfront investment can be an immediate barrier. Costs include installing IoT networks, multi-parameter sensors, data servers, AI/ML software, and reliable cloud systems. Operating costs also remain significant. A DT system requires integration of various hardware and software components from IoT devices to cloud servers, ML, and user interfaces. Inconsistent data formats or sensor calibration errors can reduce the accuracy of simulations and predictions.

Operating DT urban farms requires a skilled and interdisciplinary workforce, combining agricultural knowledge with digital literacy, data analytics, and systems management. At present, DTs are developed for specific crops, climate conditions, or farm layouts. Lack of standardization makes it difficult to integrate multiple DTs into a unified network. A breach in cloud data or IoT networks could expose sensitive operational information, disrupt farm operations, or even pose food safety risks if automated systems are compromised (Shahzad et al. 2022).

While hydroponics and aquaponics reduce soil-related pests, they can introduce other risks. Stagnant nutrient solutions may promote growth of algae or waterborne pathogens if not monitored. Mechanical failures, such as pump malfunctions or sensor breakdowns, can quickly lead to crop losses if real-time alerts are not in place.

Policy frameworks for urban vertical farming and digital agriculture are still in process. Many cities lack clear guidelines for integrating vertical farms into existing building codes, utilities, and energy grids (Clementi et al. 2024).

By acknowledging these barriers and working across disciplines, the vision of DT vertical-layered farming can go from experimental trials to urban food production strategies.

18.9 Way Forward and Recommendations

This integrated approach contributes to local and global food security. Governments, industry groups, and researchers should collaborate to create clear frameworks for systems and their interoperability. A new hybrid skillset calls for tailored vocational training, university courses, and upskilling programs. Partnerships between academic institutions, urban farms, and technology companies can help fill this gap and prepare a workforce ready to run digitally managed urban farms. Open-source models help farms in adapting DT.

Cities should view urban food production as a smart city ecosystem. Policies need to be adaptable and considerate in terms of urban spaces and infrastructures, public health and nutrition, food security, and environmental issues (Clementi et al. 2024).

The integration of plant emotions into DT systems combines plant physiology, biofeedback systems, AI, and smart agriculture technologies. Although plants may not have emotions in the human sense, they respond to environmental stimuli such as light, sound, touch, and chemical signals in measurable ways. These responses can be interpreted as indicators of plant stress, health, or satisfaction, which are sometimes metaphorically referred to as "plant emotions." To integrate these responses into a DT system, Table 18.2 presents applications in plant response, sensors, and AI that are used to detect and analyze signals (Chapin et al. 1987; Morgan and Drew 1997):

It supports plant well-being in sustainable farming. The social dimensions should not be overlooked.

Table 18.2 DT applications in plant response

Plant response type	Sensor/technology	DT application
Chlorophyll fluorescence	Fluorometer/camera	Assess photosynthesis
VOC emissions	Gas sensors	Detect pest attacks or stress
Electrical signaling	Electrodes/bioelectronic interfaces	Monitor real-time stress
Leaf temperature	Infrared thermography	Indicate water stress
Sap flow/water uptake	Sap flow meters	Evaluate transpiration and water needs
Hormonal changes (e.g., ethylene)	Biosensors	Detect ripening or stress

18.10 Conclusion

As cities worldwide witness rapid urbanization, arable land is constrained as there is a need to secure food supplies. Vertical-layered farming systems emerge as one of the promising responses to such complex challenges. Urban spaces turned as productive green ensure fresh produce is grown closer to the communities that consume it. DTs create a bridge between physical farms and their virtual counterparts. The chapter has explored the integration of enabling technologies, sensors, IoT networks, AI and ML models, cloud computing, GIS, and even blockchain in static vertical farms. These systems deliver the following benefits: stable, predictable harvests; lower resource wastage; traceable supply chains; and insights for climate-smart urban development. However, challenges such as high upfront costs, integration complexities, skill gaps, and policy uncertainties continue to limit widespread adoption. The synergy between resilient farming practices and digital tools holds vast potential to contribute directly to SDG 2: Zero Hunger and supports SDG 3: Good Health and Well Being, SDG 8: Decent Work and Economic Growth, SDG 9: Industry, Innovation and Infrastructure, SDG 11: Sustainable Cities and Communities, SDG 12: Responsible Consumption and Production, and SDG 17: Partnerships for the Goals. As cities pursue greener, healthier, and more inclusive urban features, vertical and layered farms guided by DTs stand as a vital support for the next generation of smart, self-sufficient urban food systems.

References

Buldakova TI, Suyatinov SI (2019) Assessment of the state of production system components for digital twins technology. In: Cyber-physical systems: advances in design & modelling. Springer International Publishing, Cham, pp 253–262

Chapin FS, Bloom AJ, Field CB, Waring RH (1987) Plant responses to multiple environmental factors. Bioscience 37(1):49–57

Clementi M et al (2024) GIS-based digital twin model for solar radiation mapping to support sustainable urban agriculture design. Sustainability (Switzerland) 16(15). Available at: https://doi.org/10.3390/su16156590

Dennison MS et al (2025) The role of automation and robotics in transforming hydroponics and aquaponics to large scale. Discov Sustain. Springer International Publishing Available at: https://doi.org/10.1007/s43621-025-00908-4

Food and Agricultural Organization of the United Nations (FAO) (2009) How to feed the world in 2050. Rome: Food and Agricultural Organization of the United Nations. https://www.fao.org/fileadmin/templates/wsfs/docs/expert_paper/How_to_Feed_the_World_in_2050.pdf

Goal 2: Zero Hunger – United Nations Sustainable Development United Nations. Available at: https://www.un.org/sustainabledevelopment/hunger/. Accessed 30 Jul 2025

Hu W, Zhang T, Deng X, Liu Z, Tan J (2021) Digital twin: a state-of-the-art review of its enabling technologies, applications and challenges. J Intell Manuf Spec Equip 2(1):1–34

Kalantari et al (2020) The significance of vertical farming concept in ensuring food security for high-density urban areas. Jurnal Kejuruteraan 32(1):105–111. Available at: https://doi.org/10.17576/jkukm-2020-32(1)-13

Kumari R, Kumar R (2019) Aeroponics: a review on modern agriculture technology. Indian Farmer 6(4):286–292. Available at: https://www.researchgate.net/publication/342976865

Mengi E et al (2024) A digital-twin and rapid optimization framework for optical design of indoor farming systems. Comput Mech 74(1):31–43. Available at: https://doi.org/10.1007/s00466-023-02421-9

Morgan PW, Drew MC (1997) Ethylene and plant responses to stress. Physiol Plant 100(3): 620–630

Nie J et al (2022) Artificial intelligence and digital twins in sustainable agriculture and forestry: a survey. Turk J Agric For 46(5):642–661. Available at: https://doi.org/10.55730/1300-011X.3033

Olabimpe Banke Akintuyi (2024) Vertical farming in urban environments: a review of architectural integration and food security. Open Access Res J Biol Pharm 10(2):114–126. Available at: https://doi.org/10.53022/oarjbp.2024.10.2.0017

Peladarinos N et al (2023) Enhancing smart agriculture by implementing digital twins: a comprehensive review. Sensors 23(16):1–38. Available at: https://doi.org/10.3390/s23167128

Pramanik A, Chandra B, Viswavidyalaya K (2022) Multi-layer cropping: ideal approach for better yield and increasing farm income, 3(2). Available at: https://www.researchgate.net/publication/366182512

Seyhan TG, Seyhan S (2025) Comparative analysis of deep flow and nutrient film technique in vertical farming. J Tekirdag Agri Fac 22(2):362–372. Available at: https://doi.org/10.33462/jotaf.1437386

Shahzad M et al (2022) Digital twins in built environments: an investigation of the characteristics, applications, and challenges. Buildings 12:1–19

Solanki BP et al (2024) Review on multilayer farming: a way towards farmer prosperity. Int J Environ Clim Chang 14(1):150–154. Available at: https://doi.org/10.9734/ijecc/2024/v14i13818

United Nations Department of Economic and Social Affairs (2022) UN DESA annual highlights report 2022–2023. https://desapublications.un.org/publications/un-desa-annual-highlights-report-2022-2023

Yapp EHT et al (2025) Urban farming: the challenges of hydroponic and vertical farming in Malaysia. Cogent Food Agric 11(1). Available at: https://doi.org/10.1080/23311932.2024.2448601

Zhong D, Xia Z, Zhu Y, Duan J (2023) Overview of predictive maintenance based on digital twin technology. Heliyon 9(4):e14534

Digital Twins and AI: Synergizing Intelligence for Sustainable Farm Management

19

Om Prakash Suthar, Veena Parihar, Mukesh Kumar,
Neeraj Kumar Verma, and Ashok Kumar Kashyap

Abstract

Smart and sustainable agriculture is being transformed by the merging of Digital Twin and Artificial Intelligence (AI). This chapter reviews how several agricultural technologies come together to help in managing farms intelligently. It describes the key ideas of digital twins used in agriculture and explains how AI works to boost their prediction and decision-making. This chapter covers topics such as predicting crop yields, understanding soils, and looking after livestock assets, as well as methods, actual projects, and how they fare. Experts carefully examine the main difficulties in joining data, interconnecting systems, and accepting the new system. It further introduces the concept of future-minded digital twin frameworks that can operate on their own, creating a roadmap for building dependable and eco-friendly farming with the help of intelligent systems.

O. P. Suthar (✉)
Department of Computer Engineering, Marwadi University, Rajkot, Gujarat, India

V. Parihar
Symbiosis Institute of Geoinformatics, Symbiosis International Deemed University, Pune, Maharashtra, India

M. Kumar
Department of Computer Science, IILM University, Greater Noida, Uttar Pradesh, India

N. K. Verma
Department of Data Science, Manipal University, Jaipur, Rajasthan, India

A. K. Kashyap
Department of Computer Science, Himachal Pradesh University, Shimla, India

© The Author(s), under exclusive license to Springer Nature Singapore Pte Ltd. 2026

A. L. Srivastav et al. (eds.), *Digital Twin Technology for Sustainable Agriculture*,
https://doi.org/10.1007/978-981-95-5915-2_19

Keywords

Digital Twin Technology · Artificial Intelligence · Smart Agriculture · Sustainable Farm Management · IoT in Agriculture · Predictive Analytics · Precision Farming · Real-Time Monitoring

19.1 Introduction

19.1.1 Overview of Sustainable Agriculture and Farm Management

At the beginning of the twenty-first century, sustainable agriculture remained in its early stages as a collection of generally agreed practices or model farm economy. It was not just a thought but barely beginning to manifest itself as a real possibility other than industrial agriculture. Sustainable agriculture had its champions among small farmers, environmentalists, and a small select group of agricultural scientists as opposed to conventional farming methods, the foundation of which was based on the industrial production model and was given government subsidies. These were the pioneers who were aware of the destruction caused by industrial agriculture of the late twentieth century on major natural resources such as soil and water. To this, the ecological balance means of farming everywhere in the world began to be sought, and this quest is still going on.

Traditional farming systems especially, in the US, were based on high production and efficiency based on vertical integration and mechanization. Although this model has yielded large quantities of cheap food, it has come with its own price which is associated with the poor quality of soil and water, loss of biodiversity, increased reliance on fossil fuels, and a threatening concentration of land into fewer hands that undermined the rural communities. In fact, in this way, the biological complexity and social embeddedness of farming systems were not adequately considered. It viewed farms as factories and not living ecosystems.

In contrast, sustainable agriculture is grounded in ecological principles. Various alternative models like natural, organic, biodynamic, regenerative, and holistic farming have emerged globally. Sustainable agriculture aims to meet the society's present food and textile needs without compromising the ability of future generations to meet their own. It encompasses three key pillars: environmental health, economic profitability, and social and economic equity. The aim of sustainable agriculture is to follow nature's principles for developing systems to raise crops and livestock in a way that is self-sustaining, just like nature. These practices demonstrate that farming can work in harmony with nature. Moreover, ecological soundness alone does not make a system sustainable. Economic viability is equally critical. No agriculture is sustainable if it cannot generate sufficient income to support farming families and ensure a satisfactory quality of life. Sustainable agriculture also carries social values such as targeting rich lives for families, nutritious food for everyone, etc. Sustainable practices often align with small- and medium-scale, family-run farms that operate within local food systems. These farms

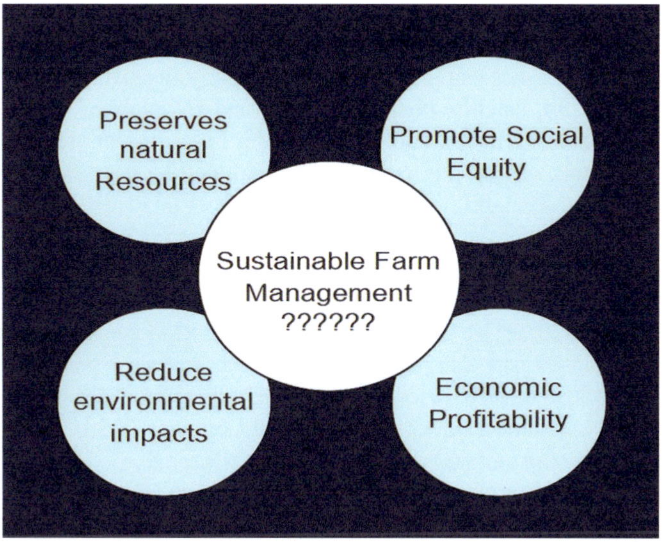

Fig. 19.1 Benefits of sustainable farming

typically engage directly with consumers through farmers' markets, community-supported agriculture, and local cooperatives. As these alternatives to industrial agriculture expand, the development of supportive markets remains a central challenge.

Sustainable agriculture emphasizes practices like crop rotation, organic farming, and efficient use of water, soil conservation, and biodiversity enhancement to overcome the problems associated with traditional farming, such as soil degradation, overuse of water, excessive use of chemical inputs, and loss of biodiversity. Currently, with the growing availability of digital tools and data-driven decision-making, there is a powerful opportunity to modernize sustainable agriculture using intelligent technologies. Digital Twin (DT) and Artificial Intelligence are now at the forefront of this transformation. They offer scalable solutions to manage resources efficiently, monitor environmental impact, and optimize yields, which enable sustainability not only from an ecological standpoint but also in terms of productivity and profitability. Figure 19.1 illustrates the broader benefits of sustainable farming.

19.1.2 Need for Intelligent Technologies in Modern Farming

The global agricultural sector is facing unprecedented challenges such as a rise in population, dwindling arable land, unpredictable weather due to climate change, and increasing demand for food security. The United Nations projects that the global population will reach 9.7 billion by 2050, necessitating a 70% increase in food production (United Nations Department of Economic and Social Affairs 2019).

Conventional methods of farming, however, are poorly placed to satisfy these increased demands without the overexploitation of natural resources.

These are urgent issues that intelligent technologies can solve. Such technologies are AI, IoT, robotics, satellite imaging, drones, and DT systems. All these are to facilitate data-driven, accurate, and effective farm management. The health of the soil, moisture levels, and crop growth can be monitored by real-time data collected by IoT sensors, and AI algorithms can analyze such data to identify pest infestations, disease outbreaks, or the most favorable time to harvest a crop. This reflects the increasing pace of the uptake of such technologies because, according to a report by MarketsandMarkets, the smart agriculture market will increase its value by USD 13.7 billion in 2020 to USD 22.0 billion in 2025 (MarketsandMarkets 2020). Moreover, McKinsey and Company emphasize that AI-based predictive analytics may save up to 30% of water used on irrigation and increase the accuracy of crop yield predictions by almost 80% (McKinsey and Company 2020). In India, efforts such as remote sensing and drone-based monitoring through the Digital Agriculture Mission by the Ministry of Agriculture reiterate how governments are proactively encouraging the use of smart agricultural systems. The technologies of DT, specifically, enable the creation of the virtual copies of farms, which include real-time data to model the behavior of crops in various conditions. This enables farmers to make proactive decisions and experiment without putting their tests to risk. The productivity of a pilot project in the Netherlands was increased by 20%, and the energy consumption was reduced by 15% (Ministry of Agriculture, and Farmers Welfare, Government of India 2021).

Given these benefits, intelligent technologies are not just useful but essential to transform modern farming into a resilient, efficient, and sustainable system. They bridge the gap between ecological responsibility and economic viability, helping build a smarter and more secure food future.

19.1.3 Scope and Objectives of the Chapter

This chapter explores the transformative potential of DT and AI technologies in enabling smart and sustainable agriculture. The specific objectives include the following:

- Presenting the principles of the sustainable development of agriculture and the digital transformation
- Explaining the architecture and operational mechanisms of DT in agriculture
- Detailing the application of various AI techniques in predictive analytics and automation
- Presenting integration models of DT and AI to build adaptive and autonomous agricultural systems
- Highlighting real-world examples, current challenges, and future directions

19.2 Related Work

Several studies have highlighted the potential of digital technologies in agriculture. Precision agriculture research has primarily focused on yield mapping, remote sensing, and variable rate application technologies. AI applications in agriculture have been explored in areas such as crop disease detection using convolutional neural networks, yield forecasting using regression and ensemble methods, and weed detection using computer vision.

19.2.1 Comparative Review—Traditional Digital Agriculture Versus Intelligent Agriculture

The evolution of agriculture technologies has moved progressively from traditional methods to digital agriculture and, more recently, to what is now termed Intelligent Agriculture (IA). The growing demand for increased productivity, resource efficiency, and climate resilience drives this shift. Understanding the distinctions between these paradigms is essential to appreciate the significance of emerging technologies like DT in agriculture. The following section presents the insights about traditional and intelligent agriculture.

19.2.1.1 Traditional Digital Agriculture

Traditional Digital Agriculture (TDA) encompasses the first wave of technology adoption in farming, which primarily involved digitizing manual tasks. It includes the use of GIS systems, remote sensing, basic sensor networks, farm management software, and satellite imagery. The core aim was data collection and monitoring, rather than advanced analytics or automation. The data usage is static and retrospective. In this process, usually, farmers collect data and analyze it manually and use insights with delay. So, the decision-making is human-driven, and technology is merely a supportive factor. The technology examples include GPS-based tractor guidance, yield monitors, weather forecasting apps, and satellite-based NDVI for crop health monitoring. The limitations for TDA are no real-time adaptation, low integration between devices, and minimal automation. According to the World Bank (2017), only 20–30% of smallholder farmers in low-income countries have access to any form of basic digital tools.

19.2.1.2 Intelligent Agriculture

Intelligent agriculture is the next frontier, integrating AI, IoT, robotics, ML, cloud computing, and DTs to create autonomous, adaptive, and predictive systems. It transforms farming into a data-intensive, real-time, and self-optimizing process. The data usage in IA is real-time, predictive, and actionable. The decision-making process is AI-/ML-driven, with minimal human intervention in repetitive or rule-based decisions. The examples include AI-based pest prediction and precision spraying, IoT-based irrigation and soil moisture sensors, drones with ML for weed

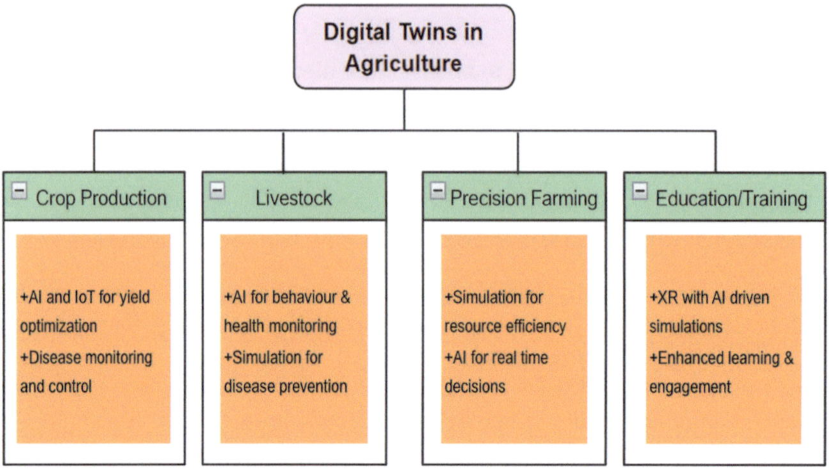

Fig. 19.2 Digital twins in agriculture

Table 19.1 TDA vs. IA

Aspect	Traditional digital agriculture	Intelligent agriculture
Decision-making	Manual, data-informed	Automated, AI-driven
Technology depth	Basic sensors, GPS, NDVI	AI, ML, DT, IoT, robotics
Data flow	Periodic, unidirectional	Real-time, bidirectional
Integration level	Low	High—multi source fusion
Adaptability	Limited	Dynamic and Adaptive
Cost efficiency	Moderate	Higher initial cost, better long-term ROI
Farmer's role	Central to all decisions	Supervisor and controller of intelligent systems

detection, and DTs replicating farm environments. Figure 19.2 represents what DTs can do in the agriculture domain.

The key features of IA are dynamic modeling and simulation, closed-loop feedback systems, and multisensory integration for holistic farm monitoring. According to FAO, there is up to 30% savings in water and fertilizer usage (FAO 2021) and 20–40% increase in crop yield due to precise decision support systems (McKinsey 2020). John Deere's connected combined harvesters now use ML algorithms to autonomously adjust threshing settings in real time, improving productivity by 10–15%.

Table 19.1 presents the comparative insights about both the approaches.

While IA offers significant advantages, its adoption remains uneven. Barriers include high initial investment, lack of technical skills, and interoperability issues across platforms and devices. The FAO (2022) notes that only 14% of farms globally use AI-based tools, highlighting the digital divide that must be bridged. The introduction of DTs represents a pivotal innovation within this intelligent ecosystem, offering real-time feedback, simulation, and enhanced precision in decision-making.

19.2.2 Digital Twins and AI in Agriculture

DT technology, initially developed for industrial systems, is now being applied to model biological and environmental systems in agriculture. Research on DTs in greenhouse management, for example, has demonstrated the benefits of real-time monitoring and simulation for microclimate control. Projects like Microsoft's FarmBeats and EU-funded SmartAgriHubs show the growing interest in integrating AI, IoT, and data platforms in agricultural innovation. Despite advancements, there remains a gap in holistic systems that integrate AI and DT to enable autonomous and sustainable farm management.

Table 19.2 presents the research work done in the field of agriculture with the integration of DTs.

The reviewed literature represents the transformative potential of DT technology in revolutionizing modern agriculture. In multiple spheres of activity, including crop production, livestock farming, precision agriculture, vertical farming, and agricultural education, DT develops as a key facilitator of data-driven decision-making, simulation, and automation. The combination of AI, IoT, XR, and genetic algorithms with the DTs proves the interdisciplinary convergence that the future-ready agricultural systems require. Research has demonstrated how DTs can enhance resource management, predicting crop yields and disease outbreaks, and personal farm management, providing a dynamic solution to threats such as changing climate, soil degradation, and food security.

Even with promising applications, a number of challenges still exist such as data integration complexities, cost barriers to small-scale farmers, and regulatory and infrastructure support. Nonetheless, the overall results of these investigations hint at a great direction toward the methodical implementation of DTs in farming, particularly in the case of compatibility with sustainability targets and precision farming operations.

19.2.3 Insights from Global Initiatives and Projects

Agricultural AI and DT integration is not a distant idea anymore but an emerging reality in different regions of the world. A number of international projects and county-scale programs are first to utilize these technologies in a bid to enhance agricultural productivity, resistance, and resilience. The section gives thematic insights of major international projects and their demonstrable effects on agriculture.

Table 19.2 Researches on integration of DT in agriculture

Authors	Technology	Contribution
Nie et al. (2022)	AI and DT	This paper investigates the role of AI and DT in promoting sustainable agriculture. It highlights AI's strength in precision farming and the challenges of data cost and examines DT's potential for real-time simulation and low-cost, high-precision farming. The authors discuss the synergy of AI and DT, current challenges, and future directions for intelligent, sustainable agricultural development.
Purcell and Neubauer (2023)	DT	This paper presents a systematic literature review of DTs in agriculture by highlighting how DT technology eliminates distinctions between state sensing, entity understanding, and physical automation through bi-directional data flows. It emphasizes real-time virtual representation combining data, modelling, and what-if simulation to enhance decision support and automation across diverse agricultural settings. It also outlines current trends and open questions and aims to raise awareness of DT's potential in digitizing agriculture.
Purcell et al. (2023)	DT	This paper reviews how DTs can address food security, land degradation, and climate change by enabling real-time data use, simulation, and decision-making. It also discusses risks and outlines a roadmap for sustainable DT adoption in agriculture.
Rogachev et al. (2022)	DT	This paper discusses the application of DTs in crop and livestock farming, outlining both opportunities and challenges. The paper emphasizes the need for systematic integration with legal, financial, and technological ecosystems to maximize their impact within Industry 4.0.
Escribà-Gelonch et al. (2024)	DT	This research work presents a review of DT applications throughout the agricultural lifecycle, which covers edaphic, phyto-technologic, postharvest, and infrastructure stages. Also highlights how real-time virtual replicas of crops enhance precision farming by optimizing irrigation, fertilization, and pest management. The discussions have been supported with case studies and address both benefits and challenges.
Shamia et al. (2023)	AI and DT	Discusses the integration of AI and DTs in precision agriculture for real-time crop monitoring, disease detection, and automated intervention. It emphasizes the application of robotic systems to the sensing of the environment and spraying specifically. It also deals with technical issues of integrating sensors, data processing, and automated decision-making schemes of future farming.

(continued)

Table 19.2 (continued)

Authors	Technology	Contribution
Pylianidis et al. (2021)	DT	The present paper explores the value and recent status of DT adoption in agriculture through a mixed-method approach. It examines 28 use cases in agriculture, contrasts with other industries and offers an incremental roadmap to wider adoption of DT in agriculture depending on complexity, benefits and readiness levels.
Dawn et al. (2023)	AI, DT and IoT	In the context of sustainable agriculture, this paper identifies the role of DT in the field of sustainable farming by uniting AI, IoT, and sensor data to form virtual farm replicas. It talks of the advantages such as higher yield and optimization of resource, particularly using low priced digital solutions, but has recognized obstacles such as high complexity of technology and cost to small scale farmers.
Neethirajan and Kemp (2021)	DT, AI and ML	The current paper will examine how DT is revolutionizing the livestock farming sector by allowing real-time tracking of animal behavior, health, and environmental status. It talks about uses of AI and simulation tools to identify the emotional state, precision farming, breeding predictions, and optimize infrastructure. The article highlights the possibility of the future and the need to carry out a cost-benefit analysis.
Nasirahmadi and Hensel (2022)	DT, IoT, AI, Big Data	This article evaluates the current condition of the DT systems in the agricultural sector, and proposes a system to be applied in soil, irrigation, robotics, farm machinery, and post-harvest processing. It emphasizes the combination of AI, IoT, and simulation to conduct real-time monitoring, data modeling, and to make better decisions to be more efficient and sustainable.
Spyrou et al. (2025)	DT, AI, Extended Reality (XR), Unity	This work proposes an AI-driven DT system within a gamified XR environment to modernize agricultural education. It demonstrates how VR with virtual educators can simulate vineyard management, support decision-making, and enhance engagement using Unity and Oculus Quest 3, based on stakeholder feedback in viticulture.
Manivasagam (2025)	DT, Sensor Integration	This paper reviews the adaptation of industrial DTs to agriculture. It highlights benefits like precision, real-time control, and self-regulating systems while discussing challenges like biological variability, data integration, and infrastructure needs. It offers strategies to address these for sustainable farming.
Ahmad (2025)	AI, DT, IoT, ML	This work explores DT systems with AI integration for hybrid crop selection and climate-resilient yield prediction. It highlights benefits in soil health, pest control, weather forecasting, and adaptive crop

(continued)

Table 19.2 (continued)

Authors	Technology	Contribution
		planning. It emphasizes challenges in data accuracy, cost, and scalability for equitable tech adoption in sustainable agriculture.
Alves et al. (2025)	DT, IoT, AI, Genetic Algorithm	This paper proposes an IoT-enabled vertical farming system using a DT and genetic algorithm to dynamically optimize RGB lighting for lettuce growth. It demonstrates reliable convergence to optimal spectral values, enabling adaptive light strategies and highlighting potential for scalability in controlled agriculture.

19.2.3.1 Microsoft FarmBeats—Enabling Data-Driven Smallholder Farming

The FarmBeats project of Microsoft is an example of how small-scale farmers can use low-cost technologies, such as AI, IoT, and edge computing, to engage in precision agriculture. The platform gathers sensors, drones, and satellite data that are then analyzed with AI models to give actionable information regarding soil moisture, plant conditions, and irrigation timetables (Chandra et al. 2020). The local edge processing of FarmBeats reduces the use of internet connectivity and cloud infrastructure, and therefore, smart farming is available even in isolated regions. India and Kenya have had field implementations that have shown up to 30% better use of water and 20–25% increase in predictable yields, thus enhancing food security and incomes of farmers (Microsoft Research 2021).

19.2.3.2 John Deere's Precision Agriculture—Intelligent Machinery and Crop Management

John Deere has been at the forefront of AI-enabled farm machinery development. Through technologies like See & Spray and Operations Center, the company integrates real-time data analytics and ML to automate functions like selective herbicide application and adaptive seeding (John Deere 2023). The See & Spray system uses computer vision and AI to detect weeds and apply herbicides only where needed, significantly reducing chemical usage by up to 77% (Gianessi 2022). These precision agriculture tools are aligned with sustainability goals and have shown notable economic returns, specifically for large-scale commercial farming operations across the US, Brazil, and Australia.

19.2.3.3 ICAR-NIASM-India—Climate-Smart Agriculture for Weather Resilience

The Indian Council of Agricultural Research–National Institute for Abiotic Stress Management (ICAR-NIASM) is driving innovation in climate-resilient farming by integrating AI and climate modeling into crop planning. Under India's National Mission for Sustainable Agriculture, NIASM employs digital twins to simulate crop responses under varying climate scenarios, helping farmers make better choices in drought-prone and flood-prone areas (Ramesh et al. 2021). Pilot implementations in Maharashtra and Andhra Pradesh indicate improved resilience planning and yield optimization by up to 18% for crops like groundnut, pearl millet, and cotton under stress conditions.

19.2.3.4 Japan's Society 5.0—National Vision for Smart Agriculture

Japan's ambitious Society 5.0 framework outlines a comprehensive integration of cyber-physical systems in all sectors, including agriculture. The Smart Agriculture Demonstration Project by the Ministry of Agriculture, Forestry and Fisheries (MAFF) uses digital twins, AI, drones, and robotics to automate rice farming in Hokkaido and Niigata. These initiatives have led to the development of autonomous tractors, AI-based disease prediction systems, and data-driven irrigation platforms. Research shows that autonomous rice cultivation can reduce labor costs by over 50% and maintain or even enhance crop quality (Ministry of Agriculture, Forestry and Fisheries (Japan) 2022). The Society 5.0 vision positions agriculture as a testbed for next-generation digital integration at a societal scale.

19.2.3.5 The Netherlands—Digital Greenhouses and Horticulture Innovation

The Netherlands, with its globally acclaimed greenhouse agriculture, is advancing its practices through the Wageningen UR's Digital Greenhouse initiative. Using AI models and digital twin frameworks, real-time simulation of microclimates inside greenhouses enables better decision-making for irrigation, ventilation, and nutrient supply. The Autonomous Greenhouse Challenge, organized by Wageningen University in collaboration with Tencent, has shown how AI-enabled greenhouse management can match or outperform human growers in terms of yield and resource use (Wageningen University and Research 2020). Dutch greenhouses yield four times more with data-driven operations than the world averages on less than 90% of water.

Significantly, these global projects do not only demonstrate the actual advantages of applying AI and DTs to agriculture but also mark a shift in paradigm toward intelligent, sustainable, and resilient farms. The effectiveness of such projects is based on the fact that they localize advanced technologies to certain agro-ecological and socio-economic conditions. These examples include smallholder farms in India and Africa and the large-scale, mechanized agriculture in the US and Japan, which proves that smart agricultural ecosystems powered by data simulation and automation have the potential to transform the world.

19.3 Key Concepts of Agri-Digital Twin

The introduction of DT into the agricultural sector is a radical change in the way the farm processes, environmental factors, and crop processes are modeled, tracked, and optimized. A DT is in essence a virtual, that is, a real-time, data-driven

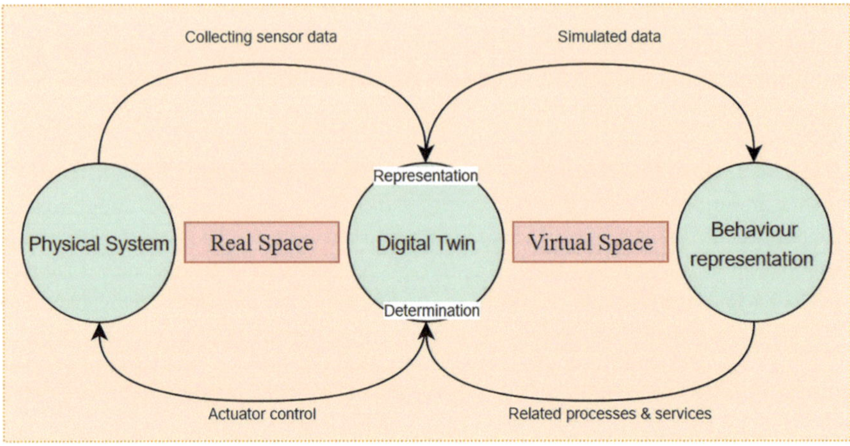

Fig. 19.3 Digital twin's working in agriculture

representation of a physical farming harvest or a prototype, be it a crop field, greenhouse, livestock system, or machine.

With the combination of AI and IoT, DTs enable farmers to simulate performance, predict anomalies, and optimize performance by using data analytics to bridge the world of the physical and technology worlds. DT represents a computerized equivalent of a farm asset that constantly adjusts its position on the basis of real-time data feeds like sensor, satellite, drone, and user feedback. This makes it possible to simulate dynamically, model predictively, and support decision making. In agriculture, the operation of DT is as shown in Fig. 19.3.

A multi-layered architecture, which combines physical systems, data acquisition, modeling and simulation engines, analytics, and user interaction interfaces, constitutes the architecture of DT in agriculture. This architecture guarantees synchronization in real time between the virtual and physical farming environments. The ultimate aim is to develop an active and smart system which can monitor, simulate, predict, and optimize the farming processes.

19.4 Implementation Framework

Deployment of AI-enabled digital twins in sustainable agriculture must be accompanied by a modular framework of implementation. These involve selecting the system topology, data sources that can be trusted, and communication channels and validating an efficient AI model development and deployment pipeline. Figure 19.4 represents the layered architecture of a digital twin in sustainable agriculture.

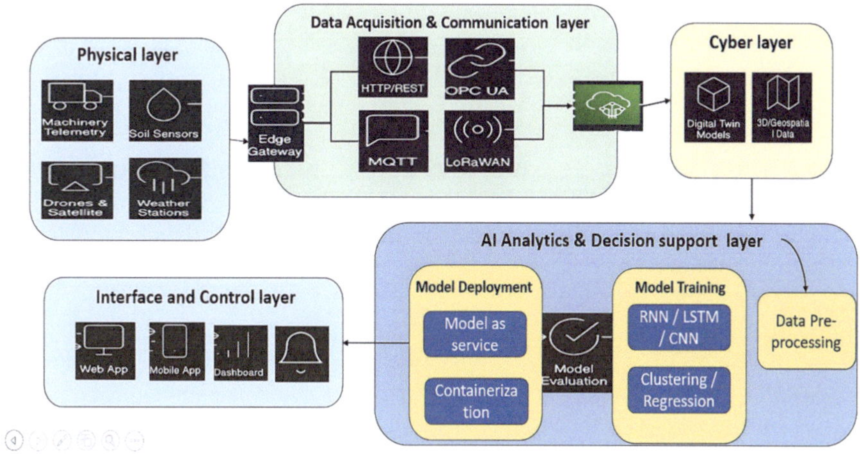

Fig. 19.4 AI-driven architecture of digital twin in sustainable agriculture

19.4.1 Architecture of AI-Driven Digital Twin System

An AI-driven digital twin system in agriculture usually has a layered architecture that comprises the following layers:

1. *Physical Layer*: It includes all the physical components, such as crops, soil, livestock, irrigation systems, drones, and machinery. The real-time data is being recorded with the help of sensors and the IoT devices embedded into it.
2. *Data Acquisition and Communication Layer*: It represents the backbone as it controls the constant data stream between the physical devices and the cyber infrastructure. Edge gadgets and gateways process information before sending it to on-premises or cloud servers.
3. *Cyber Layer (Digital Twin Layer)*: In this layer, the real-time data is utilized to generate digital representations of the physical objects. These models mirror the bio, environment, and operational status of the farm. It makes use of 3D models, geospatial data, and models that we can interact with.
4. *Central Intelligence Layer*: This is known as the AI analytics and decision support layer. It incorporates machine learning models, optimization procedures, and reasoning systems that process the digital twin's data to predict, identify anomalies, and give prescriptive suggestions.
5. *Interface and Control Layer*: This layer has dashboards, application programming interfaces, and mobile and web apps that let us see and interact with data. Agronomists and farmers are able to monitor the current states of the systems in real time and receive warnings or tips.

This architecture presents interoperability, scalability, and modularity, making it flexible to be customized in accordance with the nature of different farms in size, crops, or species of animals (Tao et al. 2021).

19.4.2 Data Sources, Sensors, and Communication Protocols

A digital twin is fed off information, and the quality thereof has a direct correlation to the performance of AI models. The following are the most common data sources and types of sensors incorporated:

- *Soil Sensors*: Detect moisture, temperature, pH, salinity, and nutrients (NPK)
- *Weather Stations*: Gather real-time information about precipitation, humidity, wind speed, and solar radiation
- *Satellite and Drone Images*: Check the health of crops (NDVI, EVI), how much of the field is covered, and signs of pests and diseases
- *Livestock Trackers*: Track position, movement, feeding patterns, and body temperature
- *Machinery Telemetry*: Monitors fuel consumption, hours worked, and planned maintenance of tractors and machinery

Standardized forms of communication are used in order to facilitate easy integration:

- *MQTT (Message Queuing Telemetry Transport)*: Lightweight protocol that is suitable for IoT devices.
- *LoRaWAN (Long-Range Wide-Area Network)*: A protocol with low power that can be used in distant fields.
- *HTTP/HTTPS/REST APIs*: Communicate over the web with cloud-based services.
- *OPC UA (Open Platform Communications Unified Architecture)*: This framework is used in a lot of industrial automation systems and makes it easier for them to work together.

Raw data are agglomerated and handled by computer resources on edge networking or cloud like AWS IoT Core, Microsoft Azure FarmBeats, or open sources like FIWARE.

19.4.3 AI Model Training and Integration Pipeline

The smartness of a digital twin relies much on the training and continual refreshment of AI models. The pipeline of AI usually has the following steps:

1. *Preprocessing of Data*: Data is cleaned, normalized, and engineered where appropriate to prepare sensor data and historical data to be used in training the model. Class imbalance can be addressed using techniques such as SMOTE in classification tasks.
2. *Choosing and Training a Model*: Select suitable algorithms based on the specific task (e.g., predicting crop yield, finding anomalies):
   - Using regression (like Random Forest or XGBoost) to figure out the yield
   - Image-based CNNs in plant disease detection
   - Using RNNs and LSTMs to predict time series (like soil moisture trends)
   - Crop zoning and soil classification using clustering (e.g., k-means and DBSCAN)
3. *Model Testing*: It is tested to ensure its performance by conventional assessments, i.e., RMSE, MAE, F1-score, and confusion matrix. Cross-validation assures the stability during various seasons or conditions on the farm.
4. *Deployment and Integration*:
   - Containerization: Portability is guaranteed through the deployment of models in the form of Docker or Kubernetes.
   - Model as a Service (MaaS): Digital twin platform may make calls to the exposed AI services when required using the RESTful APIs.
   - Continuous Learning: Retraining with new data can be used based on feedback, and the model may be adjusted to seasonal change or any other changes.
5. *Useability and Presentation* AI knowledge is delivered to the end-user as a dashboard or a custom-built mobile app through tools such as Grafana or Power BI or bespoke web interfaces.

This pipeline ensures the precision and responsiveness of AI models and their adaptability and strength to complicated environment shifts that agricultural ecosystems experience.

19.5 Applications in Sustainable Farm Management

The introduction of digital twins to an AI tech environment is likely to make a groundbreaking change in sustainable farming management. These systems have the ability to make accuracy decisions, real-time monitoring, and predictive analysis due to the capability to model the agricultural ecosystems at a virtual level. The key areas of application include crop forecasting, soil monitoring, animal management, and climate-responsive irrigation plans.

19.5.1 Crop Growth Forecasting and Yield Prediction

Crop growth modeling and crop yield forecast are the critical components of the modern farming strategy that enables farmers to make a decision regarding resource utilization, crop planning, and sustainability. Digital twins find application in the simulation of plant growth conditions based on real-time sensor data (e.g., temperature, humidity, soil moisture, and light intensity) and AI technologies, in particular, convolutional neural networks (CNNs) and recurrent neural networks (RNNs) (Li et al. 2021).

In the context of a digital twin platform, crop yield predictions are highly accurate compared to the general application of machine learning algorithms, following the analysis of historical yields, meteorological data, phenological growth cycles, and satellite images. As an example, deep learning has been effectively used to evaluate NDVI (Normalized Difference Vegetation Index) trends in the remote sensing data that are subsequently correlated to biomass predictions and their results in terms of yield (Lu et al. 2021). In addition, crop-specific digital twins enable the simulation of various agriculture conditions, which will make it possible to optimize plantations and fertilizer schedules.

19.5.2 Soil Health and Fertility Monitoring

Land, or soil, is one of the crucial but not enough assets in farming. Noninvasive soil health monitoring is possible when incorporating the digital twins with the analytics-based AI. Recent IoT sensors detect real-time variables such as pH, nutrients (N, P, K), electrical conductivity (EC), moisture content, and compaction. The data is used with AI algorithms to distinguish between soil fertility and early predictions of soil erosion or nutrient shortage (Panpatte and Jhala 2019).

There is also the possibility of advanced analytics being able to simulate nutrient cycling processes, where farmers can actually test the results of various plans of fertilization virtually, prior to them being put into practice. Support vector machines (SVMs) and random forest classifiers are commonly used in the mapping of fertility and soil classification (Singh et al. 2022). These tools make sustainability possible by reducing overfertilization, the most common degrading factor to the environment caused by nutrient spreading to the environment and the production of greenhouse gases.

19.5.3 Livestock Behavior and Productivity Optimization

In digital livestock systems, digital twins are also being developed, based on the animal profile, including parameters such as feed intake, weight gain, activity levels, reproductive status, and health records. When these computer-generated personalities are linked to AI algorithms, they can forecast health hazards and behavioral deviations and maximize feeding hours and breeding periods (Schlageter Tello et al. 2014).

An example is gait, posture, and cow-to-cow interactions that computer vision (CV) and DL models can track to determine the early signs of lameness or distress in cows. In addition, predictive models are being applied also to provide answers on ways to enhance the milk or meat quality out of the past productivity. The same realizations can also ensure the economic performance and well-being of the animals, which can be integrated into the concepts of ethical and sustainable farming (Tschopp et al. 2022).

19.5.4 Climate-Smart Irrigation and Resource Utilization

The shortage of water and climate change require water and farm resources' efficient use. Digital twins combined with AI are utilized in climate-smart agriculture through optimization of irrigation times with predictive models of evapotranspiration, precipitation predictions, and soil water balance. To get the most out of irrigation, reinforcement learning and neural networks are commonly employed to master the best strategy in the long term (Ferentinos 2018).

The involved systems are able to make real-time adjustments to water supply depending on the weather conditions and water stress of the plants, and in doing so, promote reduction of water usage without compromising the health of the crops. Further, digital twin platforms include resource management functions, which model various patterns of energy and input consumption and direct the farmer toward more eco-friendly usage. In pilot studies, there has been a reduction of up to 30% of water use without relating to the dropped crop yields with such systems (Wolfert et al. 2017).

19.6 Case Studies and Use Cases

Digital twins powered by AI are revamping Indian agriculture, as they allow precision farming, resource optimization, and help address such issues as climate variability and scarcity of resources. Two Indian case studies of smart farms based on digital twins and AI are introduced regarding real-life adoption, quantitative value, key performance indicators, lessons, and general moments. The presented examples shed light on the realistic role of digital twin technology in Indian agriculture, and they can be used by the stakeholders that seek to strengthen the sustainability and productivity of the Indian agriculture domain.

19.6.1 Case Study 1: Saagu Baagu Initiative in Telangana, India

In Telangana, the World Economic Forum-led Saagu Baagu project in the Khammam district introduced an AI-based digital twin of a chili cultivation system to farmers with the support of the state government. The system combined IoT soil sensors, drone imagery, and satellite data to develop a digital analog of chili fields, including the moisture and nutrient contents and health indices of the crop (e.g., NDVI). Alerting models that used quality testing through neural networks and yielded predictions through regression algorithms were able to provide relevant actions to farmers online through a mobile app and program. This project, which was later increased to 500,000 farmers in 10 districts by October 2023, had a digital marketplace as well, and this was meant to facilitate market access by linking farmers to buyers.

19.6.1.1 Quantified Benefits and Performance Metrics

The Saagu Baagu campaign brought considerable results, as it was reported in 2024:

- *Yield Improvement*: 21% increase in chili yields because of good use of irrigation and pesticides.
- *Income Increase*: Farmers increased their income by twofolds, and the prices at which they sold their products increased by 8% because of the AI-based quality ratings and market connections.
- *Efficiency of Resource*: A 9% reduction in pesticide use and a 5% reduction in fertilizer use were implemented to lessen the environmental impact.
- *Model Accuracy*: Neural networks reached an accuracy rate of 95% in determining the existence of crop diseases, as was supported by other competencies such as the F1 score (0.93).

These indicators were compared to those of the conventional form of farming in the area to show how the digital twin will improve productivity and sustainability, as shown in Fig. 19.5.

19.6.1.2 Learnings and Key Takeaways

- *Data Integration is Critical*: The IoT data, drone data, and satellite data were combined, requiring strong preprocessing to standardize them for joint use; therefore, standards such as MQTT or LoRaWAN should be implemented.

Fig. 19.5 AI-driven Saagu Baagu project

- *Engaging Farmers Triggers Success*: Workshops and the use of the vernacular language through mobile apps attracted the smallholder farmers to adopt, resolving digital literacy closure.
- *Infrastructure Barriers*: Low connectivity in most of the rural areas initially hindered uploading of real-time data, and hence there is a necessity concerning investment in digital infrastructure.

Key Takeaway Digital twins have the potential to dramatically increase yields and incomes for smallholder farmers, accompanied by readily available interfaces, proper data pipelines, and government-supported programs, but system upgrades are essential to scaling up.

19.6.2 Case Study 2: Cropin's SmartFarm Plus in Punjab, India

Bengaluru-based agritech startup Cropin implemented their SmartFarm Plus platform in one of the wheat and rice production areas of the Punjab state to assist it with an AI-powered digital twin solution, as shown in Fig. 19.6. The information collected by GPS-enabled smartphones, satellite images, weather stations, and IoT sensors in soil was merged into the platform to construct the virtual copies of fields, and the copied field matched soil conditions, vegetative development processes, and environmental parameters. Random Forest-based models were used to estimate yields, whereas CNN-based models were used to detect pests and deliver custom recommendations on irrigation, fertilization, and pest control. Access was provided to the farmers by mobile application in local languages in which the farmer could make real-time decisions. Carbon sequestration monitoring was also supported by the system and fits the sustainability agenda. In 2024, IndiaAI presented a case regarding the effectiveness of the mobile application for farmers.

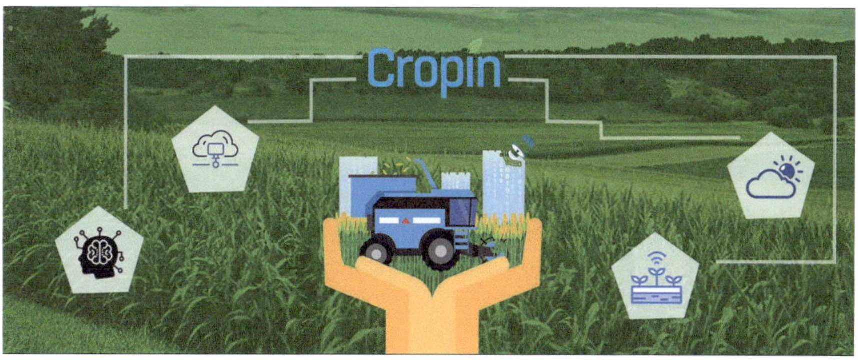

Fig. 19.6 AI-driven Cropin project

19.6.2.1 Quantified Benefits and Performance Metrics

According to the reports of 2024, Cropin SmartFarm Plus showed its quantifiable outcomes:

- *Improved Yield*: A 15–20% increase in wheat and rice yield as a result of the best use of inputs.
- *Resource Saving*: A 30% saving on the use of water by means of AI-controlled irrigation patterns, in line with the scheme Per Drop More Crop.
- *Cost Efficiency*: Reducing the cost of input to the farms (fertilizers and pesticides) by 10% using the power of accuracy.
- *Sustainability Impact*: A 12% reduction in greenhouse gas emission per farm because of reduced use of chemicals and carbon costs depending on the sequestration practice.

To determine the performance measures, RMSE = 0.9 tons/acre of yield forecast and 96% accuracy of the pest detection that was confirmed by cross-seasonal data.

19.6.2.2 Learnings and Key Takeaways

- *Utility of Data Quality*: High-quality satellite images and calibrated sensors were also important in precise digital twins which should also be regularly maintained to ensure that they do not enter into the state of data drift.
- *Scalability on Partnership*: The government schemes like PM Kisan Samman Nidhi were welcomed, yet smallholder farmers were still in need of subsidies to facilitate start-up expenditure.
- *Cultural Adaptation*: There were insights in their own languages and user-friendly interfaces that made the farmers more interested and confident.

Key Takeaway Digital twins can achieve the goal of making sustainable farming a reality due to the combination of a wide range of data sources and government funding, yet affordability and availability are critical to their use and implementation by the smallholder farmers of India.

19.7 Challenges and Limitations

Using AI-assisted digital twins in the agricultural sphere will transform the system of sustainable management of farms radically, as it will make it possible to base decisions on data and optimal resource distribution. Yet, they are accompanied by a number of issues and constraints that obstruct their further implementation and performance. Three essential areas are addressed in the section, including data availability and quality challenges, interoperability and infrastructure limits, and ethical, privacy, and adoption obstacles.

19.7.1 Data Availability and Quality Issues

The quality of data is the backbone of the proper digital twins, and AI models are quality-dependent on accurate, timely, and complete datasets. Depending on the applications, the lack of availability of data in the agriculture sector is commonly associated with disadvantaged sensor distribution, especially in small-scale or resource-limited farms. As an example, soil sensors that detect moisture, pH, or nutrients (NPK) might not be available in some remote locations, in which case their digital copies would be incomplete. Equally, satellite imagery, useful as it is to crop health indices, such as the NDVI, may not provide the necessary granularity required to make accurate farm-related decisions because of resolution challenges or clouds.

Poor model performance can be reached through inconsistent sensor calibration, noise of environmental factors, or missing data because of connection failure. In an example, inaccurate calibration of a soil moisture sensor can distort the recommendations of irrigation, minimizing the trustworthiness of the digital twin. Some of the problems can be addressed through data preprocessing methods, like normalization or outlier detection, which are, though, technically demanding and resource-intensive, thus not available to every farmer. In addition, historical data, which is vital in training powerful AI models such as Random Forest or LSTMs, may not be sufficient in new crops or places that do not have such records, hence the inability to make accurate predictions.

19.7.2 Interoperability and Infrastructure Constraints

Digital twins have a complex architecture, including five abstraction layers: physical, data acquisition, cyber, AI analytics, and interface layers, placing higher requirements on the interoperability between various devices, communication protocols, and platforms. In agriculture, attainment is, however, difficult because of disintegrated ecosystems. The device in one vendor can change the third vendor-compatible protocols, and this interferes with data integration, e.g., MQTT versus OPC UA. Consider a farmer with LoRaWAN-protocol-based soil sensors and HTTP-based weather stations and failing to collect data to a single digital twin platform.

These are worsened by infrastructural limitations, especially in rural areas where access to the internet is low. Such networks have a low bandwidth, which limits the sending of real-time data to cloud services such as AWS IoT Core or Microsoft Azure FarmBeats, therefore slowing the updating of digital twins. Edge computing may help ease the problems of latency, but implementation of edge devices is cost-intensive and therefore may not be feasible for smallholder farmers. There is also the computational intensity of the AI models, like CNNs in disease or models of clustering to define soil zoning, which cause additional strain on available infrastructure in resource-poor environments. The issue of scalability does exist, as the systems that are designed with large farms in mind might not adjust that well to the smaller farms without the added expense of customization.

19.7.3 Ethical, Privacy, and Adoption Barriers

The deployment issues and ethical or privacy concerns related to AI-powered digital twins may hinder their adoption. The information that can be gathered by means of sensors, drones, or livestock trackers may be rather sensitive, and it may refer to the location of farms, production levels, or working practices. Unless this data is encrypted or accessors are guarded, this information could be revealed to unauthorized parties, who may use it to their own advantage or disadvantage. As an example, a leakage of a cloud-hosted digital twin platform would expose the exclusive cultivation methods of a farmer.

The question of ethics in decision-making with the help of AI also appears. The excessive use of prescriptive guidelines provided by digital twins can make the farmers deskilled and less autonomous and knowledgeable about their traditional practices. There are also chances of inequities, as biased AI models, which are trained on datasets that only take into consideration large-scale farmers or even specific types of crops, can result in recommendations that benefit other farmers or even other regions.

There is a serious barrier to adoption among people with low technical literacy or the generally untrained people in the case of farmers. The interface layer can be overly complex to a nontechnical user since the dashboards and application program interfaces are complicated, and sensors, cloud subscriptions, and maintenance are expensive to invest in. Adoption is also hampered by cultural opposition to technology, particularly in rural agricultural societies. A farmer who is used to making decisions in regard to manual irrigation might not be ready to follow the AI-based advice, though a digital twin supported by real-time data could help with it.

19.8 Future Directions and Research Opportunities

The combination of AI-driven digital twins and agricultural management is destined to change the sustainable management of farms, yet there is much research and development required to realize these future possibilities. This chapter presents three areas worth following up on in the future: autonomous digital twins based on reinforcement learning, designing large-scale-capable agricultural systems, and policy, educational, and standardization requirements. These guidelines respond to technical, operational, and social problems to make way for sustainable, productive, and equal systems of farming.

19.8.1 Autonomous Digital Twins with Reinforcement Learning

The next step in agricultural intelligence is autonomous digital twins (AG farmer) that apply reinforcement learning (RL) to make systems optimize themselves. In contrast to the classical models of AI (e.g., Random Forests or CNNs) that operate on a set of fixed training data, RL-based digital twins can discover how to perform best in any farm environment based on a trial-and-error process. As an example, an RL agent may be used to optimally dynamically set irrigation schedules based on real-time soil moisture, weather forecasts, and crop growth phases to optimize yield over the lowest amount of water usage. The method is similar to real-time decision-making; the digital twins are able to respond to dynamic environments like a sudden onset of pests or a change in climate.

A research opportunity is placing RL algorithms into limited resource environments presented in the farms that can strike a trade-off between exploration (testing novel strategies) and exploitation (using the known optimal actions). Such techniques as deep RL (using both neural networks and RL) have the potential to model more complex interactions, like how to best apply fertilizer to varying types of soil. Nonetheless, such issues as high computational expenses and the necessity to train RL agents by using large amounts of real-world data should be overcome. New simulations to train the simulations through virtual environments, which include digital twins, may make physical data less relied on and allow scalable autonomy.

19.8.2 Scalable Architectures for Large-Scale Agriculture

An effective design of digital twin architectures must be modular, cost-effective, and robust in order to be applied on a large scale over vast areas (thousands of hectares) or a wide range of crops. Existing architectures, which are based on cloud platforms or edge devices, cannot be employed in localities that have poor connections or used by smallholder farmers who cannot afford high costs. Scalable solutions need to combine low-power IoT standards (such as LoRaWAN) with edge processing and hybrid cloud-edge architectures so that the data can be processed in real time and easily afforded.

The latency and cost-effective opportunities can be explored through the design of distributed architecture. As an example, federated learning may allow training AI models on farm equipment, rather than sending all the information to the cloud, ensuring data security and limiting it. The sharing of data through blockchain may improve the interoperability between farms so that they would be able to share digital twins of crops regionally. These issues are compatible with heterogeneous devices used in IoT and the computational complexity of artificial intelligence models, such as LSTMs, for time series predictions. To enable access to the digital twins by the large and small farms (i.e., democratizing access), it might be necessary to come up with the lightweight models and open-source platforms, such as FIWARE.

19.8.3 Policy, Education, and Standardization Needs

The key to the large-scale application of the AI-based digital twins in agriculture is the availability of favorable policies, affordable education, and standardized frameworks. The governments must develop policies to subsidize the implementation of sensors and digital infrastructure on the rural populations, which erases and clears the cost barriers of smallholder farmers. Education programs, which will address the needs of different literacy levels, are needed to train farmers in the usage of the digital twin interfaces, i.e., display dashboards or mobile applications. It is also relevant to standardize data format and communication protocols (such as MQTT and OPC UA) in order to simplify interactions between devices and platforms and make agritech ecosystems less fragmented.

The possible areas of research include framework development in the field of ethical AI application, ensuring the privacy of data, and avoiding bias in recommendations. As an example, one may mention a policy according to which the open data governance method should be adopted to protect farmers against theft of sensitive information, e.g., yield or location data. The digital literacy gaps may be bridged by creating training modules that are easy to understand and follow multiple languages. The new challenges are that there is a need to align international standards with domestic agricultural operations and to contribute funds to build the infrastructure. Standardization and education can be supported by the collaborative work of governments, academia, and industry and lead to equitable adoption.

The insights of this chapter are significant as they demonstrate how digital twin and artificial intelligence technologies can fundamentally change the way farms are managed—making agriculture more resilient, productive and environmentally sustainable in light of growing global food demand and climate uncertainty (McKinsey and Company 2020; Tao et al. 2021). This chapter demonstrates how intelligent systems can facilitate real-time monitoring, precise forecasting, and data-driven decision-making regarding various aspects, including soil health and market access, while also addressing the pursuit of long-term sustainability and food security for both large- and small-scale agriculture (Dawn et al. 2023; Chandra et al. 2020).

19.9 Conclusion

The integration of digital twins (DT) and artificial intelligence (AI) in sustainable agriculture signifies a pivotal moment for the agricultural industry, presenting the opportunity for data-driven, adaptive, and resilient food systems (Tao et al. 2021; Ahmad 2025). The true potential of this synergy lies not only in technical optimization but also in the widespread adoption of innovative farming for farms of all sizes. However, policymakers and innovators, to really make the world better, need to deal with important problems like high costs, lack of interoperability, and the digital divide. This will make sure that tiny farmers and marginalized communities can use the technology fairly (Dawn et al. 2023). To build trust and digital literacy in rural areas, these systems need strong policies, money spent on technological infrastructure, standardized protocols, and education programs that include everyone (Escribà-Gelonch et al. 2024; Manivasagam 2025).

We require a comprehensive research agenda focused on self-regulating digital twins, flexible architectures, and ethical frameworks for data governance and AI openness (Nie et al. 2022; Purcell and Neubauer 2023). Such government-specific yet multi-stakeholder partnership can pave the way for faster development and implementation of modular, low-cost platforms-as-a-service, which can create new opportunities for climate change adaptation, resource circularity, and on-farm innovation. By embracing technological and social aspects of these innovations, the agricultural sector can be responsive to future challenges while looking toward a more sustainable, productive, and equitable global food system (Singh et al. 2022; McKinsey and Company 2020).

References

Ahmad S (2025) AI-driven digital twin models for hybrid crop selection and climate-resilient yield prediction. Sci Res Timelines J 3(1):15–18. Available at: https://scientificrtj.com/index.php/srtj/article/view/47

Alves RG, Lima F, Guedes ÍMR, Gimenez SP (2025) Dynamic light optimization in vertical farming using an IoT-driven digital twin framework and artificial intelligence. Appl Soft Comput 174:112985. https://doi.org/10.1016/j.asoc.2025.112985

Chandra R et al (2020) FarmBeats: an IoT platform for data-driven agriculture. Microsoft Research

Dawn N, Ghosh S, Saha A, Chatterjee S, Ghosh T, Guha S, Sarkar S, Mukherjee P, Sanyal T (2023) A review on digital twins technology: a new frontier in agriculture. Artif Intell Appl 2(4): 250–262. https://doi.org/10.47852/bonviewaia3202919

Escribà-Gelonch M, Liang S, van Schalkwyk P, Fisk I, Long NVD, Hessel V (2024) Digital twins in agriculture: orchestration and applications. J Agric Food Chem 72(19):10737–10752. https://doi.org/10.1021/acs.jafc.4c01934

Ferentinos A (2018) Deep learning models for predicting irrigation needs in smart farming. Comput Electron Agric 154:287–294. https://doi.org/10.1016/j.compag.2018.09.025

FAO (2021) The state of food and agriculture 2021. Making agrifood systems more resilient to shocks and stresses. Rome, FAO. https://doi.org/10.4060/cb4476en. https://openknowledge.fao.org/server/api/core/bitstreams/fe920b40-aa2c-45e1-9ea8-8db7c2548c12/content

Gianessi L (2022) The economic benefits of precision weed control. CropLife Foundation

John Deere (2023) See & Spray™ technology overview. John Deere Official Website

Li Z, Liu J, Wang Y (2021) A digital twin driven approach for smart agriculture. IEEE Access 9: 42675–42683

Lu C, Zhang S, Li M (2021) Crop yield prediction using deep neural networks and satellite data. Comput Electron Agric 180:105876. https://doi.org/10.1016/j.compag.2020.105876

Manivasagam VS (2025) From bytes to farm: transferability of industrial digital twins in agricultural systems. J Biosyst Eng 50:130–144. https://doi.org/10.1007/s42853-025-00256-1

MarketsandMarkets (2020) Smart agriculture market by farming type, offering, application, and region - Global Forecast to 2025. Available at: https://www.marketsandmarkets.com/Market-Reports/smart-agriculture-market-239736790.html

McKinsey & Company (2020) Agriculture's connected future: how technology can yield new growth. Available at: https://www.mckinsey.com/industries/agriculture/our-insights/agricultures-connected-future-how-technology-can-yield-new-growth

Microsoft Research (2021) FarmBeats projects and pilots

Ministry of Agriculture & Farmers Welfare, Government of India (2021) Digital Agriculture Mission 2021–2025. Available at: https://agricoop.nic.in

Ministry of Agriculture, Forestry and Fisheries (Japan) (2022) Smart Agriculture Roadmap under Society 5.0

Nasirahmadi A, Hensel O (2022) Toward the next generation of digitalization in agriculture based on digital twin paradigm. Sensors (Basel) 22(2):498. https://doi.org/10.3390/s22020498

Neethirajan S, Kemp B (2021) Digital twins in livestock farming. Animals 11(4):1008. https://doi.org/10.3390/ani11041008

Nie J, Wang Y, Li Y, Chao X (2022) Artificial intelligence and digital twins in sustainable agriculture and forestry: a survey. Turk J Agric For 46(5):642–661. https://doi.org/10.55730/1300-011x.3033

Panpatte M, Jhala Y (2019) Soil fertility and management. In: Soil science: an introduction. Springer, pp 135–160

Purcell W, Neubauer T (2023) Digital twins in agriculture: a state-of-the-art review. Smart Agricultural Technology 3:100094. https://doi.org/10.1016/j.atech.2022.100094

Purcell W, Neubauer T, Mallinger K (2023) Digital twins in agriculture: challenges and opportunities for environmental sustainability. Curr Opin Environ Sustain 61:101252. https://doi.org/10.1016/j.cosust.2022.101252

Pylianidis C, Osinga S, Athanasiadis IN (2021) Introducing digital twins to agriculture. Comput Electron Agric 184:105942. https://doi.org/10.1016/j.compag.2020.105942

Ramesh K et al (2021) Climate smart agriculture in India. ICAR-NIASM

Rogachev AF, Skiter NN, Ketko NV, Simonov AB, Makarevich IV (2022) Digital twins as a tool for systemic integration of innovative digital technologies in agriculture. IOP Conf Ser Earth Environ Sci 1069(1):012042. https://doi.org/10.1088/1755-1315/1069/1/012042

Schlageter Tello A et al (2014) Automated lameness detection in dairy cattle using a combined approach of gait analysis and machine learning. J Dairy Sci 97(9):5592–5601

Shamia D, Suganyadevi S, Satheeswaran V, Balasamy K (2023) Digital twins in precision agriculture monitoring using artificial intelligence. In: Digital twin for smart manufacturing. Elsevier, pp 243–265. https://doi.org/10.1016/b978-0-323-99205-3.00004-3

Singh SJ et al (2022) Machine learning techniques for soil fertility mapping: a case study. Environ Monit Assess 194(3). Article 3.A

Spyrou O, Ariza-Sentís M, Vélez S (2025) Enhancing education in agriculture via XR-based digital twins: a novel approach for the next generation. Appl Syst Innov 8(2):38. https://doi.org/10.3390/asi8020038

Tao Y, Jin M, Zhang L (2021) Digital twin and AI integration in smart agriculture. IEEE Internet Things J 8(16):12950–12960. https://doi.org/10.1109/JIOT.2021.3050849

Tschopp JH, Stoyanov M, Anastasov K (2022) Digital twins in animal farming: enhancing welfare and productivity. Smart Agric Technol 3:100045

United Nations Department of Economic and Social Affairs (2019) World population prospects 2019. Available at: https://population.un.org/wpp/

Wageningen University & Research (2020) Autonomous greenhouse challenge report. Available at: https://www.wur.nl/en

World Bank Annual Report (2017) (English) Washington, D.C.: World Bank Group. http://documents.worldbank.org/curated/en/143021506909711004

Wolfert M, Ge L, Verdouw C, Bogaardt M-J (2017) Big data in smart farming—a review. Agric Syst 153:69–80. https://doi.org/10.1016/j.agsy.2017.01.023

Digital Twin Implementation in Agriculture: Strategies Leveraging AI and IoT

20

Pardeep Kumar, Gobind Kumar, and Navneet Kaur

Abstract

Digital twin technology is transforming the agriculture industry by designing virtual versions of real-life farming systems that can optimize farming operations through real-time monitoring and decision-making. This chapter provides a detailed discussion on the implementation of digital twins with the assistance of artificial intelligence (AI) and the Internet of Things (IoT), in the agricultural sector. The building block integrates IoT sensor networks to gather precise data, cloud and edge computing to compute data in real time, and AI/machine learning models to derive predictive analytics. The most important aspects are extremely precise soil moisture sensors (98%), environmental tracking tools, and livestock health gadgets. Cyber-security concerns, sensor deployments, data integration tools, and standardization are also the issues addressed in the chapter. According to the latest studies, the potential benefits are significant: AI irrigation can reduce water use by 18–25% and crop yields by 11–20%, detect pests more accurately than 92%, and accurate fertilizers can lead to a 25–30% reduction in resources. Problems such as technical complexity, economic barriers, and resistance to adoption are resolved with suggestions on how they can be removed. The future

P. Kumar (✉)
Computer Science, Government College for Women, Sirsa, Haryana, India

G. Kumar
Government Polytechnic College, Sirsa, Haryana, India
e-mail: govind166kumar@gmail.com

N. Kaur
Department of Computer Science and Engineering, Chaudhary Devi Lal University, Sirsa, Haryana, India
e-mail: nkaur5156@gmail.com

© The Author(s), under exclusive license to Springer Nature Singapore Pte Ltd. 2026

A. L. Srivastav et al. (eds.), *Digital Twin Technology for Sustainable Agriculture*,
https://doi.org/10.1007/978-981-95-5915-2_20

perspective shows the introduction of AI, 5G, and quantum computing to support sustainable change in agriculture.

Keywords

Digital twins · Agriculture · Artificial intelligence · Internet of Things · Precision farming · Smart agriculture · Implementation strategies

20.1 Introduction

Digital twin technology promises to become the paradigm of modern agriculture whereby real-life agricultural operations can be optimized by the use of virtual models of the physical agricultural system. Agricultural digital twins form virtual models of individual farms that have promising opportunities to improve productivity and efficiency and reduce energy consumption and wastes. AI, Internet of Things (IoT), and digital twin technology can bring together a level of data-driven decision-making that can maximize resource utilization, productivity, and mindfulness of sustainable cultivation (Nasirahmadi and Hensel 2022).

20.1.1 Overview of Digital Twin Concept in Agriculture

The digital twin architecture of agricultural systems consists of a few interrelated meta-classes that collaborate with each other to develop detailed virtual models of farms, as shown in Fig. 20.1. Such components are environmental monitoring sensors to monitor atmospheric conditions, such as temperature, humidity, wind, solar radiation, and precipitation, which are important data of the crop growth model and irrigation scheduling. Sensors are installed on equipment and infrastructure to track the status and performance of agricultural equipment, irrigation systems,

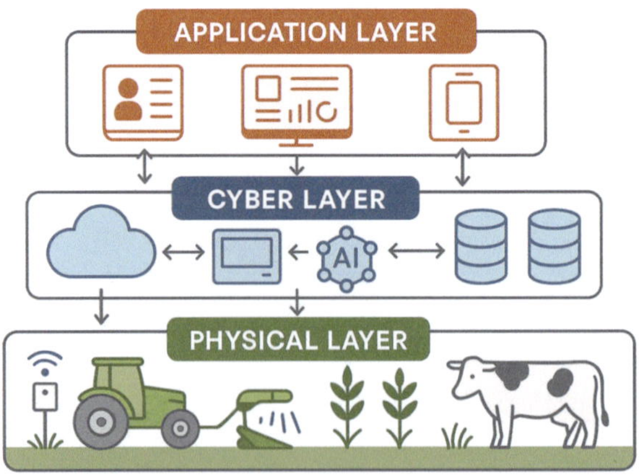

Fig. 20.1 Digital twin architecture for agriculture

storage facilities, and other farm infrastructure, providing predictive maintenance and optimizing equipment use (Escribà-Gelonch et al. 2024).

Digital twin technology in agriculture is made to be adaptable to integrate various datasets and large data models, which makes it possible to roll out end-to-end integration of agricultural data. The technology has become one of the foundations of an intelligent crop platform capable of resource-efficient decision-making and recommendations in agronomy (Ngo et al. 2018).

20.1.2 Importance of Implementation Strategies

Applying digital twins systematically to agriculture is essential to solve the complicated problems modern agriculture experiences. Against a backdrop of an expanding world population that is estimated to reach 9.7 billion in 2050, agricultural systems would have to produce and multiply food provision by 70% yet mitigate environmental footprints and adapt to climate change in the process. Initial applications of digital twin technology are yielding positive performance improvements of 15–20% crop yields, 25% reduction in resource use, and 30% reduction in pest- and disease-related crop losses. The agriculture innovation agenda formulated by the United States Department of Agriculture (USDA) has highlighted the revolutionary role of digital technologies and has requested to raise the agronomic output by 40% and reduce the environmental impact by 50% by 2050. The barriers to implementation, such as technical, economic, and adoption, should be discussed as well as interoperability and standardization of the agricultural systems (Semin et al. 2023).

20.1.3 Role of AI and IoT in Enabling Digital Twins

The main enabling technologies to agricultural digital twins are AI and IoT, which provide the data, tools, and computing power to develop models of farming systems that accurately represent the subsystems within those systems. The machine learning and deep learning variant of AI are utilized in the IoT infrastructure that in combination could be of great assistance in making intelligent decisions at any level of a farming system accrediting various factors and nonintuitive behavioral patterns. This combination has facilitated accuracy farming where farmers are able to know and monitor their crops closely and utilize resources depending on the real-time data gathered. (Sodhi and Jamwal 2024).

20.2 Key Components of Implementation

The use of the agricultural digital twins is quite possible, and three core functional technologies that encompassed the data collection, cloud and edge, and AI/ML to prepare the models to forecast the result are IoT tools, cloud, and edge tools. The combined factors allow massive scale monitoring, smart analysis, and automatic decision-making in the high-precision agriculture of today.

20.2.1 IoT Devices and Sensors for Data Acquisition

Modern implementations of the digital twins of agriculture rely on large networks of IoT sensors that can measure a wide range of various environmental and operational factors. Smart sensor provides us with optimal information about water and crop water requirements, and the mechanically controlled water supply systems can help farmers optimize irrigation. IoT sensors allow them to measure soil parameters such as the temperature of soil, the level of NPK (nitrogen, phosphorus, and potassium), and soil moisture with an extraordinary level of accuracy (Bouhachlaf et al. 2023).

Soil monitoring sensors are sophisticated soil monitors that detect the key parameters such as water-to- soil ratio, soil pH, temperature, and fertilizer reserves (Sun et al. 2016). Such sensors are installed all over the agricultural regions to ensure complete coverage and quality of data (Pechlivani et al. 2023). In new use cases, a level of accuracy of over 98% and a very low standard deviation rate are obtained with soil moisture sensors (Muthmainnah et al. 2024). Environmental monitoring system captures atmospheric data such as temperature, humidity, wind speed, sun radiation, and precipitation that are measured by the weather station and other environmental sensors (Idbella et al. 2020). Additionally, livestock health monitoring employs IoT sensors to track livestock behavior and health indicators, which allow predicting disease outbreaks in early times and optimization in animal welfare.

20.2.2 Cloud and Edge Computing for Real-Time Data Processing

The deployment of DTs in agriculture needs solid computation infrastructure that has the ability to process big quantities of heterogeneous data en route. The complex data analysis in agriculture will require computing power and scalability which are available on cloud-based platforms (Tholhappiyan et al. 2023). NoSQL databases such as MongoDB can be very well suited to store agricultural big data because of the approach in storing real-world big data (high temporal resolution). The edge computing technology enables the processing of information to occur at the edge and reduce the time wastage when making important decisions in a critical situation. By integrating edge computing devices, real-time processing and detection of agricultural data are possible, with some systems having a low-level processing time of 17 ms. This type of combination of cloud and edge computing can be easily scaled and is responsive in real time simultaneously, with the implementation of a digital twin (Andriulo et al. 2024).

20.2.3 AI/ML Models for Predictive Analytics

The analytical core of agricultural digital twins consists of artificial intelligence and machine learning models that can predict with raw sensor data to provide actionable insights. Artificial intelligence, machine learning, and deep learning algorithms extract insights about actions out of uncooked agricultural data. The XGBoost algorithms achieved a superior coefficient of determination, as shown in Eq. (20.1), when predicting evapotranspiration using fewer meteorological variables.

$$R^2 = 0.97 \tag{20.1}$$

Machine learning approaches on crop yield prediction combine the effect of the environment, nutrient concentration, and crop composition to form comprehensive descriptions of crops. The superiority of polynomial regression models in the prediction of the nutrient content, especially with predicting the percentage of nitrogen, has been reported. The model achieved high detection performance, with the accuracy reported in Eq. (20.2). Digital twin applications in pest management identify a variable rate in the accuracy of the classification performance of 88.01% for the training set (Eq. 20.3) and 85.73% for the testing set (Eq. 20.4) (Dai et al. 2023).

$$\text{Accuracy} = 92\% \tag{20.2}$$

$$\text{Accuracy train} = 88.01\% \tag{20.3}$$

$$\text{Accuracy test} = 85.73\% \tag{20.4}$$

20.3 Integration of IoT and Sensor Networks

The agricultural digital twins rely on the complex sensor networks that could record real-time operational and environmental data. Agricultural IoT sensor specifications and cost analysis are shown in Table 20.1. This convergence involves a variety of sensors to monitor soil conditions, weather conditions, and the state of livestock, reliable communication protocol to aid data delivery, and integration and fusion of various sources of information, typically of different nature, existing in farming settings.

20.3.1 Types of Sensors for Agriculture (Soil, Weather, and Livestock Health)

Agricultural digital twin applications can use various types of sensors to track other activities that are related to farming. Precision agriculture is based on soil sensors

Table 20.1 Agricultural IoT sensor specifications and cost analysis for digital twin implementation

Sensor type	Parameters measured	Accuracy	Communication protocol	Estimated cost (USD)
Soil moisture sensor (YL-69)	Soil moisture content	98.71%	Analog/digital	\$5–15
Temperature sensor (DHT11)	Air temperature, humidity	99.67%	Digital (DHT protocol)	\$3–8
pH sensor	Soil pH levels	±0.1 pH units	Analog	\$25–50
NPK sensor	Nitrogen, Phosphorus, Potassium	±5% for each nutrient	Digital/I2C	\$150–300
Weather station	Temperature, humidity, wind, radiation	±0.5Â°C, ±3% RH	WiFi/LoRa	\$200–500
Livestock health monitor	Heart rate, temperature, activity	±2 BPM, ±0.2Â°C	Bluetooth/WiFi	\$100–250
Camera/vision sensor	Crop health, pest detection	>92% detection rate	Ethernet/WiFi	\$500–2000
LoRa communication module	Data transmission	99.9% packet delivery	LoRa	\$20–40

that oversee crucial parameters that have direct impacts on the growth and development of crops.

The most popular actual soil monitoring systems use various kinds of sensors with highly accurate readings such as moisture sensors, temperature sensors, and pH sensors (Sivakumar et al. 2023). Weather stations combine a range of sensors, including those measuring temperature, moisture, wind velocity, solar radiation, and precipitation. Investigational wearable sensors monitor and track vital signs, behavior, health status as well as energy expenditure on livestock in real time. Such systems make it possible to control animal welfare on an ongoing basis and identify health problems at an early stage and contribute to better management and productivity of livestock.

20.3.2 Communication Protocols and Data Transmission Methods

Data transmission in agriculture–IoT networks needs to be managed in a way that ensures effective communications across the networks despite the challenges in their rural areas (Kiv et al. 2022). As shown in Table 20.1, IoT systems applied to agriculture use numerous wireless protocols: LoRa, WiFi, and cellular networks to send data back to the central processing devices using wireless transmissions. The communication protocol to use can be either limited by the range that is needed or restricted by the amount of power it can allow to be used or restricted by the frequency of data that needs to be carried across the communication protocol (Wilberforce and Mwebaze 2025). Sophisticated agrarian systems use technologies such as Hive, Flume, and Sqoop to collect and synchronize data on various sensors and databases. Such programs make the data exchanges between distributed sensor networks and central processing systems more reliable (Li and Zhang 2020).

20.3.3 Data Fusion Techniques

The use of data fusion in agriculture implies uniting two or more sources of information that reside in different heterogeneous sources to generate insights into the agricultural systems (Barbedo 2022). Fusion of agricultural data involves the combination of multiple resources of data such as satellite imagery information, sensors on the ground, weather stations, and historic information. Approaches to spatial data infrastructure make it possible to combine geospatial data with both time series. Farming data are often spatial, time-series, multifaceted, heterogeneous, nonuniform, and enormously large.

20.4 AI-Driven Decision-Making in Digital Twins

AI makes the raw agricultural information into usable intelligence via prediction modeling and optimization algorithms. Using IA systems, farming can be proactive because they can predict crop production and cattle health, identify unusual events pointing to disease or outbreak of pests, as well as improve the effectiveness and efficiency of irrigation and fertilizer application strategies.

20.4.1 Predictive Modeling for Crop Yield and Livestock Health

The principal concept of the digital twin in agriculture is the application of predictive AI-based models to handle the received sensor data to provide farm management information that can be performed. The models have included different sets of data to forecast the crop yields, to facilitate the optimization of resources, and to keep livestock healthy as never before.

Crop Yield Prediction Models Powerful machine learning algorithms like Random Forest fine-tuned using genetic algorithms lead to crop yield forecasting over 85% better than other algorithms. The neural net algorithm and big data analysis tools are fairly efficient and indicate greater yield prediction rates, by 15–20%, than the existing methods. The models consider a large number of environmental features, dosage of nutrients, and crop profile that can be used to come up with detailed simulation of crop traits (Ghazvini et al. 2024).

Livestock Health Monitoring The wearable sensors of the digital twin type will monitor the nutritional condition and the behavioral patterns of animals under the influence of AI algorithms. IUMENTA framework will take advantage of the EnergyTag system to assist real-time energy costs monitoring to facilitate ongoing update and personalization of energy balance digital twins. They describe a disease of livestock health and behavior change with over 90% accuracy (Youssef et al. 2024).

20.4.2 Anomaly Detection for Disease and Pest Outbreaks

The AI-based anomaly detection systems can be provided with valuable early warning capabilities in disease and pest control of an agricultural digital twin.

Pest Detection Systems Pest management of a digital twin system is able to achieve 92% pest detection and less than 10% error rates in crop growth indicators. The predictive pest populations are genetically optimized random forest algorithms with 88.01% training and 85.73% test accuracy. These systems allow correcting the situation in time, which may result in the minimization of pesticide use and the ongoing control of pests (Dai et al. 2024).

Disease Monitoring The AI-driven disease-detecting systems will be in a position to process the multispectral image and sensor values of the disease symptoms far in advance of the visual manifestations of the disease. The deep learning models and the YOLOv8 frameworks have outstanding performance in the field of detection of abnormalities with an increased accuracy of more than 92. With such an innovation in the methods of detection at an early age, preventive types of treatment can be adopted, and this would go a long way in crop saving (Li et al. 2025).

20.4.3 Optimization Algorithms for Irrigation and Fertilization

AI optimization algorithms embedded in digital twins facilitate precise management of resources that enhance the extent to which crops can yield and lessen the impact on the environment. Such algorithms consider real-time sensor data, weather forecasts, and requirements toward crops to optimize irrigation and fertilizing schedules.

Irrigation Optimization AI and IoT enable smart irrigation systems to optimize water use by an 18% water consumption decrease and 11% crop growth increase by using sensor information and data analysis. The optimal time to irrigate is forecasted by machine learning models using soil moisture, weather forecasts, and crop water requirement. Through these systems, water saving of 20% could be achieved without altering or decreasing crop quality (Mekonen 2025).

Fertilization Management Machine learning systems can be used to control fertilizers where their systems evaluate the levels of nutrients in the soil, crop development, and climatic conditions to guide how to apply fertilizers. Polynomial regression models are more superior when predicting the requirements of the nutrients, especially in nitrogen management. Such systems provide controlled fertilizer application, which saves 30% on fertilizer amount and ensures an optimum nutrition of crops (Shi et al. 2025).

20.5 Data Collection, Storage, and Analysis

Agricultural data that may prove highly complex to manage in bulk volumes and of different types demands the use of powerful big data infrastructure enabling the processing of such sources. The discussion in this section will be on the architecture of storage and analysis of agricultural datasets, including preprocessing pipelines, to guarantee data quality, consistency, and reliability to optimize the performance of digital twin systems.

20.5.1 Big Data Architectures in Agriculture

Agricultural digital twins produce terabytes of heterogeneous data that need to have custom big data architecture to meet the needs of processing and analysis. The agricultural datasets can be described as the big data applications regarding their volume, variety, velocity, and veracity (Ngo et al. 2018, 2019).

Distributed Storage Systems Distributed storage systems are designed to process extreme sets of heterogeneous data, using Hadoop and NoSQL databases. The MongoDB NoSQL databases are the most applicable when dealing with real-world big data, especially in agriculture, because of their flexibility and high performance (Ngo et al. 2019).

Data Warehouse Implementation Agricultural data warehouses offer data integration of high-performance and data-flexible schema design, multi-dataset integration, data science support, security, and governance, made at the continental level. Such systems would empower query management and historical and real-time agricultural data analysis (Ngo et al. 2020).

20.5.2 Data Cleaning, Preprocessing, and Visualization

Data preprocessing is important in the overall accuracy and reliability of agricultural digital twin systems. It is common to find inconsistent, missing, and noisy values in the agricultural data, and these need to be dealt with prior to analysis, as shown in Fig. 20.2.

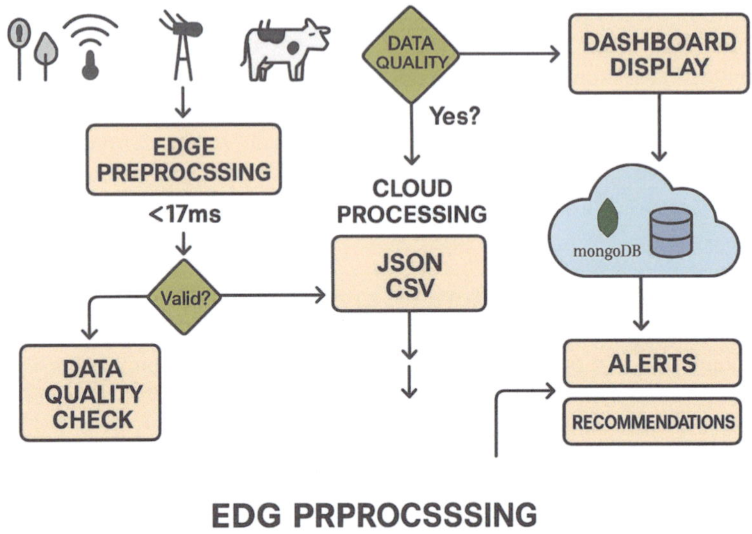

Fig. 20.2 Data processing pipeline

Data Quality Management We propose a data quality management technique based on advanced data cleaning techniques making use of ERA5-Land reanalysis data and deep learning to identify and correct the errors in our agricultural sensor data. Such approaches provide R2 = 0.96 and RMSE = 0.04 in scaling temperature estimates, which guarantee quality data to be used later (Hachimi et al. 2022).

Preprocessing Pipelines Flexible data preprocessing pipelines will combine data across different sources such as real-time sensors, devices in the IoT domain, reanalysis, and nonprocessed files. The pipelines convert data into standard formats and management of missing values, including preparation to be used by AI and machine learning algorithms.

Visualization Systems The highly developed visualization software presents information in interactive dashboards and tools that allow farmers and researchers to browse through fine-grained information. Real-time monitoring, tendency analysis, and decision support will also be possible through the help of user-friendly graphical user interfaces through the systems.

20.5.3 Handling Heterogeneous Agricultural Datasets

The syntheses of heterogeneous data in different forms, temporal peculiarities, and semantics are the problems that motorized agricultural digital twins will have to solve. Such a diversity is a major issue to data integration and analysis.

Data Integration Frameworks Agricultural data integration systems use constellation schema frameworks that allow them to be open and accommodative to absorb a large spectrum of datasets and big data paradigms. These structures assist in solving the problem of incorporating nonstandardized information provided by different sources of information regarding agriculture (Ngo and Kechadi 2020).

Semantic Standardization Standardized data with standard interfaces and protocols allow communication between heterogeneous data sources and allow interoperability between such data. These platforms will take care of the problem of datasets in different agricultural fields being created based on different ontologies, modalities, or forms (López-Morales et al. 2020).

20.6 Framework for Digital Twin Development in Farms

Introduction of digital twin technology should be systematic in taking into account security, technical, and operational factors in the development process. In this environment, stakeholders are able to monitor the planning to deployment phases, and it has been able to guarantee interoperability, cyber-security, and sustainable implementation strategies that maximize the agricultural operation to return on investment.

20.6.1 Steps from Planning to Deployment

The technical, operational, and economic aspects should be the sources of systematization of development and implementation of the digital twin in the agricultural environment. The framework will be implemented in a coherent way to ensure they are adopted and maintained in the long term.

Planning and Assessment Phase The undertakings during this phase include evaluation of the existing infrastructure, information, and also access to resources. Every farmer should pay close attention to their own operational needs and must be capable of easily adopting digital twin technologies (Xie et al. 2024).

System Design and Architecture Digital twin architecture is composed of various interconnected layers that encompass physical layer (crops, infrastructure, and equipment), virtual layer (semantic mapping and simulating environments), and the synchronization layer (communication systems). It uses an IoT infrastructure comprising nine layers resembling the FIWARE characteristics to support the bidirectional communication flow round the clock (Fernández et al. 2025).

Implementation and Integration It is an implementation stage that entails implementation of IoT sensors, communication network, as well as integrating data processing systems. The stepwise implementation of the process in three stages, including monitoring and data collection, decision-making and virtual simulation, and autonomous operation with the model refinement based on the feedback, provides the ease of transition and adoption (Reyes et al. 2022).

Testing and Validation Verification testing tests the system thoroughly to ensure its performance, accuracy, and reliability, prior to being fully implemented. The effectiveness in a specific after-prototype setting is proven through experimental demonstration.

20.6.2 Interoperability and Standardization

One of the most pressing issues in the implementation of agricultural digital twins is interoperability, specifically with the number of different systems interacting to promote agricultural practices extending on a vast landscape. Standardization is aimed at developing shared protocols and interfaces that facilitate broad compatibility to allow interconnection of distinct platforms and vendors (Falcão et al. 2022).

Industry Standards The implementation of agricultural digital twin is aligned to accepted industry standards such as IEC 62443 and cyber-security in Agricultural Automation Control Systems (AACS). The implementation of the interoperable digital twin systems is offered by Open Digital Twin Platform (ODTP) standards (Kristen et al. 2021).

Data Standards To guarantee that farming data can be shared and used correctly across systems, the FAIR (Findable, Accessible, Interoperable, and Reusable) principles of data are implemented to guarantee agricultural data can be shared and effectively used across systems. Homogeneous data sources can be integrated with the help of standardized information formats and semantic modeling (Pan et al. 2023).

Communication Protocols Communication protocol standardization will make IoT devices and systems compatible with one another. Interoperable spatial data infrastructure standards (OGC, Open Geospatial Consortium) are relevant in the agricultural context (Bordogna et al. 2016).

20.6.3 Cyber-Security and Privacy Considerations

Deployment of digital twins in the agricultural sector creates tremendous security and privacy problems that need to be solved in order to make the system secure and trusted by the farmers. Agricultural systems grow weak to cyber-threats as they involve the usage of connected devices and data exchange (Hazrati et al. 2022).

Security Frameworks Broad cyber-security frameworks like IEC 62443 offer systematic ways of securing an agricultural Internet of Things. Such frameworks deal with security risks such as data breach, unauthorized access to information, and the manipulation of a system (Al Asif et al. 2021).

Privacy Protection Privacy-enhancing solutions are provided to the data-sharing issue on the part of farmers due to the risk of data misuse. The frameworks enable the sharing of the data in a secure way and provides protection of sensitive farm data against unauthorized access and commercial use (Rahaman et al. 2024).

Data Governance Well-defined data governance establishes the policy of data collection, storage, sharing, and utilization and complies with laws including GDPR without jeopardizing the integrity and safety of agricultural data.

20.7 Case Examples

The case studies throughout the world indicate that agricultural digital twins have the potential to deliver physical value through an improved efficacy of irrigation, livestock health regulation, and enhanced productivity, as illustrated in Fig. 20.3.

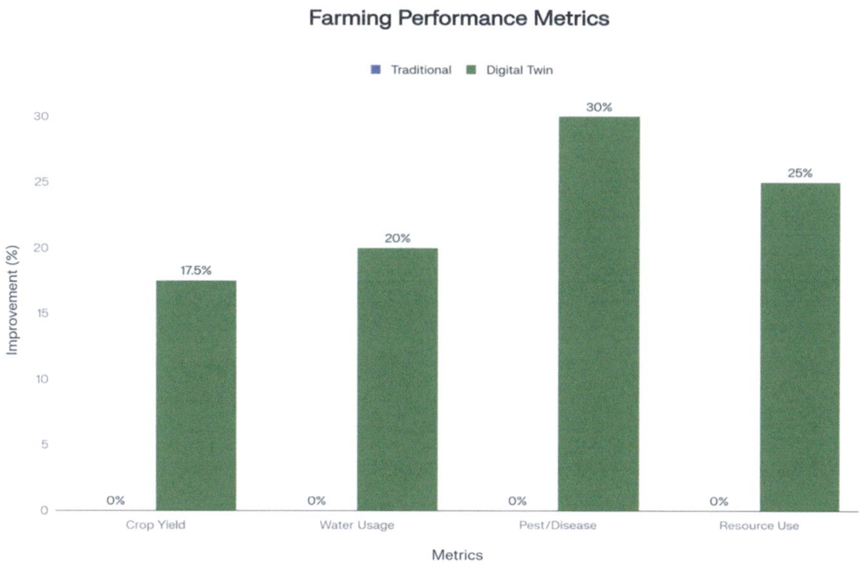

Fig. 20.3 Case study results: Comparison performance improvements achieved through digital twin implementation vs traditional farming methods in agriculture

20.7.1 AI-Powered Irrigation Management System

Examples of digital twin applications in irrigation management show a lot of potential in optimizing the use of water without having to sacrifice and/or increase crop yield. Innovations in AI-powered irrigation systems can be described as one of the most successful digital twin implementations in agriculture.

System Architecture This system is based on IoT-enabled soil moisture monitoring sensors that measure real-time soil moisture status, soil temperature, humidity conditions, and atmospheric conditions (Varathaiah et al. 2024).

Performance Results Results of water efficiency and crop productivity are high after implementation. With AI-based irrigation systems, 18% is saved in the use of water, and 11% growth in crops is obtained over conventional irrigation methods. The system results in 20% water savings with the maintenance of optimum levels of moisture in soils (Shi et al. 2025).

Technology Integration XGBoost ML models are used to schedule the irrigation management and showed strong predictive ability, with the performance given in Eq. (20.5). Streaming data analytics allows automatic irrigation command that responds instantaneously due to a latency of less than 17 ms. The combination of weather and soil moisture prediction allows advanced irrigation control to adapt to the changing environmental factors (Hachimi et al. 2022).

$$R^2 = 0.97 \tag{20.5}$$

20.7.2 Livestock Digital Health Tracking

The digital twin of the livestock involves the combination of wearable sensors, environmental monitors, and behavioral analysis algorithms to develop a complete virtual representation of the animal. The IUMENTA framework will use specialized tools to build and simulate digital twins of animals that integrate the energy balance models and the ability to monitor the animal in real time (Arulmozhi et al. 2024). The system achieved a high monitoring accuracy as reported in Eq. (20.6) on both the detection of wellness abnormalities and behavior change. Disease symptoms and health concerns are identified early enough and can be intervened with the help of real-time analysis.

$$\text{Accuracy} = 92\% \tag{20.6}$$

20.8 Challenges and Solutions

Despite the high potential of the digital twin, integration has faced certain obstacles to widespread implementation in the agricultural industry. These problems include technical complexity, financial constraints, and refusal to adopt technology by traditional agricultural operations. Awareness of the presence of the above barriers and development of effective mitigation programs to facilitate accelerated agricultural transformation are crucial.

20.8.1 Technical, Economic, and Adoption Barriers

Use of the application of digital twin in agriculture faces numerous obstacles that vary in intensity between the technical issues and the cost constraint to the adoption problems. Learning about these challenges will be crucial toward the establishment of mitigation strategies to roll-out.

Technical Challenges Digital twin systems in agriculture should be able to process diverse and heterogeneous data on different sources but in real time in terms of analysis and processing capabilities. There is a technical challenge to the interconnection of the IoT devices, AI algorithms, and cloud computing infrastructure. Additional technical challenges are data quality problems, the limited reliability of sensors, and a communication network in rural regions (Kashliev and Tracy 2024).

Economic Barriers Economic barriers are the initial high cost of the digital twin infrastructure such as sensors, communication systems, and software platforms, which are the obstacles to many farmers. Financial problems are associated with the limited availability of high-quality data, costly monitoring instruments, and other costs in maintaining devices, especially among the small-scale farmers. Calculating the return on investment in digital twin implementations is rather complex, and to calculate it as an economic good is also rather hard (Xie et al. 2024).

Adoption Barriers Farmers have little technical skills, do not prefer to adopt technology easily, and are worried about privacy and security of data. With regard to the availability of digital twin technologies, the digital divide between various regions and farm sizes is an issue (Adenyi et al. 2024).

20.8.2 Possible Mitigation Strategies

The problems associated with the deployment of digital twins can be resolved only with the use of complex strategies that meet technological, financial, and social challenges. To address the stakeholders of agriculture, proper mitigation policies need to be context-based on the specifics and constraints of the stakeholders.

Technical Solutions The technical issues can be overcome by coming up with cost effective and user-friendly digital twin platforms that can be utilized to enhance access. Such digital and Internet of Things infrastructure needs to be promoted by investment especially to the rural exclaves, so that it becomes acceptable to the masses. The complexity of integration in the form of data format and communication protocols and system interface standardization may help to improve interoperability (Bhat et al. 2021).

Economic Mitigation Government subsidies can also reduce the economic cost of digital twin technology and can subsidize it to farmers. Raising modular-scale systems makes incremental implementation (spreading the costs over a time period) possible. The economic benefits of implementing digital twins can be explained by demonstrating a viable benefit of investment (return on investment) in pilot projects and case studies (Mekonen 2025).

Adoption Strategies The mass trainings and technical assistance may generate the local capacity and increase the trust of the farmers in digital technologies. The ease of use and acceptability can be enhanced by the development of mobile environments and simplified user-friendly interfaces to suit the needs of the farmers (Kaur et al. 2022).

Policy Support The agricultural sector requires digital regulation which will promote the uptake of digital agriculture by upgrading farmers and securing their interests. Cooperation and standardization protection within the industry can accelerate the process of establishing cross-compatible systems and reduce their implementation costs (Marrano et al. 2025).

20.9 Conclusion and Future Scope

One of the avenues that will help implement the sustainable food production systems that can tackle the global security challenges is the digital twin. This is a discussion of one of the strategic approaches of smart agriculture adoption and also an analysis of the revolutionary capability of new emerging technologies such as advanced AI, 5G connectivity, and quantum computing to transform agricultural practices.

20.9.1 Roadmap for Smart Agriculture Adoption

The future of the principle of the digital twin in agriculture is the necessity to organize a joint work in order to eliminate the existing limitations and apply the new technologies which have appeared to design more sophisticated and accessible systems.

Short-Term Objectives (1–3 Years) Investment in the development of digital twin applications that are easy and less costly to operate by farmers who do not have a deep understanding of technology, scaling up pilot projects to demonstrate their effectiveness, and working out the standards.

Medium-Term Goals (3–7 Years) The sophistication of digital twins and addressing privacy challenges and cyber-security concerns by introducing complementary frameworks will be enhanced by adding novel AI characteristics and an ecosystem-wide strategy (Kristen et al. 2021).

Long-Term Vision (7–15 Years) Digital twin technology will be mature enough to enable complete autonomy in farming systems and will be able to optimize agriculture on planetary scales thanks to global climate monitoring and high simulation capacities. (Tzachor et al. 2022).

20.9.2 Role of Emerging Technologies

The already emerging interaction between digital twins and agriculture will surely be enriched with the adoption of new technologies that enhance capabilities and performance. The technologies will address the current deficiencies and elaborate more on the opportunities of optimization of agriculture.

Artificial Intelligence Advancement Large language models and other cutting-edge computer vision products designed as artificial intelligence models in the future will offer an analytical improvement to the agricultural digital twins in terms of making real-time decisions with lower latency (Nasirahmadi and Hensel 2022; Agarwal et al. 2024).

Extended Reality (XR) Technologies AR, VR, and MR technologies will offer immersive interfaces to enable farmers to see and feel the relationship between the complex data and how management decisions would affect the results (Fiaz et al. 2024).

Blockchain and Distributed Systems The blockchain will improve the security, traceability, and trust of agricultural digital twin systems, as well as facilitating secure data-sharing among the members of the ecosystem (Aliyu and Liu 2023; Paulus and Leiding 2023).

Quantum Computing Complex agricultural computational problems such as weather forecasting and crop modeling will be optimized with the help of quantum algorithms.

5G and Advanced Connectivity The network will allow real-time data transfer and processing with the support of advanced applications and continuous connection of agro-distributed systems, which is possible using 5G networks (Zeng et al. 2025).

The application of digital twin in agriculture is one of the characteristics and a breakthrough to the achievement of the concept of food security in the world and sustainability of the environment. Correct response to the most frequent problems and an attentive integration of the new technology will contribute to the digital twin approach to revolutionize the agricultural production and create more resilient and productive sustainable agricultural systems. It is also worthy to mention that despite the field of agriculture seeming bright in future with the application of huge biotechnology, there are numerous stakeholders including farmers, technology providers, policymakers, and researchers who should be on the roadmap to realize this so that it is completely achieved.

References

Adenyi AO, Okolo CA, Olorunsogo T, Babawarun O, Adenyi AO, Okolo CA, Olorunsogo T, Babawarun O (2024) Leveraging big data and analytics for enhanced public health decision-making: a global review. GSC Adv Res Rev 18(2):450–456

Agarwal K, Ananthanarayanan S, Srinivasan S (2024) Enhancing IoT based plant health monitoring through advanced human plant interaction using large language models and mobile applications. arXiv preprint arXiv:2409.15910

Al Asif MR, Hasan KF, Islam MZ, Khondoker R (2021) STRIDE-based cyber security threat modeling for IoT-enabled precision agriculture systems. In: 2021 3rd international conference on sustainable technologies for industry 4.0 (STI). IEEE, pp 1–6

Aliyu AA, Liu J (2023) Blockchain-based smart farm security framework for the internet of things. Sensors 23(18):7992

Andriulo FC, Fiore M, Mongiello M, Traversa E, Zizzo V (2024) Edge computing and cloud computing for internet of things: a review. Informatics 11(4):71. MDPI

Arulmozhi E, Deb NC, Tamrakar N, Kang DY, Kang MY, Kook J, Basak JK, Kim HT (2024) From reality to virtuality: revolutionizing livestock farming through digital twins. Agriculture 14(12): 2231

Barbedo JGA (2022) Data fusion in agriculture: resolving ambiguities and closing data gaps. Sensors 22(6):2285

Bhat SA, Huang NF, Sofi IB, Sultan M (2021) Agriculture-food supply chain management based on blockchain and IoT: a narrative on enterprise blockchain interoperability. Agriculture 12(1):40

Bordogna G, Kliment T, Frigerio L, Brivio PA, Crema A, Stroppiana D, Boschetti M, Sterlacchini S (2016) A spatial data infrastructure integrating multisource heterogeneous geospatial data and time series: a study case in agriculture. ISPRS Int J Geo Inf 5(5):73

Bouhachlaf L, Benslimane O, El Hajjaji S (2023) Monitoring soil elements for irrigation management using internet of things (IoT) sensors. World Water Policy 9(4):756–766

Dai R, Kannampallil T, Kim S, Thornton V, Bierut L, Lu C (2023) Detecting mental disorders with wearables: a large cohort study. In: Proceedings of the 8th ACM/IEEE conference on internet of things design and implementation. Association for Computing Machinery, New York, pp 39–51

Dai M, Shen Y, Li X, Liu J, Zhang S, Miao H (2024) Digital twin system of pest management driven by data and model fusion. Agriculture 14(7):1099

Escribà-Gelonch M, Liang S, van Schalkwyk P, Fisk I, Long NVD, Hessel V (2024) Digital twins in agriculture: orchestration and applications. J Agric Food Chem 72(19):10737–10752

Falcão R, Matar R, Rauch B (2022) Using I4. 0 digital twins in agriculture. In: European conference on software architecture. Springer International Publishing, Cham, pp 483–498
Fernández R, Navas E, Rodríguez-Nieto D, Rodríguez-González AA, Emmi L (2025) DigiHortiRobot: an AI-driven digital twin architecture for hydroponic greenhouse horticulture with dual-arm robotic automation. Futur Internet 17(8):347
Fiaz F, Sajjad SM, Iqbal Z, Yousaf M, Muhammad Z (2024) Metassi: a framework for personal data protection, enhanced cybersecurity and privacy in metaverse virtual reality platforms. Futur Internet 16(5):176
Ghazvini A, Sharef NM, Balasundram SK, Lee LS (2024) A concentration prediction-based crop digital twin using nutrient co-existence and composition in regression algorithms. Appl Sci 14(8):3383
Hachimi CE, Belaqziz S, Khabba S, Sebbar B, Dhiba D, Chehbouni A (2022) Smart weather data management based on artificial intelligence and big data analytics for precision agriculture. Agriculture 13(1):95
Hazrati M, Dara R, Kaur J (2022) On-farm data security: practical recommendations for securing farm data. Front Sustain Food Syst 6:884187
Idbella M, Iadaresta M, Gagliarde G, Mennella A, Mazzoleni S, Bonanomi G (2020) Agrilogger: a new wireless sensor for monitoring agrometeorological data in areas lacking communication networks. Sensors 20(6):1589
Kashliev A, Tracy K (2024) A provenance-aware approach to big data workflow management in heterogeneous cloud environments. In: 2024 9th international conference on cloud computing and big data analytics (ICCCBDA). IEEE, pp 113–119
Kaur J, Hazrati Fard SM, Amiri-Zarandi M, Dara R (2022) Protecting farmers' data privacy and confidentiality: recommendations and considerations. Front Sustain Food Syst 6:903230
Kiv D, Allabadi G, Kaplan B, Kravets R (2022) Smol: sensing soil moisture using LoRa. In: Proceedings of the 1st ACM workshop on no power and low power internet-of-things. The Association for Computing Machinery, New York, pp 21–27
Kristen E, Kloibhofer R, Díaz VH, Castillejo P (2021) Security assessment of agriculture IoT (Aiot) applications. Appl Sci 11(13):5841
Li Y, Zhang D (2020) Hadoop-based university ideological and political big data platform design and behavior pattern mining. In: 2020 international conference on advance in ambient computing and intelligence (ICAACI). IEEE, pp 47–51
Li H, Chen H, Liu J, Zhang Q, Liu T, Zhang X, Li Y, Qian Y, Zou X (2025) Deep learning-based detection and digital twin implementation of beak deformities in caged layer chickens. Agriculture 15(11):1170
López-Morales JA, Martínez JA, Skarmeta AF (2020) Digital transformation of agriculture through the use of an interoperable platform. Sensors 20(4):1153
Marrano A, Cabugos L, Hafner A, Kapoor B, McNamara J, O'Donnell M, Reiser L, Tello-Ruiz MK, Zhang H, Staton M (2025) A teaching and training framework to promote findable, accessible, interoperable, and reusable data generation in agriculture. Database *2025*:baaf034
Mekonen BM (2025) Integrating AI and remote sensing in precision agriculture for advancing sustainable irrigation monitoring and management in Ethiopia. Am J Artif Intell 9:22–29
Muthmainnah M, Mulyadi MF, Tazi I, Mulyono A, Hananto FS, Chamidah N (2024) Development of an automated monitoring system for soil moisture and temperature in smart agriculture to enhance lettuce farming productivity based on IoT. Multidiscip Sci J 6(11):2024233–2024233
Nasirahmadi A, Hensel O (2022) Toward the next generation of digitalization in agriculture based on digital twin paradigm. Sensors 22(2):498
Ngo VM, Kechadi MT (2020) Crop knowledge discovery based on agricultural big data integration. In: ICMLSC'20: proceedings of the 4th international conference on machine learning and soft computing; 2020; Haiphong City. Association for Computing Machinery, New York. [online]
Ngo VM, Le-Khac NA, Kechadi M (2018) An efficient data warehouse for crop yield prediction. arXiv preprint arXiv:1807.00035

Ngo VM, Le-Khac NA, Kechadi MT (2019) Designing and implementing data warehouse for agricultural big data. In: International conference on big data. Springer International Publishing, Cham, pp 1–17

Ngo VM, Le-Khac NA, Kechadi MT (2020) Data warehouse and decision support on integrated crop big data. Int J Bus Process Integr Manag 10(1):17–28

Pan Y, Sun J, Yu H, Bai G, Ge Y, Luck J, Awada T (2023) Transforming agriculture with intelligent data management and insights. In: 2023 IEEE international conference on big data

Paulus S, Leiding B (2023) Can distributed ledgers help to overcome the need of labeled data for agricultural machine learning tasks? Plant Phenomics 5:0070

Pechlivani EM, Papadimitriou A, Pemas S, Ntinas G, Tzovaras D (2023) IoT-based agro-toolbox for soil analysis and environmental monitoring. Micromachines 14(9):1698

Rahaman M, Lin CY, Pappachan P, Gupta BB, Hsu CH (2024) Privacy-centric AI and IoT solutions for smart rural farm monitoring and control. Sensors 24(13):4157

Reyes YA, Abbasi R, Martinez P (2022) Digital twinning of hydroponic grow beds in intelligent aquaponic systems. Sensors (Basel) 22(19):7393

Semin A, Mironov D, Kislitskiy M, Zasypkin A, Ivanov V (2023) Improving the theoretical and methodological framework for implementing digital twin technology in various sectors of agriculture. Emerging Sci J 7(4):1100–1115

Shi Y, Chen X, Shao W, Huang Y, Yu X (2025) Hexapod robotic agricultural crop growth monitoring system: applications of digital twins and energy optimization. In: Second international conference on electronics, electrical, and control system (EECS 2025), vol 13659. SPIE, pp 381–392

Sivakumar VG, Baskar VV, Vadivel M, Vimal SP, Murugan S (2023) IoT and GIS integration for real-time monitoring of soil health and nutrient status. In: In 2023 international conference on self sustainable artificial intelligence systems (ICSSAS). IEEE, pp 1265–1270

Sodhi GK, Jamwal P (2024) Smart farming: harnessing the power of IoT for agricultural transformation. In: In 2024 5th international conference on recent trends in computer science and technology (ICRTCST). IEEE, pp 406–412

Sun S, Rappaport TS, Thomas TA, Ghosh A, Nguyen HC, Kovacs IZ, Rodriguez I, Koymen O, Partyka A (2016) Investigation of prediction accuracy, sensitivity, and parameter stability of large-scale propagation path loss models for 5G wireless communications. IEEE Trans Veh Technol 65(5):2843–2860

Tholhappiyan T, Sankar S, Selvakumar V, Robert PD (2023) Agriculture monitoring system with efficient resource management using IoT. In: In 2023 second international conference on augmented intelligence and sustainable systems (ICAISS). IEEE, pp 1628–1633

Tzachor A, Richards CE, Jeen S (2022) Transforming agrifood production systems and supply chains with digital twins. NPJ Sci Food 6(1):47

Varathaiah D, Yogarayan S, Mae AE, Min PP (2024) IoT-driven soil moisture monitoring for optimizing irrigation. In: In 2024 international conference on intelligent cybernetics technology & applications (ICICyTA). IEEE, pp 1066–1071

Wilberforce N, Mwebaze J (2025) A framework for IoT-enabled smart agriculture. arXiv preprint arXiv:2501.17875

Xie Q, Zhong Y, Ariannejad M, Budi STD, Zeyu C, Shuyi C (2024) Recent advances in the digital twin system for smart farming: a five-year review. In: In 2024 international conference on computing innovation, intelligence, technologies and education (CIITE). IEEE, pp 1–6

Youssef A, Vodorezova K, Aarts Y, Agbeti WE, Palstra AP, Foekema E, Aguilar L, Torres RDS, Grübel J (2024) IUMENTA: a generic framework for animal digital twins within the open digital twin platform. arXiv preprint arXiv:2411.10466

Zeng L, An Y, Zhou H, Luo Q, Lin Y, Zhang Z (2025) A hybrid machine learning-based data-centric cybersecurity detection in the 5G-enabled IoT. Secur Priv 8(2):e472

21 Digital Twins for Sustainable Agriculture: Advancing Productivity, Resilience, and Equity

Anita Rani Mehta, Pardeep Kumar, Sanjay Tyagi, Kanwal Garg, and Shalini Aggarwal

Abstract

The application of digital twin (DT) technology has a significant impact on agriculture, offering an interactive and data-driven digital representation of agricultural cropping systems. This approach enhances productivity, optimizes the efficient use of resources, and improves climate resilience in crop production, animal production, aquaponics, and supply chain efficiency. This chapter emphasizes that DT technology has advanced precision agriculture by enabling predictive analytics, real-time monitoring in an environment, and scenario simulation. Evidence indicates significant enhancements of 10–20% in yield, more than 30% irrigation water saving, as well as improved crop disease diagnosis and greater accountability in food supply chains. Nonetheless, the current research also identifies significant challenges in the adoption of DT technology, including elevated implementation costs, connectivity issues, cybersecurity concerns, and technological disparities between advanced and limited resources in agricultural systems. The results indicate that developing the concept of integration demands policy support, enhancing capacity, and creating affordable, portable, and accessible DT depictions of various agricultural situations. The study indicates that DTs are not merely a technological advancement, but they are also a crucial facilitator of sustainable and climate-resilient agricultural cropping systems. This innovative technology promotes proactive decision-making and innovates

A. R. Mehta (✉) · P. Kumar · S. Tyagi · K. Garg
Department of Computer Science and Applications, Kurukshetra University, Kurukshetra, Haryana, India
e-mail: anita_mehta.dcsa@kuk.ac.in; pmittal@kuk.ac.in; sanjaytyagi@kuk.ac.in; gargkanwal@kuk.ac.in

S. Aggarwal
S.U.S. Govt. College, Matak Majri (Indri), Karnal, Haryana, India

© The Author(s), under exclusive license to Springer Nature Singapore Pte Ltd. 2026

A. L. Srivastav et al. (eds.), *Digital Twin Technology for Sustainable Agriculture*,
https://doi.org/10.1007/978-981-95-5915-2_21

resource distribution, ultimately enhancing food security for the expanding global population. Moreover, collaborative solutions that integrate innovations and equality are necessary to achieve this promise and make DTs available to farmers across different sizes and regions.

Keyword

Digital Twin · Sustainable Farming · Agriculture 4.0 · Cyber-Physical System · Food Supply Chain

21.1 Introduction

21.1.1 Background and Context

Agriculture is the most significant component in the survival of humans and socio-economic platforms among small and marginal farmers. The agricultural sector is currently facing several critical challenges, including a burgeoning population, climate change, and depletion of natural resources; consequently, there is an immediate requirement to enhance the sustainability of agricultural cropping systems. According to the estimates of the United Nations (UN), the world's food demand may increase by 60% by 2050, which may result in further depletion of natural resources, including arable land, soil quality, and freshwater. The answer to such a need is for agriculture to become more efficient, resilient, and sustainable with the least possible environmental impacts. The recent years brought a paradigm shift in the agricultural sphere, which might be called Agriculture 4.0 because advanced technologies are used to transform the industry, including the Artificial Intelligence (AI), Internet of Things (IoT), Cloud Computing, Robotics, and Remote Sensing (Tsolakis et al. 2022; Verdouw et al. 2021). Among these developments, DT technology has emerged as a disruptive framework poised to transform the way agricultural systems are established, monitored, enhanced, and maintained.

The DT technology is a computerized model of a physical object, process, or system that constantly updates real-time data using sensors, the cyber-physical, and IoT representations (Tuegel et al. 2011). This DT replicates the real-life version of the object, allowing for the study of its behavior in an advanced manner through simulations, analysis, prediction, and optimization (Jones et al. 2020). Applied to agriculture, DTs enable stakeholders to construct virtual models of farms, greenhouses, livestock, aquaponics systems, and food supply chains, providing data-driven evidence to support sustainable decision-making (Purcell and Neubauer 2023).

21.1.2 Importance of Sustainability in Agriculture

Sustainability in agriculture is defined in terms of environmental, social, and economic impacts. In the environment, it aims to reduce the application of resources (water, energy, fertilizers, and pesticides), preserve biodiversity, and minimize carbon emissions. Socially, it emphasizes equity, food security, and the well-being of farmers. Economically, it necessitates the enhancement of productivity and profitability, in addition to the long-term sustainability of the agricultural system. DT technology focuses on four key goals: efficiency, predictive analytics, sustainability monitoring, and resilience building, as it makes direct contributions toward each of them. By allowing sustainability to be incorporated into digital design and management, DT technologies fit into the general agenda of sustainable agriculture advocated by the United Nations Sustainable Development Goals (SDGs), as well as Goal 2 (Zero Hunger), Goal 12 (Responsible Consumption and Production), and Goal 13 (Climate Action).

21.1.3 Emergence of DT in Agriculture

DT solutions in agriculture have a relatively short history, yet they are growing at an incredibly rapid pace in several areas, including crop production, greenhouse horticulture, smart irrigation and water management, livestock farming, aquaponics and hydroponics, and food supply chains. The IoT Analytics (2023) report and Straits Research (2023) report both identify agriculture as one of the emerging markets where DT is expected to be utilized, as it can lead to sustainable food systems.

21.1.4 Research Problem and Significance

Although there have been some encouraging advances, several issues related to the implementation of DT technology in agriculture remain. High prices and technical complexity are particularly challenging for smallholder farmers. The absence of compatibility across DT systems in the agricultural sectors is a significant challenge (Jones et al. 2020). Data security and privacy-related concerns are raised because of the use of IoT and cloud frameworks (Kalyani et al. 2023). The challenges of scalability and interoperability of integrating DTs with legacy systems in the agricultural domain are significant (Nasirahmadi and Hensel 2022).

This chapter answers key questions as follows:

(i) What are the possible areas where DT technology can be used to promote sustainability in thc agriculture sector?
(ii) Which are the most excellent DT applications and success stories within the agriculture sector?
(iii) What are the limitations, gaps, and challenges to use DT in sustainable agriculture?

(iv) What are the future prospects and possibilities to improve DT-based agricultural systems?

By addressing these questions, this study contributes to the academic and practical understanding of DTs as enablers of sustainable agriculture, providing insights for researchers, policymakers, and industry stakeholders.

21.1.5 Research Objectives

The major objectives of this chapter are to:

(i) Discuss the development and the situation of DT technology in the agriculture sector in the context of sustainability
(ii) Identify and classify DT applications in the areas of crop production, irrigation, livestock, aquaponics, and supply chains
(iii) Examine the advantages, issues, and constraints to the use of DTs in sustainable agriculture
(iv) Cite future avenues of research and policy reforms to increase adoption, scalability, and inclusiveness

21.1.6 Structure of the Chapter

This study is structured as follows:

Section 21.2 performs a comprehensive Literature Review, which is a review of the theoretical basis, development, and use of DT technology in agriculture, as well as gaps in the available studies. In Sect. 21.3, i.e., Research Methodology, the approach used is explained, which involves a systematic literature review and thematic synthesis. Section 21.4 discusses the utilization of DTs within the framework of sustainable agriculture, introducing several potential use cases across various agricultural sectors. Section 21.5 presents a Critical Discussion of the results, outlining the benefits, limitations, and ramifications of the use of DT. The conclusion is presented in Sect. 21.6, and it provides a summary of the findings, the contributions, and future directions.

21.2 Literature Review

21.2.1 Concept and Evolution of DTs

The idea of the DT was initially presented by Michael Grieves in the very beginning of the 2000s in the context of Product Lifecycle Management (PLM). Closer to the present work is the formalization of the idea in aerospace engineering, where a technology roadmap was proposed, in which DTs simulated and predicted the behavior of space vehicles in real time (Tuegel et al. 2011). A DT is understood to be defined as a virtual copy of a physical object, process, or system that can continually be updated with real-world measurement data, imitating, monitoring, and optimizing it (Jones et al. 2020; Tao et al. 2019).

At its core, a DT integrates three essential components:

- Physical Entity: In real life system (e.g., greenhouse, livestock, and irrigation system)
- Virtual Model: A digital version, which replicates the behavior, structure, and processes of an entity
- Data Connection: IoT sensors, network, and cloud-edge infrastructures capable of real-time data exchanges

Traditional areas of application of early DT were in the manufacturing and aerospace sectors. As an example, the aircraft industry utilized DTs to foretell the fatigue in structures, optimize the maintenance schedule, and prolong the life of the airframes (Tuegel et al. 2011). In a similar fashion, DTs are now at the heart of Industry 4.0 as enablers of cyber-physical systems (CPS) and of smart factories, as well as predictive maintenance (Tao et al. 2019). As technology advanced, their applications in healthcare, logistics, construction sites, energy systems, and farming were complemented by other applications for handling and mitigating complex, data-intensive systems (Nasirahmadi and Hensel 2022).

21.2.2 DT in Agriculture 4.0

The agricultural sector has had lower levels of digital adoption compared to other sectors, such as manufacturing. Nonetheless, with the introduction of Agriculture 4.0 (where IoT, AI, Robotics, and Data Analytics are used in agriculture), the ground has been transformed into a greenhouse where DT can flourish (Tsolakis et al. 2022; Verdouw et al. 2021). The IoT Analytics (2023) report identifies agriculture as one of the most promising emerging markets for digital transformation implementation, with growth driven by sustainability, resource optimization, and global food security demands. The DTs are unlike previous monitoring systems, as they provide a closed-loop modeling and sensing combined with decision-making, allowing farmers and stakeholders to not only observe but also simulate what the future could hold.

- *Role of IoT and Sensors*: IoT serves as a basis of agricultural DTs. The use of sensors in soil, irrigation, livestock collars, drones, and aquaponics units has enabled the live streaming of data, which is then processed into digital models (Pylianidis et al. 2021). These are temperature, humidity, the pH of soil, moisture levels, nutrient levels, crop maturity, and animal well-being indicators.
- *Integration with Cloud, Fog, and Edge Computing*: The performance of the agricultural DTs is highly dependent on the data processing architectures

considered. Cloud computing is considered centralized computation, while fog and edge computing are characterized by local and low-latency analyses, which are significant for time-efficient applications such as irrigation control and disease detection (Kalyani et al. 2023). The hybrid Cloud–Fog–Edge architecture enables scalability and resilience in deploying DT in agricultural environments with limited connectivity, particularly in rural areas.

- *AI and Predictive Modeling*: Algorithms such as AI incorporated into DTs offer predictive mechanisms of estimating crop yields, predicting outbreaks of pests, and animal behavior (Nasirahmadi and Hensel 2022). DTs increase the precision of their models with the aid of neural networks to discover unseen patterns across large sets of agricultural data.
- *CPS and Agriculture 4.0*: DT in agriculture can be viewed as a cyber-physical connection between actual farms and their DTs. This alignment enables simulation-driven optimization to test the effects of fertilizer strategies or irrigation schedules in the computational space before implementing them in the field (Tsolakis et al. 2022). In short, DTs will represent the dream of sustainable, data-driven, and precision-oriented agriculture because they enable predictive decision-making, risk mitigation, and system-level optimization.

21.2.3 Applications Across Agricultural Domains

21.2.3.1 Crop Production and Greenhouse Horticulture

DTs may be most usefully applied in the initial agricultural activities, crop production, and in greenhouse management. DT applications are suitable in controlled-environment agriculture, where greenhouses and plant factories are a perfect environment to test DT deployment, although there is also much progress to be made (Slob et al. 2023). DT of a greenhouse combines the data collected by the sensors that read the temperature, humidity, concentration of carbon dioxide, the moisture in the soil, and the nutrients. It provides growers with the opportunity to develop virtual models of various environmental conditions and their impact on plant growth (Howard et al. 2022). As another example, DTs were created with the help of virtual reality, thus providing farmers with the opportunity to interact with virtual greenhouse environments, test their intervention, and train staff on how to manage crops (Slob et al. 2023).

Another important use case is plant growth modeling. By integrating machine learning models with real-time environmental data, DTs can emulate the physiology of plants and forecast yields under various management alternatives (Zhang et al. 2023). This is especially essential in precision horticulture, where small changes in environmental conditions may result in pronounced yields and quality effects. The applications of DTs in sustainable plant factories, as digital modeling aids in optimizing the lighting systems, nutrient cycles, and energy usage, have also been discussed in the literature (Liu et al. 2023). Likewise, a rapid optimization framework utilizing DT was presented to optimize the design of indoor farms with the potential to save energy and concurrently achieve optimal growth in plants (Mengi et al. 2024).

21.2.3.2 Smart Irrigation and Soil Management

The role of water scarcity is one of the most influential sustainability issues threatening agriculture today. In the field of irrigation systems, DTs can be implemented to provide a data-driven solution for water flow, soil moisture, and crop water needs (Barkakoti and Joshi 2023). The present study involves the creation of an IoT-DT-based irrigation system that works on real-time soil data to increase water efficiency. Their results showed a vast improvement in water use efficiency and stability of crop yields (Manocha et al. 2024). In this regard, DTs are used in predictive irrigation, including irrigation based on predictive tools such as weather forecasts, soil conditions, and evapotranspiration models, to prevent the wastage of water (Barkakoti and Joshi 2023).

DTs enhance the management of nutrient and soil health. Monitoring the soil (pH, organic matter, and nutrient availability) is crucial for DTs, as it allows them to provide farmers with tests to check the viability of their fertilization plan before it is implemented in the soil. This limit is to prevent over-fertilization, soil erosion, and groundwater contamination. DT-based irrigation can help ensure climate resilience by conserving water resources, mitigating drought risks, and preserving soil quality over the long term. However, challenges still exist in areas without IoT infrastructure or reliable connectivity, a necessity for implementing the real-time aspect of DT (Kalyani et al. 2023).

21.2.3.3 Livestock Farming

Applications successfully implemented on DTs in livestock farming focus on animal welfare, productivity, and health tracking. Neethirajan and Kemp (2021) note that DTs may be created at the animal or herd level, incorporating data on animals that wear biometric sensors, GPS collars, and video surveillance equipment among others. These digital doubles facilitate the tracking of health indicators in these animals, including body temperature, heart rate, feeding habits, and movements. The DTs can simulate the effect of feeding on animal growth and milk production, thereby optimizing feed management policies. A major strong side of DTs in farming livestock is their capability to combine all data sources: barn, open pastures, and supply chains into a single decision-support system. The limitation, however, lies in the fact that it is a very expensive project on a wide scale in deploying IoT sensors, as well as the ethical issues surrounding monitoring and animal data privacy. However, livestock diseases are a perfect representation of the One Health concept, which combines animal welfare, human health, and environmental sustainability.

21.2.3.4 Aquaponics, Hydroponics, and Aquaculture

Hydroponics and aquaponics are new closed-loop agricultural systems, where considerable focus is required to resource efficiency and optimization of agricultural systems. DTs have been utilized to model water quality and nutrient cycling as well as fish crop interactions. The researchers proposed a DT model of an aquaponics factory as a power system, examining the dynamics of the power system and possible opportunities to optimize it. Their work also emphasized the ability of DTs to solve problems such as nutrient balance and minimizing energy use (Xu et al. 2023). Hydroponic grow beds are developed using DTs so that the growth and water circulation of plants could be monitored and optimized in real-time environmental conditions to increase the input use efficiency (Yanes et al. 2022). In aquaculture, DTs have been applied to recreate fish farming conditions, model fish growth rates, predict disease outbreaks, and monitor water quality changes (Zhao et al. 2022). More recently, DT-enabled decision support systems are used to improve the urban aquaponics systems by the incorporation of IoT and AI, where people can now make correct decisions based on the simulated predictions performed with the help of AI (Ghandar et al. 2021). Nonetheless, the major setback is that there are no standard DT models that can be used in the widely varying number of existing aquaponic/hydroponic systems worldwide.

21.2.3.5 Food Supply Chains and Logistics

DTs can be used in postharvest management as well as food supply chains beyond the use in the farm. Supply chains today are fragile to shocks such as those induced by climate change, pandemics, and geopolitical strains; therefore, resilience and visibility are required (Defraeye et al. 2021). Using a DT-based gray causal modeling (GCM) method, the food supply chain's resilience and sustainability have been significantly improved (Singh et al. 2023). They showed the importance of being able to model logistic scenarios and predict the risks of logistic activities and calculate the most effective resource utilization strategies using DTs.

The application of DTs is studied to examine how the COVID-19 pandemic affected food supply chains, offering guidelines to increase resilience (Burgos and Ivanov 2021). Other use cases include monitoring fruit life cycles in the postharvest stage using DT simulations, logistics of perishable goods in cold chains, and transparent traceability of food by integrating blockchain with DTs (Shoji et al. 2022; Defraeye et al. 2021). Sustainability-wise, DT-controlled supply chains

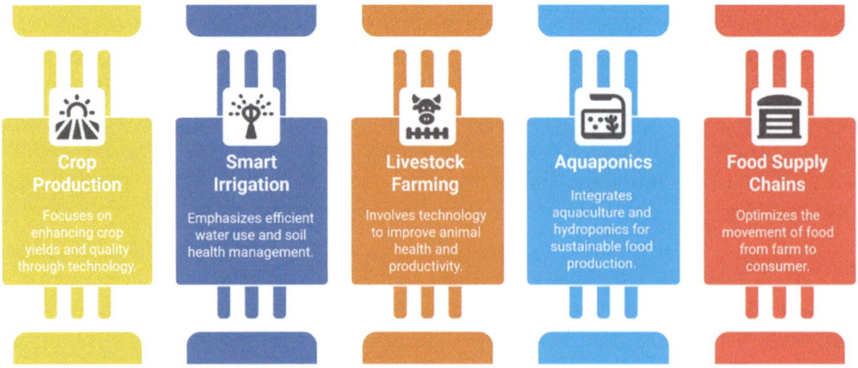

Fig. 21.1 Agricultural technology applications

minimize wastage of food, energy, and greenhouse gases and enhance traceability and trusting relations throughout the farm-to-fork supply chain. Figure 21.1 demonstrates the applications of agricultural technology.

21.2.4 State-of-the-Art Reviews

DT usage in agriculture is relatively new, although it is quite advanced in manufacturing, aerospace, and healthcare. Nevertheless, reviews and meta-analyses are emerging as valuable sources of systematized information on the maturity, typology, and promise of DTs in agriculture and food systems.

21.2.4.1 Classification of DTs

Several authors have attempted to categorize DTs into various groups based on the maturity level, integration abilities, and functionalities. In this context, static digital models that do not include automated data exchange are referred to as digital models, and digital objects wherein alterations in the physical apparatus will cause the digital representation to refresh, but not the other way around, are referred to as digital shadows. However, more complete models in which modifications done to one side of a model (physical or digital) are reflected in the other side are categorized as DTs (Kritzinger et al. 2018).

This framework can be more applicable to agriculture, and it is notable that many existing solutions, including greenhouse monitoring systems, still operate at the level of only a digital shadow, rather than an interactive twin (Purcell and Neubauer 2023). The analysis of DT is conducted according to phases in the lifecycle (design, operation, maintenance, and decommissioning) and by identifying the enablers of DT through technologies such as IoT, AI, and cloud infrastructures (Jones et al. 2020). It is clear from their systematic review that interoperability standards are crucial for making strides in DT applications across multiple industries, including agriculture. More recently, the DTs and CPS are differentiated by observing that although CPS focuses on real-time observation and control, DTs expand on this model by adding the elements of predictive simulation, lifecycle control, and feedback (Tao et al. 2019). These definitions reflect the reality that, as of the latest report, many agricultural DTs are still in their infancy, acting as CPS or digital shadows, rather than full-fledged DTs.

21.2.4.2 Reviews in Agricultural Context

Agriculture-oriented reviews of DT applications have begun to appear in recent years. The majority of DTs in agriculture are experimental or pilot projects, and the scale is limited to some commercial applications. They analyzed that greenhouse horticulture and supply chain management have the most developed DT implementations, but livestock and aquaponics are not such well-studied areas yet (Purcell and Neubauer 2023). In the greenhouse horticultural industry, DTs state that greenhouse DTs provide advanced crop modeling, climate modeling, and energy optimization (Ariesen-Verschuur et al. 2022). UV also emphasized the importance of virtual reality and human-to-machine interaction to enhance usability for farmers. The surveyed DT paradigms are applicable to the livestock business and argued that uniform guidelines and economical IoT integrations are required to help render DTs reachable to small and intermediate farmers (Nasirahmadi and Hensel 2022). The review by Singh et al. (2023) and Defraeye et al. (2021) examined food supply chains to clarify that DTs play a crucial role in supporting resilience, openness, and provisioning of the circular economy. These reviews suggest that the adoption of DT in supply chains is progressing at a faster pace compared to farm-level implementations, as logistics networks are already highly digitalized. In the case of aquaponics and hydroponics, Xu et al. (2023) and Yanes et al. (2022) carried out a review on DT opportunities, with the modeling of nutrient balance, water reuse, and fish–plant integration as the noted areas where DTs have the capacity to increase the resilience and sustainability of the cropping systems.

21.2.5 Gaps and Research Opportunities

Although the body of research is expanding to advocate for the promise of DT technology in agriculture, significant gaps and drawbacks remain. Recognizing these gaps is crucial in informing future research and achieving the necessary change to transition DTs from experimental projects to mainstream agricultural practice.

21.2.5.1 Technological Gaps

A particularly large problem is the immaturity of the enabling technologies of agricultural DTs.

- *IoT Limitations*: Most of the applications of DTs in agriculture are too dependent on the use of IoT sensors. Still, there are limitations due to their high cost, energy constraints, and connectivity issues in rural areas (Pylianidis et al. 2021). In the developing world, DT benefits are inaccessible to smallholder farmers, primarily due to inadequate infrastructure.
- *Computational Needs*: DTs necessitate high-performance computing (HPC), cloud, and occasionally edge/fog computing to work with live data. This presents a challenge in places that lack the digital infrastructure (Kalyani et al. 2023).
- *Standardization Deficit*: There is a lack of a universal framework or agreement on standards for agricultural DT development. This poses interoperability difficulties, as models created to support one type of farm or region will rarely be adapted to support another farm region (Jones et al. 2020).

21.2.5.2 Data-Related Gaps

Agricultural DTs rely on quality and up-to-date data, but a number of problems still exist:

- *Lack of Data and Bits of Data*: Data in the agricultural context is highly dispersed compared to that in the manufacturing industry, where systems are a little more standardized. Model applicability is restricted by data heterogeneity in dirt types, climate, crops, and livestock breed (Nasirahmadi and Hensel 2022).
- *Uncertainty and Incompleteness*: Sensor breakdowns or senseless values and noisy data produce the imprecision of the DT models (Xu et al. 2023).
- *Data Ownership and Security*: Farmers believe that their control over data is often ignored because large companies that offer digital services may have monopolistic situations in relation to agricultural data (Verdouw et al. 2021).

21.2.5.3 Economic and Adoption Gaps

Socio-economic factors are hindering the implementation of DT due to high costs, complexity, and a lack of awareness. These are described as follows:

- *Expensive Implementation*: Implementation costs of IoT, AI, and DT systems are high, and specific farmers (subsidized or collective infrastructure) will not be subsidized to implement them.
- *Digital Literacy Gap*: Most farmers lack the technical expertise to operate DT systems, creating a usability barrier (Purcell and Neubauer 2023).
- *Uncertainty of Return on Investment (ROI)*: Small farms will generally not be in a position to efficiently compute the economic benefits linked to the implementation of DT, and investing in untested technologies is unsafe.

21.2.5.4 Research Gaps

Literature lists several under-researched areas where research is sorely needed:

- *Livestock and Aquaponics*: Crop-based DTs (so far mainly in greenhouses) attracted interest. Still, so far, only very few studies have been implemented in the fields of livestock welfare monitoring, aquaponics, and integrated fish–plant systems.
- *Scalability and Inclusion*: DT applications are being written to run on large-scale farms traditionally located in wealthy nations. The distinction is that smaller, similar DTs are not yet being made to be cheap and modular, thereby not benefiting the smallholder farmer in the developing world.
- *Lifecycle Integration*: DTs in agriculture so far primarily deal with real-time control and monitoring. Further studies are required to scale DTs throughout the lifecycle of agricultural systems, such as managing food harvesting, food processing, and the circular economy.
- *The Human–Machine Interaction*: The decision-making process of farmers is not necessarily thoroughly implemented into DT systems. Farming-oriented design and participatory development of DT should be researched (Ariesen-Verschuur et al. 2022).

Based on the identified gaps, future research directions would include the following: One of the major research possibilities is to incorporate sustainability

Table 21.1 Comprehensive view of the contributions, gaps, and focus area

S. No.	Author (s) and year	Domain	Contribution	Gaps
1.	Neethirajan and Kemp (2021)	Livestock Farming	Investigates the application of DT technology to enhance livestock farming practices and animal welfare.	Limited focus on the practical challenges and long-term monitoring of livestock through DTs.
2.	Pylianidis et al. (2021)	DT in Agriculture	Introduces the concept of DTs to agriculture, highlighting key areas of application and future potential.	Limited exploration on the scalability and cost of implementing DT win systems on small farms.
3.	Yanes et al. (2022)	Hydroponic and Aquaponic Systems	Discusses Digital Twinning of hydroponic grow beds within intelligent aquaponic systems, enhancing production optimization.	More research is required to explore cost-effectiveness and long-term sustainability of aquaponic systems using DTs.
4.	Shoji et al. (2022)	Postharvest Supply Chains	Uses DT technology to track the postharvest journey of imported fruits from packhouses to retail stores.	Need for models that can predict and optimize postharvest shelf life across diverse environmental conditions.
5.	Nasirahmadi and Hensel (2022)	Digitalization in Agriculture	Explores the potential of DT for advancing agricultural digitalization and its role in smart farming paradigms.	Insufficient focus on the practical application in diverse agricultural contexts and the integration of real-time environmental data.
6.	Tsolakis et al. (2022)	Drone Applications in Digital Farming	Uses a cyber-physical approach to analyze drone applications in agriculture, focusing on sustainability in farming 4.0.	Further work is needed to address the high costs of drone technology and how to effectively scale it to smallholder farming operations.
7.	Zhao et al. (2022)	Marine and Aquaculture Systems	Focuses on the application of DTs for rapid damage detection in aquaculture net panels, improving operational efficiency.	Limited application to broader aquaculture systems; further research needed on scaling and real-time operational integration.
8.	Zhang et al. (2023)	Plant factory systems	Researches the design and implementation of a DT system for mapping multi-environmental variables in plant factories.	Need for improvements in handling complex environmental variables and scaling the models for large-scale facilities.

(continued)

Table 21.1 (continued)

S. No.	Author (s) and year	Domain	Contribution	Gaps
9.	Kalyani et al. (2023)	Smart Agriculture in Cloud-Fog-Edge Infrastructure	Proposes the deployment of DT systems in agriculture using cloud, fog, and edge computing to improve data processing and system responsiveness	More practical validation needed to assess the feasibility of large-scale deployments in rural areas with limited infrastructure.
10.	Slob et al. (2023)	Virtual Reality and Greenhouses	Focuses on Virtual Reality-based DT models for greenhouses, particularly in terms of human interaction with the system.	The study does not fully integrate crop-specific needs and environmental factors into the VR-based models.
11.	Manocha et al. (2024)	Smart Irrigation Systems	Proposes a smart irrigation approach inspired by DT technology for optimizing water usage in agriculture.	Need for real-world field data to assess the effectiveness of these systems in different climates and agricultural conditions.
12.	Mengi et al. (2024)	Indoor Farming Systems	Uses DT technology and rapid optimization for the design of indoor farming systems, improving efficiency.	Integration with existing urban farming practices is still an open challenge.

measurements into DT assessments. Table 21.1 represents a comprehensive view of the contributions, gaps, and focus areas in the realm of DT technology in agriculture and related domains.

21.3 Research Methodology

This study is predominantly qualitative and exploratory, employing a systematic literature review (SLR) as its technique, supplemented with a conceptual synthesis approach. Considering that DT applications in agriculture are innovative, a systematic literature review (SLR) facilitates the integration of disparate material, delineates the current status of the field, and identifies gaps and prospective research avenues.

21.3.1 Research Design

This study is intended to be a multi-stage study, and it will comprise:

- SLR: The search, screening, and synthesis of peer-reviewed scholarly literature, industry reports, and conference proceedings regarding applications of DT in agriculture
- Conceptual Synthesis: Multidisciplinary (computer science, agronomy, supply chain management, and sustainability studies) synthesis of a holistic model of DT application in agriculture
- Thematic Analysis: Grouping applications by greenhouse horticulture, irrigation management, livestock farming, aquaponics, and supply chains and how they relate to sustainable agriculture

The design is suitable since the area of DT in agriculture remains a developing research field with fewer empirical case studies than theoretical and technical work.

21.3.2 Data Sources

To ensure comprehensiveness, various databases and sources were involved, including Web of Science, Scopus, IEEE Xplore, SpringerLink, ScienceDirect, Wiley Online Library, MDPI, industry reports, conference papers, and government/industry reports. This multisource approach was both industry- and scholarly-relevant.

21.3.3 Search Strategy

A combination of Boolean operators and key strings was used to search. The main keywords were: DT AND Agriculture, DT AND Sustainable Farming, Smart Agriculture AND CPS, IoT AND DT AND Food Supply Chain, and Precision Agriculture AND Simulation Models. The publications dated after 2010–2025 were only searched, as indicated by their name, because it is the period during which DT has started to become a mainstream concept in the sphere of agriculture.

21.3.4 Inclusion and Exclusion Criteria

The subsequent criteria were employed to ensure relevance and quality:

Inclusion Criteria:

(a) Conference papers, books, and peer-reviewed journal articles.
(b) Industry reports with data on the trends in DTA implementation.
(c) Articles that specifically comment on the application of DT in agriculture, food systems, or supply chains.
(d) Papers that would relate explicitly to the development of DT, such as enabling technologies (IoT, AI, and cloud/edge computing), are also sought.

Exclusion Criteria:

(a) Not available in English studies
(b) Articles on general use of IoT/AI in agriculture without DT connection
(c) Non-scientific publications, blogs, or advertising media

21.3.5 Screening Process

The PRISMA model (Preferred Reporting Items in Systematic Reviews and Meta-Analysis) was used to screen. There were 512 original records in the scholarly and industrial databases, and 168 articles were screened by sifting through the titles and abstracts (Fig. 21.2). The full-text review was conducted to identify the direct relevance of DT in agriculture. A subsample of 92 studies was found, out of which 35 high-quality sources (academic and industry) were ultimately selected to undergo further analysis and synthesis. This methodology was designed to ensure reliability, transparency, and reproducibility of the review.

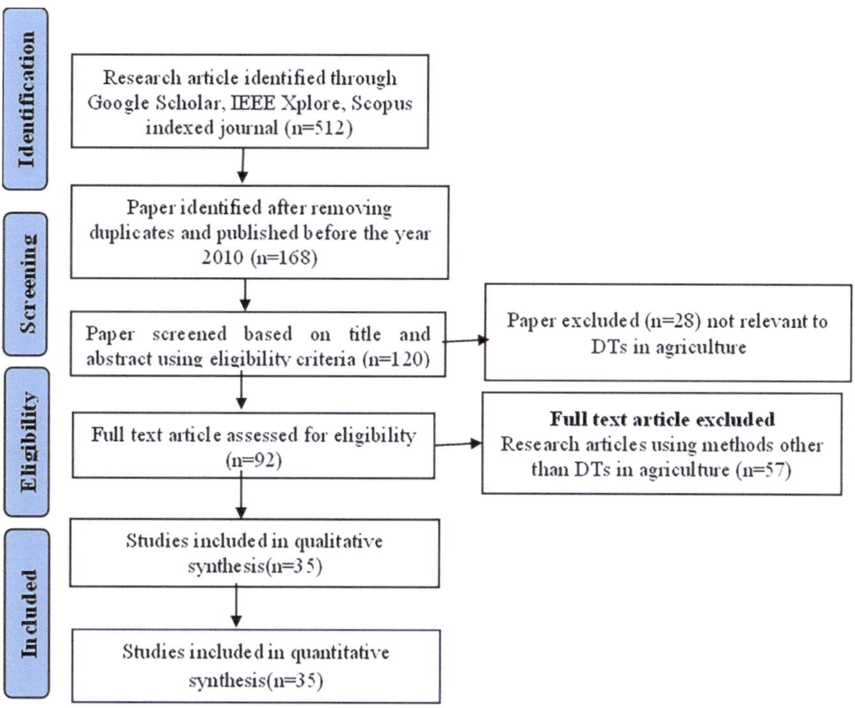

Fig. 21.2 PRISMA guidelines of the search process

21.3.6 Limitations of Methodology

Despite the study's careful design, there are some limitations. As a publication bias, the more successful the outcomes of applying DT, the more likely the studies are to be published and influence the data. The vast majority of agricultural DT research is conceptual or pilot in nature, lacking large-scale empirical studies. The materials were confined to the English language, and this may have overlooked local developments in Asia, Africa, and Latin America. As this field is rapidly evolving, there is a possibility that some of the more recent developments are not yet reflected in peer-reviewed literature.

21.4 DT Applications in Sustainable Agriculture

DT is becoming recognized as a disruptive technology in agriculture, enabling the industry to transition toward sustainability, resilience, and data-driven decision-making. By creating simulated versions of real-life farming systems, DTs can be utilized to control, forecast, and optimize agricultural processes in real time. Its use in five vital areas is reviewed in the following subsections: greenhouse horticulture, irrigation and soil management, livestock monitoring, aquaponics/hydroponics, and food supply chains.

21.4.1 DT in Greenhouse Horticulture

Greenhouses are exceptionally managed places, which is why they are one of the earliest to implement DT technology. To investigate the relationships between crops and microclimate, as well as pests and their variability, digital reconstruction of greenhouse ecosystems is conducted using IoT devices, sensor networks, and AI analytics. It is widely used for monitoring temperature, carbon dioxide levels, humidity, and light intensity. It is also used to anticipate ideal circumstances and improve crop yield through virtual testing of alternative irrigation, nutrient, and energy plans. DTs help farmers decrease their water and fertilizer use, reduce dependence on pesticides, and optimize energy use in heating/cooling. Research has also revealed that predictive analytics based on DT can enhance yields by 15–30% (Koumpouros et al. 2023). Despite these advantages, it faces several challenges, including the high costs of installing IoT-enabled greenhouses, incompatibility with existing equipment, and security issues related to real-time data.

21.4.2 Accuracy in Soil and Irrigation Control

One of the most important agricultural issues is water shortage. The DTs operate with a simulated soil model that estimates moisture content, nutrient availability, and crop water demand, ensuring proper irrigation scheduling. It is widely used in weather forecasting and predicting water demand during the crop lifecycle. It helps in the installation of soil sensors in conjunction with DTs to achieve irrigation efficiency and mitigate groundwater loss through giving farmers simulation-based irrigation programs. It also assists in reducing water use by up to 40% without compromising yield (Gong et al. 2025). It improves soil health by reducing over-irrigation and leaching of fertilizers and adds value to climate-smart agriculture by mitigating resource overexploitation. However, on the other hand, it requires a strong connection in rural regions. There is also a requirement for technological literacy and training of farmers to facilitate adoption. Data privacy is another concern when sensor data is gathered and processed through third-party systems.

21.4.3 DT in Livestock Monitoring

In livestock farming, DT is applied to combine wearable sensors, health sensors, and behavior sensors to produce digital equivalents of animals or livestock. It is used for real-time monitoring of body temperature, activity, and food habits. It is helpful in predictive modeling of health parameters to detect diseases at their early stages. It is used to simulate optimal feed and breeding strategy. It enhances the well-being of animals by providing active healthcare. Additionally, it improves productivity by maximizing nutrition and minimizing mortality rates, while also reducing the over-use of antibiotics to facilitate precise veterinary treatment. Despite its several benefits, it also presents some challenges. There are ethical issues in collecting animal data. Smart collars and wearable sensors are costly. There is a lack of interoperability between platforms across different livestock systems.

21.4.4 Aquaponics and Hydroponics

Hydroponics and aquaponics are controlled environment farming systems becoming important to urban agriculture. DTs mimic the dynamics of the water–nutrient cycle, fish health (in aquaponics), and plant growth to optimize resource use. They are utilized in the simulation of nutrient compartmentation in hydroponic systems and to monitor pH, dissolved oxygen, and water circulation. They help maximize the symbiosis between fish and plants in aquaponics. They ensure the cultivation of food throughout the year in urban areas with limited arable land and enable closed-loop circular economies. There is up to 70% reduction in water use compared to traditional agriculture. However, the initial capital cost is high and requires a constant power supply. There is also a requirement for complex AI algorithms to control sophisticated interactions in aquaponics systems.

21.4.5 Food Supply Chain and Logistics

The food supply chain suffers from significant inefficiencies, resulting in nearly one-third of food being wasted annually. DTs are being used to improve supply chain sustainability, logistics, and traceability. It can be applied to develop online solutions that create digital representations of food supply chains, allowing for the tracking of movement, quality, and storage. Blockchain can be integrated for transparency in food origin and certification. It enhances traceability, thereby reducing the risk of contamination and food fraud. It helps to reduce supply–demand disjuncture and wastage while increasing consumer confidence in sustainable sources. However, there is a requirement for globally standardized data formats. Another significant challenge is the cybersecurity vulnerabilities in blockchain.

21.5 Results and Discussion

DT technology, which is the focus of the reviewed studies in the agricultural sector, has a revolutionary potential to the industry. Nevertheless, the potential is significant, yet, there are a number of major concerns that should be discussed to make sure that the DT technologies will be efficiently integrated in a variety of agrifood settings, especially the ones that involve smallholder farmers, the development of policies, and the following research.

Among the main obstacles to the successful implementation of DT technology in agriculture, scalability is a major issue among smallholder farmers. A lot of the reviewed papers emphasize the complex and expensive character of DT applications that can be expensive in terms of financial resources, high-tech infrastructure, and well-trained staff. As an illustration, the systems, which operate with cloud–fog–edge infrastructure (Kalyani et al. 2023) or real-time data acquisition with the sensors (Verdouw et al. 2021), are made to be used by a large-scale operation. Smallholder farmers who are usually working with limited resources encounter high challenges in applying such high-tech solutions.

Barkakoti and Joshi (2023) and Neethirajan and Kemp (2021) observe that although the advantages of DT technology are evident, especially in its application to efficiently irrigate land or provide better livestock management, small-scale farmers will have a hard time taking advantage of these technologies due to high start-up expenses and technical demands. On top of that, very few models that can be modified to be cost-effective and easy to use in smaller operations have been demonstrated (Pylianidis et al. 2021). Therefore, additional research and development is required to downsize DT applications and make them more affordable to smallholder farmers, potentially by introducing low-cost sensors, mobile-based applications, and less complicated decision-support systems.

DT technologies to be implemented successfully in agriculture will be impossible without substantial policy interventions, especially the subsidies and training of farmers. The accelerated technological change, as reported in other studies (Defraeye et al. 2021; Ghandar et al. 2021), is a challenge to policymakers since it is important not only to have equitable access to these technologies but also to organize the incorporation of the latter into the existing agricultural systems.

Among the major suggestions that can be made based on the reviewed studies is that government subsidies or other forms of finances should reduce the cost of entry to the smallholder farmers. Otherwise, the DT technology can continue to be centralized to big, commercially oriented, agricultural businesses. In the absence of proper policy frameworks, as noted by IoT Analytics (2023), the digital gap between large- and small-scale farmers may only widen, and the influence of these technologies will be constrained.

Capacity-building and training is also very important. A large number of smallholder farmers do not have the technical capacity to learn and apply sophisticated digital systems. Kalyani et al. (2023) assert that, to be effective, DTs do not only need farmers to have access to the technology but also capability to interpret and respond to the data produced. Therefore, the policies should focus on the aspects of training to provide farmers with the necessary skills on how to operate DT technologies. Partnerships with schools, technology players, and NGOs might assist in providing this much needed training so that farmers can have the power to make the right choice on embracing DT systems in their business.

The DT technology in agriculture also has a bright future, yet it also offers a number of research opportunities that will allow resolving the challenges described above. The development of Artificial Intelligence (AI) at the edge, as mentioned in numerous papers (Barkakoti and Joshi 2023; Purcell and Neubauer 2023), will become one of the key factors in the further development of the capabilities of DTs in agriculture. Edge AI allows real-time data processing in place, eliminating the use of centralized cloud systems and accelerating and enhancing the quality of decision-making. It is especially significant to smallholder farmers whose access to the internet can be limited because edge computing does not require constant access to the cloud.

More studies of how edge AI is integrated into DT systems have the potential to make such technologies stronger and more available to farmers living in remote locations. Moreover, it is important that farmers are engaged in the design, development, and implementation of DT solutions. To make the solutions to match their unique needs and environments, a closer involvement in the development process can be achieved. Research works by Howard et al. (2022), Yanes et al. (2022), and others emphasize the necessity of taking local knowledge and practices into account during agricultural technology designing. The next thing that future research should develop is to seek ways of involving the farmers in the co-designing of these systems so that the systems are not only practical but effective as well.

The other topic that needs to be investigated more urgently is the standardization of DT in agriculture in the future. As observed in the works by Verdouw et al. (2021) and Jones et al. (2020), the lack of standardization in data formats, communication protocols, and system architectures is a major obstacle to interoperability and scalability of DTs. A set of standardized DT applications would enable a more

Table 21.2 Applications, benefits, and challenges of DTs in agricultural sustainability

Application domain	Key applications	Benefits	Challenges	References
Crop management	Soil health monitoring, crop growth simulation, precision irrigation	15–30% yield improvement, 40% water savings, and optimized fertilizer use	High sensor costs, lack of interoperability	Koumpouros et al. (2023), Manocha et al. (2024), Gong et al. (2025)
Livestock farming	Health monitoring, feed optimization, disease prediction	Improved animal welfare, reduced antibiotic use, productivity gains	Data integration complexity, cybersecurity concerns	Jones et al. (2020), Nasirahmadi and Hensel (2022)
Aquaponics and controlled environment	DT-driven greenhouse and aquaponics management	Real-time monitoring, energy efficiency, reduced input waste	High computational requirements, reliance on stable connectivity	Tao et al. (2019), Xu et al. (2023)
Supply chain and logistics	Farm-to-fork traceability, demand forecasting, logistics optimization	Transparency, reduced food loss, improved market access	Adoption gap for smallholders, need for policy frameworks	Tsolakis et al. (2023), Singh et al. (2023)
Climate resilience and sustainability	Climate impact simulations, carbon footprint assessment, circular farming	Supports SDGs, resilient food systems, resource efficiency	Unequal access to technology, need for global standards	Defraeye et al. (2021), Kalyani et al. (2023)

effortless integration of various systems and sources of data and would also lower the complexity and cost of implementation. Moreover, this kind of standardization might help to establish cooperation among technology suppliers, agricultural interest groups, and policy-makers and achieve more effective and consistent solutions. The applications of DTs in different agricultural sectors are highlighted in Table 21.2.

21.6 Conclusion and Future Directions

This research is the first one to ensure that the progress in DT technology in agriculture is integrated and analyzed in a complex manner, presenting the applications, issues, and perspectives of the technology. It explains the optimizing nature of DTs in irrigation, crop management, and supply chain logistics, especially in the fresh produce and livestock farming industry. As good as these advantages are, other issues such as scalability to smallholder farmers and supportive policy are also important barriers to adoption.

The major policy implication of this research is that the governments and industry players need to work together in facilitating the adoption of DTs. This consists of providing subsidy to make DT technology cheaper to smallholder farmers and investing in training programs to impart the knowledge to the farmers on how to operate these sophisticated systems. In the absence of these interventions, the digital divide between big and small-scale farmers will increase, restricting the extensive application of DTs in the agricultural sector.

The next line of research should consider incorporating DTs with the new technologies such as AI and blockchain. AI may be used to boost the predictive capabilities of DTs, so that real-time information would be used to make more accurate decisions, and blockchain would provide security and transparency of the data, especially in the agricultural supply chains. Besides, the edge computing technology of DT systems will enable real-time data processing in the remote regions without relying on cloud infrastructure and make systems more efficient.

The other avenue that is promising is the co-design approach where farmers are involved in the development of DT solutions that meet their requirements. This will make the technology viable and contextual. Through solving these problems and exploiting these future possibilities, DTs have the ability to realize their full potential of transforming agriculture to enhance productivity, sustainability, and resiliency in various agricultural environments.

References

Ariesen-Verschuur N, Verdouw C, Tekinerdogan B (2022) Digital twins in greenhouse horticulture: a review. Comput Electron Agric 199:107183. https://doi.org/10.1016/j.compag.2022.107183

Barkakoti C, Joshi S (2023, March) Advancement of digital twin in irrigation and smart farming. In: 2023 International Conference on Sustainable Computing and Data Communication Systems (ICSCDS). IEEE, pp 1245–1250. https://doi.org/10.1109/ICSCDS56580.2023.10104641

Burgos D, Ivanov D (2021) Food retail supply chain resilience and the COVID-19 pandemic: a digital twin-based impact analysis and improvement directions. Transp Res Part E Logist Transp Rev 152:102412. https://doi.org/10.1016/j.tre.2021.102412

Defraeye T, Shrivastava C, Berry T, Verboven P, Onwude D, Schudel S, Bühlmann A, Cronje P, Rossi RM (2021) Digital twins are coming: will we need them in supply chains of fresh horticultural produce? Trends Food Sci Technol 109:245–258. https://doi.org/10.1016/j.tifs.2021.01.025

Ghandar A, Ahmed A, Zulfiqar S, Hua Z, Hanai M, Theodoropoulos G (2021) A decision support system for urban agriculture using digital twin: a case study with aquaponics. IEEE Access 9: 35691–35708. https://doi.org/10.1109/ACCESS.2021.3061722

Gong J, Chen J, Chen Z, Chi H, Liu J (2025, March) Overview of the application progress of digital twins in agriculture. In: Proceedings of the 2025 International Conference on Digital Management and Information Technology, pp 295–303. https://doi.org/10.1145/3736426.3736473

Howard DA, Ma Z, Jørgensen BN (2022, October) A case study of digital twin for greenhouse horticulture production flow. In: 2022 IEEE 2nd international conference on digital twins and parallel intelligence (DTPI). IEEE, pp 1–6. https://doi.org/10.1109/DTPI55838.2022.9998914

IoT Analytics. Digital twin market: analyzing growth and emerging trends. 2023. Available online: https://iot-analytics.com/digital-twin-market-analyzing-growth-emerging-trends/. Accessed on 21 April 2025

Jones D, Snider C, Nassehi A, Yon J, Hicks B (2020) Characterizing the digital twin: a systematic literature review. CIRP J Manuf Sci Technol 29:36–52. https://doi.org/10.1016/j.cirpj.2020.02.002

Kalyani Y, Bermeo NV, Collier R (2023) Digital twin deployment for smart agriculture in cloud-fog-edge infrastructure. Int J Parallel Emergent Distrib Syst 38(6):461–476. https://doi.org/10.1080/17445760.2023.2235653

Koumpouros Y, Georgoulas A, Kremmyda G (eds) (2023) Transcending humanitarian engineering strategies for sustainable futures. IGI Global

Kritzinger W, Karner M, Traar G, Henjes J, Sihn W (2018) Digital twin in manufacturing: a categorical literature review and classification. Ifac-PapersOnline 51(11):1016–1022. https://doi.org/10.1016/j.ifacol.2018.08.474

Liu J, Wang L, Wang Y, Xu S, Liu Y (2023) Research on the interface of sustainable plant factory based on digital twin. Sustainability 15(6):5010. https://doi.org/10.3390/su15065010

Manocha A, Sood SK, Bhatia M (2024) IoT-digital twin-inspired smart irrigation approach for optimal water utilization. Sustain Comput Inform Syst 41:100947. https://doi.org/10.1016/j.suscom.2023.100947

Mengi E, Becker CJ, Sedky M, Yu SY, Zohdi TI (2024) A digital-twin and rapid optimization framework for optical design of indoor farming systems. Comput Mech 74(1):31–43. https://doi.org/10.1007/s00466-023-02421-9

Nasirahmadi A, Hensel O (2022) Toward the next generation of digitalization in agriculture based on digital twin paradigm. Sensors 22(2):498. https://doi.org/10.3390/s22020498

Neethirajan S, Kemp B (2021) Digital twins in livestock farming. Animals 11(4):1008. https://doi.org/10.3390/ani11041008

Purcell W, Neubauer T (2023) Digital twins in agriculture: a state-of-the-art review. Smart Agricultural Technology 3:100094. https://doi.org/10.1016/j.atech.2022.100094

Pylianidis C, Osinga S, Athanasiadis IN (2021) Introducing digital twins to agriculture. Comput Electron Agric 184:105942. https://doi.org/10.1016/j.compag.2020.105942

Shoji K, Schudel S, Onwude D, Shrivastava C, Defraeye T (2022) Mapping the postharvest life of imported fruits from packhouse to retail stores using physics-based digital twins. Resour Conserv Recycl 176:105914. https://doi.org/10.1016/j.resconrec.2021.105914

Singh G, Rajesh R, Daultani Y, Misra SC (2023) Resilience and sustainability enhancements in food supply chains using digital twin technology: a grey causal modelling (GCM) approach. Comput Ind Eng 179:109172. https://doi.org/10.1016/j.cie.2023.109172

Slob N, Hurst W, Van de Zedde R, Tekinerdogan B (2023) Virtual reality-based digital twins for greenhouses: a focus on human interaction. Comput Electron Agric 208:107815. https://doi.org/10.1016/j.compag.2023.107815

Straitresearch (2023) Digital Twin Market. Available online: https://straitsresearch.com/report/digital-twin-market. Accessed on 1 May 2025

Tao F, Qi Q, Wang L, Nee AYC (2019) Digital twins and cyber–physical systems toward smart manufacturing and industry 4.0: correlation and comparison. Engineering 5(4):653–661. https://doi.org/10.1016/j.eng.2019.01.014

Tsolakis N, Bechtsis D, Vasileiadis G, Menexes I, Bochtis DD (2022) Sustainability in the digital farming era: a cyber-physical analysis approach for drone applications in agriculture 4.0. In: Information and communication technologies for agriculture-theme IV: actions. Springer International Publishing, Cham, pp 29–53. https://doi.org/10.1007/978-3-030-84156-0_2

Tsolakis N, Harrington TS, Srai JS (2023) Digital supply network design: a circular economy 4.0 decision-making system for real-world challenges. Prod Plan Control 34(10):941–966. https://doi.org/10.1080/09537287.2021.1980907

Tuegel EJ, Ingraffea AR, Eason TG, Spottswood SM (2011) Reengineering aircraft structural life prediction using a digital twin. Int J Aerosp Eng 2011(1):154798. https://doi.org/10.1155/2011/154798

Verdouw C, Tekinerdogan B, Beulens A, Wolfert S (2021) Digital twins in smart farming. Agric Syst 189:103046. https://doi.org/10.1016/j.agsy.2020.103046

Xu L, Yu H, Qin H, Chai Y, Yan N, Li D, Chen Y (2023) Digital twin for aquaponics factory: analysis, opportunities, and research challenges. IEEE Trans Ind Inform 20(4):5060–5073. https://doi.org/10.1109/TII.2023.3331135

Yanes AR, Abbasi R, Martinez P, Rafiq A (2022) Digital twinning of hydroponic grow beds in intelligent aquaponic systems. Sensors 22(19):7393. https://doi.org/10.3390/s22197393

Zhang Z, Zhu Z, Gao G, Qu D, Zhong J, Jia D, Du X, Yang X, Pan S (2023) Design and research of digital twin system for multi-environmental variable mapping in plant factory. Comput Electron Agric 213:108243. https://doi.org/10.1016/j.compag.2023.108243

Zhao YP, Lian L, Bi CW, Xu Z (2022) Digital twin for rapid damage detection of a fixed net panel in the sea. Comput Electron Agric 200:107247. https://doi.org/10.1016/j.compag.2022.107247

MIX
Papier aus verantwortungsvollen Quellen
Paper from responsible sources
FSC® C105338

If you have any concerns about our products,
you can contact us on
ProductSafety@springernature.com

In case Publisher is established outside the EU,
the EU authorized representative is:
Springer Nature Customer Service Center GmbH
Europaplatz 3, 69115 Heidelberg, Germany

Printed by Libri Plureos GmbH
in Hamburg, Germany